Pathogen-Derived Immunomodulatory Molecules

ADVANCES IN EXPERIMENTAL MEDICINE AND BIOLOGY

Recent Volumes in this Series

Volume 659
HOT TOPICS IN INFECTION AND IMMUNITY IN CHILDREN VI
Edited by Adam Finn, Nigel Curtis, and Andrew J. Pollard

Volume 660
PARAOXONASES IN INFLAMMATION, INFECTION, AND TOXICOLOGY
Edited by Srinu Reddy

Volume 661
MEMBRANE RECEPTORS, CHANNELS AND TRANSPORTERS IN PULMONARY CIRCULATION
Edited by Jason X. –J. Yuan, and Jeremy P.T. Ward

Volume 662
OXYGEN TRANSPORT TO TISSUE XXXI
Edited by Duane F. Bruley and Eiji Takahasi

Volume 663
STRUCTURE AND FUNCTION OF THE NEURAL CELLADHESION MOLECULE NCAM
Edited by Vladimir Berezin

Volume 664
RETINAL DEGENERATIVE DISEASES
Edited by Robert E. Anderson, Joe G. Hollyfield, and Matthew M. LaVail

Volume 665
FORKHEAD TRANSCRIPTION FACTORS
Edited by Kenneth Maiese

Volume 666
PATHOGEN-DERIVED IMMUNOMODULATORY MOLECULES
Edited by Padraic G. Fallon

A Continuation Order Plan is available for this series. A continuation order will bring delivery of each new volume immediately upon publication. Volumes are billed only upon actual shipment. For further information please contact the publisher.

I

Pathogen-Derived Immunomodulatory Molecules

Edited by

Padraic G. Fallon, BSc, MSc, PhD
Inflammation and Immunity Group, Institute of Molecular Medicine, Trinity College, Dublin, Ireland

Springer Science+Business Media, LLC

Landes Bioscience

Springer Science+Business Media, LLC
Landes Bioscience

Printed in the USA.

Springer Science+Business Media, LLC, 233 Spring Street, New York, New York 10013, USA
http://www.springer.com

Please address all inquiries to the publishers:
Landes Bioscience, 1002 West Avenue, Austin, Texas 78701, USA
Phone: 512/ 637 6050; FAX: 512/ 637 6079
http://www.landesbioscience.com

The chapters in this book are available in the Madame Curie Bioscience Database.
http://www.landesbioscience.com/curie

Pathogen-Derived Immunomodulatory Molecules, edited by Padraic G. Fallon. Landes Bioscience / Springer Science+Business Media, LLC dual imprint / Springer series: Advances in Experimental Medicine and Biology.

ISBN: 978-1-4419-1600-6

While the authors, editors and publisher believe that drug selection and dosage and the specifications and usage of equipment and devices, as set forth in this book, are in accord with current recommendations and practice at the time of publication, they make no warranty, expressed or implied, with respect to material described in this book. In view of the ongoing research, equipment development, changes in governmental regulations and the rapid accumulation of information relating to the biomedical sciences, the reader is urged to carefully review and evaluate the information provided herein.

Library of Congress Cataloging-in-Publication Data

Pathogen derived immunomodulatory molecules / [edited by] Padraic G. Fallon.
p. ; cm. -- (Molecular biology intelligence unit)
Includes bibliographical references and index.
ISBN 978-1-4419-1600-6
1. Pathogenic microorganisms. 2. Biological response modifiers. I. Fallon, Padraic G. (Padraic Gerard), 1966- II. Series: Molecular biology intelligence unit (Unnumbered : 2003)
[DNLM: 1. Immunologic Factors. 2. Microbiological Phenomena--immunology. QW 568 P297 2009]
QR175.P374 2009
579'.165--dc22

2009024025

DEDICATION

This book is dedicated to the Fallon ladies:
Rosie, Caitríona and Niamh.

PREFACE

Pathogen-Derived Immunomodulatory Molecules is a book title that may require some explanation. Pathogens that are present today have evolved following a long association with man and have developed unique strategies that have been optimized by natural selection to subvert the host immunity. As we approach the 200th anniversary of Charles Darwin's birth, it is appropriate to appreciate that Darwin recognized that pathogens (infections) play a significant and potent role in natural selection, encompassed by the concept "infection begets natural selection". This book therefore examines the molecules that pathogens produce, which can modulate or usurp the functions of the immune system.

The idea of using molecules from pathogens as a therapeutic is an ancient concept in medicine. Such a strategy is exemplified by vaccination, with pathogen molecules employed to induce protective immunity against the given or related species of pathogen. The following chapters explore the concept of using pathogen-derived immune modulating molecules as a therapy. In doing so, they may provide the drug cabinet of the future for treating a spectrum of unrelated disease. Herein, a range of immune modulating molecules or strategies from various pathogens is examined in one volume.

The intention of the book was to have chapters addressing immunomodulating molecules from different pathogens. The range of pathogens considered includes bacteria (chapters by Williams, van Strijp and Rooijakkers), viruses (chapters by Bowie, McFadden), protozoan parasites (Aliberti), helminths (Harnett, Fallon), fungi (Sorrell) and parasitic ticks (Anguita). Chapters also address specific immunomodulatory molecules or strategies. The diversity of aspects addressed in the book is highlighted by Lucas and colleagues review of the 'saga' of viral serine proteinase inhibitors, with a focus on Serp-1, the first new generation of pathogen immunomodulatory molecule currently in clinical trials. While Elliott and Weinstock have contributed a provocative chapter exploring the use of live parasitic helminth infections as a therapeutic strategy for immune-mediated diseases; indeed trials have already been completed for such an approach. With respect to pathogens usurping an immune pathway, Alcami and colleagues here reviewed the growing number of pathogens that have evolved a range of molecules that can modify many aspects of the chemokine system.

This book is timely due to the need to expand the horizons of conventional drug discovery. A trend in the biopharmaceutical pipeline of fewer drugs to market is illustrated by USA FDA in 2007 approving the lowest number of new molecular entities since 1983. As the drug discovery and development industry broadens its search for new drugs to less traditional strategies, this book will be a reference to the potential for exploiting pathogen as a source of the anti-inflammatory drugs of the future.

Finally, I hope that this book whets the appetite for the reader, whether in academia or industry, to explore opportunities for exploiting pathogens for the discovery of new processes in immunobiology and, ultimately, for development of new therapies for human inflammatory diseases.

Padraic G. Fallon, BSc, MSc, PhD

ABOUT THE EDITOR...

PADRAIC G. FALLON is Stokes Professor of Translational Immunology, School of Medicine, Trinity College Dublin, Ireland. Formally he was a Wellcome Trust Fellow, in the Department of Pathology, University of Cambridge, UK. He is currently the head of the Inflammation and Immunity Research Group, Institute of Molecular Medicine, St James's Hospital, Trinity College Dublin. Using both animal and cellular models, Professor Fallon's research addresses the mechanisms of modulation of human inflammatory diseases by pathogens, with a focus on parasitic helminths. The objective of the research is to develop novel therapeutic strategies for the treatment of inflammatory or autoimmune diseases. Allergic lung and skin inflammation and inflammatory bowel disease are the principal human diseases investigated by the Fallon Inflammation and Immunity Research Group. Since 1998, pathogen-derived immunomodulatory molecules have been a central theme in Professor Fallon's research.

PARTICIPANTS

Poom Adisakwattana
Inflammation and Immunity Group
Institute of Molecular Medicine
Trinity College Dublin
Dublin
Ireland

Antonio Alcami
Centro de Biología Molecular Severo Ochoa
Consejo Superior de Investigaciones Científicas
Universidad Autónoma de Madrid
Madrid
Spain
and
Deparment of Medicine
University of Cambridge
Addenbrooke's Hospital
Cambridge
UK

Julio Aliberti
Division of Molecular Immunology
Cincinnati Children's Hospital Medical Center and University of Cincinnati College of Medicine
Cincinnati, Ohio
USA

Juan Anguita
Department of Veterinary and Animal Sciences
University of Massachusetts Amherst
Amherst, Massachusetts
USA

Mee Y. Bartee
Department of Medicine
and
Department of Molecular Genetics and Microbiology
University of Florida
Gainesville, Florida
USA

Erik Biessen
University of Maastricht
Maastricht
The Netherlands

Ilze Bot
University of Maastricht
Maastricht
The Netherlands

Andrew G. Bowie
School of Biochemistry and Immunology
Trinity College Dublin
Dublin
Ireland

Sharon C.A. Chen
Centre for Infectious Diseases and Microbiology and Westmead Millennium Institute
University of Sydney
and
Department of Infectious Diseases
Westmead Hospital
Sydney
Australia

Alexander Christov
University of Oklahoma
Oklahoma City, Oklahoma
USA

Erbin Dai
Department of Medicine
University of Florida
Gainesville, Florida
USA
and
Robarts Research Institute
University of Western Ontario
London, Ontario
Canada

Jennifer A. Davids
Department of Medicine
and Department of Molecular
Genetics and Microbiology
University of Florida
Gainesville, Florida
USA

David S. Donaldson
Department of Cellular and Molecular
Medicine
School of Medicine Sciences
University of Bristol
Bristol
UK

David E. Elliott
Division of Gastroenterology
and Hepatology
Department of Internal Medicine
University of Iowa
Roy J. and Lucille A. Carver
College of Medicine
Iowa City, Iowa
USA

Padraic G. Fallon
Inflammation and Immunity Group
Institute of Molecular Medicine
Trinity College
Dublin
Ireland

Margaret M. Harnett
Division of Inflammation
Infection and Immunity
University of Glasgow
Glasgow
UK

William Harnett
Strathclyde Institute of Pharmacy
and Biomedical Sciences
University of Strathclyde
Glasgow
UK

Barry Harrington
School of Biochemistry
and Immunology
Trinity College Dublin
Dublin
Ireland

Mark Hatton
McMaster University
Hamilton, Ontario
Canada

Ilse Jongerius
Medical Microbiology
University Medical Center Utrecht
Utrecht
The Netherlands

Ignacio J. Juncadella
Department of Veterinary
and Animal Sciences
University of Massachusetts Amherst
Amherst, Massachusetts
USA

Liying Liu
Department of Medicine
University of Florida
Gainesville, Florida
USA
and
Robarts Research Institute
University of Western Ontario
London, Ontario
Canada

David A. Lomas
Cambridge Institute for Medical Research
University of Cambridge
Cambridge
UK

Alexandra Lucas
Department of Medicine
and
Department of Molecular Genetics and Microbiology
University of Florida
Gainesville, Florida
USA
and
Robarts Research Institute
University of Western Ontario
London, Ontario
Canada

Colin Macaulay
Viron Therapeutics, Inc
London, Ontario
Canada

Fabiana S. Machado
Division of Molecular Immunology
Cincinnati Children's Hospital Medical Center
and
University of Cincinnati College of Medicine
Cincinnati, Ohio
USA

Grant McFadden
Department of Medicine
and
Department of Molecular Genetics and Microbiology
University of Florida
Gainesville, Florida
USA
and
Robarts Research Institute
University of Western Ontario
London, Ontario
Canada

Leslie Miller
University of Maastricht
Maastricht
The Netherlands

Richard Moyer
Department of Molecular Genetics and Microbiology
University of Florida
Gainesville, Florida
USA

Orla Mulhern
School of Biochemistry and Immunology
Trinity College Dublin
Dublin
Ireland

Ganesh Munuswamy-Ramunujam
Department of Medicine
and
Department of Molecular Genetics and Microbiology
University of Florida
Gainesville, Florida
USA

Hendrik J. Nel
Inflammation and Immunity Group
Institute of Molecular Medicine
Trinity College Dublin
Dublin
Ireland

Mark Poznansky
Robarts Research Institute
University of Western Ontario
London, Ontario
Canada

Masmudur M. Rahman
Department of Molecular Genetics and Microbiology
College of Medicine
University of Florida
Gainesville, Florida
USA

Sanjay Ram
Division of Infectious Diseases and Immunology
University of Massachusetts Medical School
Worcester, Massachusetts
USA

Jakob Richardson
University of Oklahoma
Oklahoma City, Oklahoma
USA

Mary Richardson
Department of Molecular Genetics and Microbiology
University of Florida
Gainesville, Florida
USA

Suzan Rooijakkers
Medical Microbiology
University Medical Center Utrecht
Utrecht
The Netherlands

Margarida Saraiva
Life and Health Sciences Research Institute
School of Health Sciences
University of Minho
Braga
Portugal

Sean P. Saunders
Inflammation and Immunity Group
Institute of Molecular Medicine
Trinity College Dublin
Dublin
Ireland

Tania C. Sorrell
Centre for Infectious Diseases and Microbiology and Westmead Millennium Institute
University of Sydney
Sydney
Australia

Collin Sullivan
Department of Molecular Genetics and Microbiology
University of Florida
Gainesville, Florida
USA

Jos A.G. van Strijp
Medical Microbiology
University Medical Center Utrecht
Utrecht
The Netherlands

Karin Ellen Veldkamp
Medical Microbiology
University Medical Center Utrecht
Utrecht
The Netherlands

Kasinath Viswanathan
Oregon Health Sciences University
Portland, Oregon
USA

Hao Wang
Oregon Health Sciences University
Portland, Oregon
USA

Joel V. Weinstock
Division of Gastroenterology and Hepatology
Department of Internal Medicine
Tufts Medical Center
Boston, Massachusetts
USA

Neil A. Williams
Department of Cellular and Molecular Medicine
School of Medicine Sciences
University of Bristol
Bristol
UK

Robert Zhong
University of Maastricht
Maastricht
The Netherlands

CONTENTS

Chapter 1

Bacterial Toxins as Immunomodulators

David S. Donaldson and Neil A. Williams*

Abstract

Bacterial toxins are the causative agent of pathology in a variety of diseases. Although not always the primary target of these toxins, many have been shown to have potent immunomodulatory effects, for example, inducing immune responses to co-administered antigens and suppressing activation of immune cells. These abilities of bacterial toxins can be harnessed and used in a therapeutic manner, such as in vaccination or the treatment of autoimmune diseases. Furthermore, the ability of toxins to gain entry to cells can be used in novel bacterial toxin based immuno-therapies in order to deliver antigens into MHC Class I processing pathways. Whether the immunomodulatory properties of these toxins arose in order to enhance bacterial survival within hosts, to aid spread within the population or is pure serendipity, it is interesting to think that these same toxins potentially hold the key to preventing or treating human disease.

Introduction

Since the discovery that the disease associated with *Corynebacterium diphtheriae* (diphtheria) was caused by a secreted toxin,[1] many other toxins that are the causative agent of the pathology in bacterial diseases have been isolated (Table 1). Although the primary target involved in the induction of pathology by these toxins is often not the immune system, many of them also appear to have potent immunomodulatory effects in vivo and in vitro. Furthermore, these effects have the potential to be harnessed and applied in a therapeutic manner. In this review, the immunomodulatory effects and potential applications of a variety of bacterial toxins, including the Anthrax toxins, the *Helicobacter pylori* vacuolating cytotoxin, Clostridial toxins, toxins produced by *Bordetella pertussis*, listerolysin O and the Cholera-like enterotoxins, will be discussed.

Toxins Secreted by Bacillus Anthracis

Edema Toxin (EdTx) and Lethal Toxin (LeTx) are two toxins with immunomodulatory activity that are produced by *B. anthracis*, the cause of the disease anthrax. Both toxins are composed of a heptameric complex of protective antigen (PA) bound to either edema factor (EF) or lethal factor (LF).[2-4] The heptameric complex of PA is responsible for receptor binding and cellular entry,[5] whereas toxicity is associated with both EF, a calmodulin-dependent adenylate cyclase that induces increases in cytosolic cAMP[6] and LF, a metalloprotease that cleaves mitogen-activated protein kinase kinases (MAPKK).[7,8]

Many studies have shown that LeTx harbours a potent immunomodulatory capacity. LeTx induces apoptosis of murine macrophages and human monocytes; however, they may need to be activated before this occurs.[9-12] Additionally, LeTx inhibits chemotaxis and the expression of pro-inflammatory cytokines and chemokines in stimulated macrophages, by preventing nuclear factor (NF)-κB and interferon regulatory factor (IRF)-3 activation, and diminishes their T-cell

*Corresponding Author: Neil A. Williams—Department of Cellular and Molecular Medicine, School of Medicine Sciences, University of Bristol, University Walk, Bristol, BS8 1TD, UK. Email: neil.a.williams@bristol.ac.uk

Pathogen-Derived Immunomodulatory Molecules, edited by Padraic G. Fallon.

Table 1. Common bacterial diseases associated with toxin production

Bacteria	Toxin(s)	Disease
Bacillis anthracis	Edema toxin	Anthrax
	Lethal toxin	
Bacillus cereus	Cereulide	Food poisoning
Bordetella pertussis	Petussis toxin	Whooping cough
	Adenyate cyclase toxin	
Clostridium botulinum	Neurotoxins	Botulism
Clostridium difficle	Toxin A	Antibiotic assoicated
	Toxin B	Diarrhoea
Clostridium tetani	Tetanus toxin	Tetanus
Corynebacterium diphtheriae	Diphtheria toxin	Diphtheria
Escherichia coli	Heat-labile	Traveller's diarrhoea
	Enterotoxin	
Listeria monocytogenes	Listeriolysin O	Listeriosis
Pseudomonas aeruginosa	Exotoxin A	Pneumonia
Shigella dysenteriae	Shiga toxin	Dysentery
Vibrio cholerae	Cholera toxin	Cholera

stimulatory capacity.[12-17] Macrophage production of nitric oxide (NO) is also reduced by LeTx,[14] potentially through its ability to cleave neural NO synthase (nNOS).[18] The activity of LeTx is not limited to monocytes/macrophages, as it also induces the apoptosis of dendritic cells (DC)[19] and inhibits LPS-induced activation of murine and human DC, associated with reduced cytokine expression, lack of costimulatory molecule up-regulation and an impaired ability to stimulate $CD4^+$ T-cells.[20,21] The ability of B-cells to proliferate and produce antibodies in response to a variety of stimuli is impaired by LeTx[22] as is the proliferation of $CD4^+$ T-cells, due in part to a reduction in both interleukin (IL)-2 secretion and NF-κB activation.[23,24]

EdTx has been shown to act as a mucosal adjuvant, enhancing both systemic and mucosal antibody responses to ovalbumin (OVA) and inducing a mixed T helper $(T_H)1/T_H2$ OVA-specific $CD4^+$ T response following intra-nasal co-administration,[25] potentially related to its ability to induce AP-1 and C/EBP-β expression in murine macrophages.[26] In contrast, EdTx inhibits murine $CD4^+$ T-cell proliferation[23] and the production of IL-12 and tumour necrosis factor (TNF)-α by murine DC in response to *B. anthracis* spores.[27] Additionally, both LeTx and EdTx have been shown to inhibit human neutrophil chemotaxis[17,28] and superoxide production.[29,30]

The ability of PA to deliver LF into the cytosol has been exploited in order to deliver antigens into MHC class I processing pathways and stimulate cytotoxic T-lymphocyte (CTL) responses.[31] A detoxified version of LF (LFn) has also been used to deliver conjugated antigens into the cytosol of antigen presenting cells (APC), resulting in the induction of both $CD4^+$ and CTL responses.[32-34]

H. pylori Vacuolating Cytotoxin

H. pylori, a Gram-negative bacteria implicated in the pathogenesis of peptic ulcer disease and gastric adenocarcinoma,[35] secretes a toxin, referred to as the vacuolating cytotoxin (VacA), capable of inducing the formation of large vacuoles in epithelial cells.[36] Although not required for infection, secretion of VacA confers a selective advantage to expressing strains[37] and administration of VacA alone can result in gastric injury.[38] Additionally, VacA has been shown to induce the activation of signalling molecules[39] and to have many immunomodulatory activities.

VacA interacts with CD18 on human $CD4^+$ T-cells[40] resulting in inhibited proliferation, primarily by preventing nuclear factor of activated T-cells (NFAT) nuclear translocation and the up-regulation in IL-2 secretion associated with this.[41-43] Interestingly, this effect of VacA on $CD4^+$ T-cells confers the ablity to resist primary HIV-1 infection.[44] VacA also induces cycloxagenase (COX)-2 expression in neutrophils and macrophages,[42] inhibits MHC Class II presentation in B-cells[45] and inhibits phagosome maturation in macrophages.[46] Mast cell migration and degranulation is inhibited by VacA, whearas secretion of TNF-α, macrophage inflammatory protein (MIP)-1α, IL-1β, IL-6, IL-10 and IL-13 is enhanced.[47,48]

Toxins Produced by *Clostridium* Species

Clostridium difficile, a common cause of antibiotic associated diarrhoea,[49] produces two toxins with known immunomodulatory properties, referred to as toxin A and B (TcdA and TcdB).[50] Both toxins display glucosyltransferase activity, resulting in the inactivation of Rho, Rac and Cdc42 within target cells.[51,52] Both TcdA and TcdB are capable of inducing colitis similar to that induced by *C. difficile*, suggesting that they play an important role in pathology.[53]

Both TcdA and TcdB induce IL-1β, TNF-α, IL-6 and IL-8 secretion from human monocytes.[54,55] TcdA also inhibits human neutrophil chemotaxis and oxidative activity,[56] induces apoptosis in lamina propria macrophages and peripheral blood T-cells[57] and stimulates IL-1 production by mouse macrophages.[58] Rat peritoneal mast cells produce TNF-α in response to TcdA, whearas mast cell degranulation in response to a nematode infection is inhibited.[59] A similar capacity has been attributed to TcdB, which can inhibit FcεRI-mediated and polyclonally-stimulated degranulation by rat basophilic leukaemia cells and mast cells.[60,61] Conversely, both TcdA and B induce the degranulation of human mast cells and induce them to produce prostaglandin (PG) E_2 and IL-8.[62]

Other toxins produced by Clostridium species have been shown to have immunomodulatory activities. The Type D toxin produced by *Clostridium botulism* inhibits the production of TNF-α by LPS stimulated human monocytes.[63] Additionally, the *Clostridium perfingens* delta toxin has been shown to be cytotoxic to rabbit macrophages, but not to thymocytes.[64]

Toxins Produced by *B. pertussis*

B. pertussis is the cause of the acute respiratory illness referred to as Whooping Cough.[65] *B. pertussis* produces a variety of toxins that are associated with its pathogenicity including the pertussis toxin (PT) and the adenylate cyclase toxin (CyaA), both of which have been shown to have potent immunomodulatory activities. PT displays ADP-ribosyltransferase activity[66,67] and is associated with the systemic symptoms of *B. pertussis* infection.[68] Furthermore, PT appears to enhance colonisation of *B. pertussis*[69] CyaA is a calmodulin-dependent adenylate cyclase.[70,71] The receptor for CyaA is CD11b/CD18 ($\alpha_M\beta_2$ integrin), restricting its immunomodulatory effects to monocytes/macrophages, neutrophils and DC.[72] Mice challenged with CyaA-deficient strains of *B. pertussis* have greatly reduced bacterial loads, suggesting that CyaA is important for colonisation.[73]

PT can act as an adjuvant, inducing systemic and mucosal antibody responses characterised by high levels of immunoglobulin (Ig)G, IgA and IgE[74-76] and enhanced delayed type hypersensitivity reactions[77] to co-administered antigens. Furthermore, immunisation of keyhole limpet hemocyanin (KLH) with PT resulted in the generation of a mixed T_H1/T_H2 response, with increased KLH-specific secretion of IL-4, IL-5 and interferon (IFN)-γ.[78] Recent evidence has also suggested that PT can induce the differentiation of T_H17 cells.[79,80] Administration of PT exacerbates the autoimmune disease experimental autoimmune encephalomyelitis (EAE),[81] and it has become common practice to include injection with PT as an inherent part of protocols to trigger a number of autoimmune disease models including EAE and experimental autoimmune uveitis (EAU).[82,83] The reasons for its inclusion are largely supposed to be due to its ability to reduce the competence of the blood brain barrier,[84] however, the ability of PT to exacerbate autoimmune disease may also be related to its ability to prevent activated $CD4^+$ T-cell deletion[85] and reduce the number of $CD4^+CD25^+Foxp3^+$ regulatory T-cells (Treg).[86,87] Interestingly, adoptive transfer of PT-stimulated murine bone marrow-derived (BM)DC can substitute for PT in the induction of EAU, suggesting that PT-mediated modulation of DC may also have a role in the promotion of autoimmune disease.[88] Conversely, the B-subunit of PT has been shown to inhibit EAE induction[89] and strains of *B. pertussis* deficient in PT show enhanced responses to *B. pertussis* antigens, suggesting that PT down-regulates this response.[90]

PT induces the maturation of monocyte-derived (MD)DC, resulting in the up-regulation of CD80, CD86, CD83 and MHC Class II[91,92] and enhances the production of IL-12, TNF-α, IL-8 and IL-10, resulting in the generation of T_H1 responses.[91,93,94] However, LPS-induced IL-12 and TNF-α production by MDDC can be inhibited by PT.[92] This effect is limited to MDDC as PT appears to have no effect on human plasmacytoid DC.[94] PT promotes *B. pertussis* infection by suppressing alveolar macrophage function,[95] potentially through its ability to inhibit macrophage chemotaxis and migration and impair Fc mediated phagocytosis.[96] PT also inhibits LPS-induced NO production[97] and alters the responsiveness of macrophages to cytokine stimulation.[98] Despite this, PT prevents the apoptosis of BM macrophages,[99] induces the expression of CD80 and CD86[78] and enhances LPS-induced secretion of TNF-α.[97] PT-deficient strains of B. *pertussis* induce higher levels of IL-6 and IL-10 production from BM mast cells, suggesting that PT has a role in down-regulating this response also.[100]

PT inhibits the chemotaxis of neutrophils[69,101,102] and lymphocytes.[103] PT has been shown to act as a T-cell mitogen,[104] and induces the proliferation of both naïve and memory $CD4^+$ T-cells.[105] This proliferation is associated with increased NFAT and NF-κB nuclear translocation and up-regulated CD40L, CD69 and CD25 expression.[105] B-cell survival in vitro is also impaired by PT[106] and adoptively transferred PT-treated B-cells fail to migrate to lymph nodes or into the white pulp of the spleen.[107] Despite this, administration of PT enhances the expression of CD80 and CD86 on splenic B-cells.[78]

CyaA also acts as both a systemic and mucosal adjuvant. Immunisation of mice with CyaA and OVA induces anti-OVA IgG and IgA responses.[108,109] Furthermore, co-administration of CyaA and KLH enhances IgG_1, but not IgG_{2a}, production and results in the generation of $CD4^+$ T-cells which secrete high levels of IL-5 and IL-10 upon restimulation, suggesting that it biases responses towards a T_H2/Tr1 phenotype.[110] Conversely, immunisation of mice with CyaA conjugated to a viral peptide has been shown to induce the generation of T_H1 responses.[111]

CyaA has multiple effects on cells of the immune system. Neutrophil function and chemotaxis is impaired by CyaA.[112-114] Furthermore, CyaA induces the apoptosis of macrophages[115,116] and alters their response to LPS stimulation, resulting in increased IL-10 and IL-6 and reduced IL-12 and TNF-α secretion.[110,117] In addition, CyaA inhibits the oxidative response of human monocytes and inhibits LPS-induced production of TNF-α.[118,119] CyaA also inhibits the ability of murine macrophages to induce $CD4^+$ T-cell proliferation and induces the differentiation of T_H17 cells by inducing PGE_2 production.[120] CyaA has similar effects on DC. CyaA suppresses LPS-induced TNF-α and IL-12 production by BMDC whilst increasing their production of IL-10 and IL-6.[110,117,121] Furthermore, LPS-induced CD40 and ICAM-1 up-regulation was suppressed, whereas expression of CD80 and MHC Class II remained up-regulated.[110,117] CyaA also has similar effects on human MDDC, inducing increased expression of CD80, CD86, CD83 and MHC class II, increasing their ability to stimulate $CD4^+$ T-cells, whilst inhibiting LPS-induced IL-12 production.[92,122]

Both PT and CyaA have been used to deliver viral and tumour peptides into cells in order to prime CTL responses. PT mediated delivery of a HIV gp120 into the MHC Class I pathway was able to induce the generation of CTL responses in vitro and in vivo.[123] CyaA mediated peptide delivery can result in the stimulation of both the MHC Class I and II pathways, resulting in the activation of both CTL and $CD4^+$ T-cell responses.[124-129] The ability of CyaA to target antigen specifically to APC through it's binding to CD11b/CD18 has meant that huge interest has been generated in its potential use in human anti-tumour vaccines.[72]

Listeria monocytogenes Listeriolysin O

Listeriolysin O (LLO) is a pore-forming cholesterol-dependent cytolysin that is produced by *L. monocytogenes* and mediates its escape from phagosomes.[130,131] In addition to this, LLO has also been shown to have immunomodulatory activities. LLO induces the apoptosis of lymphocytes and DC[132,133] whilst retaining the ability to induce IFN-γ, IL-1α, IL-6, IL-10, IL-12 and TNF-α production from murine splenocytes, predominantly from macrophages and NK cells.[134-136] LLO also induces degranulation and leukotriene formation by neutrophils.[137] LLO-treated APC also lose the ability to process antigen, without altering MHC class II expression.[138] LLO-induced increases in IFN-γ and IL-12 skews the immune responses toward a T_H1 phenotype and accordingly LLO has been used to suppress both the sensitization and effector phase of a murine model of allergy.[139,140]

The ability of LLO to mediate escape from phagosomes has been utilized in order to deliver antigens into the MHC class I pathway. *Escherichia coli* expressing LLO and antigens have been shown to successfully deliver these antigens into the MHC class I presentation pathways of murine BMDC and macrophages and human MDDC, subsequently stimulating CTL responses.[141-143] Furthermore, vaccination with *E. coli* expressing LLO and OVA has been shown to impair the suppressive ability of $CD4^+CD25^+$ Treg cells.[144] In order to minimise the risk associated with administering live bacteria, liposomes containing LLO and OVA and translational fusions of LLO and antigens have been shown to be effective at inducing CTL responses in vitro and in vivo.[145-147]

The Cholera-Like Enterotoxins

The *E. coli* heat-labile enterotoxin (Etx) and the related Cholera Toxin (Ctx), produced by enterotoxigenic *E. coli* and *Vibrio cholerae* respectively, are major virulence factors associated with traveller's diarrhoea and cholera. Both toxins are composed of a single A subunit (CtxA/EtxA) and five B subunits arranged in a pentameric ring (CtxB/EtxB), are highly homologous at the amino acid level (82%)[148] and have near identical 3D structures.[149,150] Both toxins bind the cell surface receptor ganglioside G_{M1}, which is found ubiquitously on mammalian cells and the ganglioside G_{D1b}.[151] Etx can also interact with asialo-G_{M1}, lactosylcerimide and certain galactoproteins, however with much lower affinity.[152-155]

Both Ctx and Etx are potent immunogens when administered either parenterally or mucosally. Both induce strong mucosal (IgA) and systemic (IgG) antibody responses against themselves,[156,157] and can induce similar responses to co-administered antigens when administered mucosally.[156,157] Responses to co-administered antigens induced by Ctx are long-term and can persist for up to 2 years in mice.[158] Although many similarities exist between the two toxins, Ctx is associated with the generation of a predominantly T_H2 response,[159] whereas Etx induces a more balanced T_H1/T_H2 response to co-administered antigens.[160] Furthermore, Etx can induce IgA responses and the differentiation of IL-5 secreting T-cells in IL-$4^{-/-}$ mice, whereas Ctx cannot.[161] The basis of these differences appears to lie within the B subunit as chimeras of CtxA with EtxB and EtxA with CtxB display the adjuvant properties of their respective B subunits.[162]

Ctx enhances HLA-DR expression on human B-cells, enhances their stimulatory capacity[163] and promotes their differentiation into plasma cells.[164] Both Ctx and Etx induce the up-regulation of CD86 on macrophages, DC and B-cells, increasing their T-cell stimulatory capacity.[165,166] Furthermore, the induction of antibody responses has been linked to this increase in CD86 expression.[166,167] Etx induces the up-regulation of CD80 and down-regulation of CD40 on human MDDC and impairs their ability to present antigens.[168] Ctx also induces the maturation of human MDDC, by increasing PGE_2 and NO production,[169] whilst retaining the capacity to suppress LPS-induced cytokine expression.[170] Ctx-matured MDDC have the ability to stimulate $CD4^+$ T-cells and induce their differentiation into T_H2 cells.[170,171] However, Ctx inhibits anti-CD3 induced PBMC CD40L expression and proliferation.[172] Sub-cutaneous administration of Ctx with KLH in mice has been shown to result in the generation of IL-10-secreting $CD4^+$ T-cells, with properties similar to those of Tr1 cells.[173] Furthermore, Ctx was able to induce higher levels of IL-10 production by BMDC in response to LPS whilst inhibiting expression of ICAM-1 and CD40.[173]

Originally it was thought that the ability of Ctx and Etx to act as mucosal adjuvants was linked to the ADP ribosyltransferase activity of the A subunit as a mutant lacking enzymatic activity was ineffective as an adjuvant when administered orally.[174] However, the development of other nontoxic mutant forms of Ctx and Etx has shown that loss of ADP ribosyltransferase activity is not associated with a loss of adjuvant activity.[175-177] Furthermore, recombinantly produced CtxB and EtxB appear to retain some adjuvant activity. Intranasal administration of EtxB with herpes simplex virus (HSV)-1 glycoproteins in the absence of Etx can induce immunity to HSV-1.[178,179] EtxB has also been shown to induce immune responses to influenza virus haemagluttinin (HA) and OVA.[180,181] Intraperitoneal injection of CtxB with bovine serum albumin (BSA) resulted in increased systemic and mucosal anti-BSA antibodies.[182] Since CtxB is a poor mucosal adjuvant alone,[183] mucosal immunisation with CtxB-antigen conjugates, presumed to increase stability, have also been shown to be effective at inducing immunity to the conjugated antigens.[184]

Both CtxB and EtxB have been shown to have effects on a variety of cells in vitro. They alter cytokine production from monocytes/macrophages and inhibit pro-inflammatory cytokine production after LPS stimulation.[182,185-188] Pre-incubation of splenic APC with CtxB has been shown to confer the APC with the ability to induce increased IL-4 production from a $CD4^+$ T-cell clone.[189] Both CtxB and EtxB are capable of modulating antigen processing by macrophages, resulting in either decreased[190] or increased[191-193] presentation. CtxB can also induce the maturation of DC, manifested by increased MHC Class II and CD86 expression and increased secretion of IL-12.[194] EtxB and CtxB can also induce polyclonal activation of B-cells, associated with an up-regulation of MHC class II, CD80, CD86, CD40 and CD25 expression on the cell surface.[195-197] The mechanism by which this occurs appears to involve the activation of the mitogen activated protein (MAP) kinase and the extracellular signal related kinase (Erk).[198] Despite this, CtxB and EtxB inhibit anti-mouse IgM and LPS induced proliferation of B-cells.[199,200] CtxB has also been shown to enhance IL-4 mediated IgG_1 and TGF-β_1 mediated IgA isotype switching in murine B-cells.[201,202] Both CtxB and EtxB are capable of inducing apoptosis in $CD8^+$ T-cells in vitro[203,204] and appear to have differential effects on $CD4^+$ T-cells depending on their phenotype. Activation-induced proliferation of naïve $CD4^+$ T-cells can be suppressed with CtxB or EtxB[199,200] whereas CtxB can enhance the proliferation of antigen-experienced $CD4^+$ T-cells.[205] Intra-nasal administration of EtxB with antigen also enhances antigen-specific $CD4^+$ T-cell proliferation.[206] The underlying reason for this differential effect is unknown, however it has been reported that cross-linking of G_{M1} by CtxB can substitute for costimulation following T-cell activation, which could ultimately lead to increased proliferation.[207]

Although it has been demonstrated that EtxB or CtxB retain adjuvant activity and can stimulate immune responses to co-administered antigens, immunisation with EtxB and CtxB has also been associated with the induction of peripheral tolerance and protection from autoimmune disease development. Orally administered CtxB-antigen conjugates have been shown to down regulate both cell mediated and humoral immunity to the conjugated antigens and protect from T_H1 mediated autoimmune disease development, such as EAE, Type 1 diabetes (T1D) and collagen induced arthritis (CIA)[208-213] and T_H2 mediated diseases, such as allergy.[214] Intra-nasal administration of EtxB alone has also been shown to protect from T1D and CIA development in animal models.[215,216] Furthermore, intra-nasal administration of an EtxB-OVA conjugate has been shown to suppress the induction of OVA-specific IgE production.[217] Recent studies have identified that mucosal administration of CtxB-antigen conjugates appears to modulate a population of $CD4^+CD25^+Foxp3^+$ Treg in a transforming growth factor (TGF)-β_1-dependent manner and these are likely to underlie its ability to induce protection from both autoimmune disease and allergy.[218-220]

The ability of EtxB to mediate binding and entry to the cells has been exploited for other purposes. Binding of EtxB has been shown to enhance presentation of Epstein-Barr virus latent

Table 2. Other bacterial toxins displaying immunomodulatory activities

Bacterial Species	**Toxin Name**	**Immunomodulatory Activities**
Bacillus cereus	Cereulide	Inhibits NK cell cytotoxicity and cytokine production[227]
Corynebacterium diphtheriae	Diphtheria toxin	Delivery of peptides into MHC class I pathways[228,229]
Escherichia coli	Type II enterotoxins (LT-IIa/LT-IIb)	Mucosal and systemic adjuvant[230]
		Activates splenic DC in vitro[230]
		Induces apoptosis of $CD8^+$ T-cells and inhibits $CD4^+$ T-cell proliferation[231]
		Inhibits LPS-induced NF-κB activation and cytokine secretion in monocytes[232]
		B subunits interact with TLR2 and can activate monocytes[233]
Haemophilus ducreyi	Cytolethal distending toxin	Induces lymphocyte and DC death in vitro[234,235]
		Inhibits B-cell proliferation and antibody production[234]
Mycobacterium ulcerans	Mycolactone	Inhibits human DC maturation and chemotaxis[236]
		Inhibits murine macrophage TNF-α production in response to whole bacteria[237]
Pasteurella multocida	*P. multocida* toxin	Prevents human DC migration[238]
		Activates human and mouse DC[239]
		Inhibits LPS-induced IL-12 production from human DC[239]
Pseudomonas aeruginosa	Exotoxin A	Delivery of peptides into MHC class I and II pathways[240,241]
		Acts as a T-cell mitogen[242,243]
		A subunit inhibits human PBMC proliferation and differentiation[244]
		Alters human monocyte cell surface protein expression[244]
		Inhibits B-cell responses to T-dependent and T-independent antigens[245]
		Induces Mast cell apoptosis[246]
Shigella dysenteriae/ Escherichia coli	Shiga toxin/ Shiga-like toxin	Induction of cytokine secretion by monocytes/ macrophage[247-249]
		Prevents neutrophil apoptosis[250]
		Delivery of peptides into MHC class I pathways[251-253]
		Inhibits phagocytosis and superoxide production by human neutrophils[254]
		Treatment of B-cell lymphomas[255]

membrane proteins on lymphoblastoid cells by inducing their internalisation, rendering the cells more susceptible to killing by CTL.[221] EtxB has also been used to successfully deliver a peptide into the nucleus that suppresses HSV-1 replication.[222] Furthermore, conjugation of EtxB to an OVA peptide was able to deliver the peptide into the MHC class I processing pathway in DC, facilitating cross-presentation and suggesting that it may be used to generate CTL responses against tumour-derived antigens.[223] DNA vaccination of mice with a construct containing carcinoembryonic antigen, an antigen expressed on a number of different malignancies, fused to EtxB has been shown to induce CTL responses and reduce tumour growth.[224] An alternative vaccine strategy uses only the A1 fragment of Ctx conjugated to two staphylococcal protein A component Ds (CTA1-DD) targeting it to Ig-expressing B-cells.[167] Following intra-venous or intra-nasal administration of CTA1-DD with KLH, KLH-specific systemic and mucosal antibody responses were successfully induced. Linking antigens to CtxA2 and then combining this with CtxB has also been shown to be an effective way of inducing systemic and mucosal antibody responses to the conjugated antigens following mucosal delivery.[225]

Concluding Remarks

Within this review, the known immunomodulatory effects of the most commonly studied bacterial toxins have been presented. Many other bacterial toxins also have immunomodulatory activities, some of which are summarised in Table 2. Understanding the abilities of these toxins to modulate the immune system not only aids our understanding of the pathogenesis of disease and treatment of disease, it also allows the discovery of novel applications of these toxins, such as their use in vaccination to unrelated pathogens or in the treatment of immunopathological diseases. For example, oral administration of the Beçhet's disease specific peptide from heat shock protein 60 conjugated to CtxB has been shown to suppress relapsing uveitis in a Phase I/II clinical trial.[226] The immunomodulatory properties of bacterial toxins appear to have evolved in order to enhance bacterial survival within hosts, ultimately causing the death of many, however, these same toxins potentially hold the key to saving many lives.

References

1. Roux E, Yersin A. Contribution a l'etude de la diphtherie. Annales de l'Institut Pasteur 1888; 2:629-661.
2. Smith H. The basis of immunity to anthrax. Proc R Soc Med 1958; 51(5):375-377.
3. Stanley JL, Smith H. Purification of factor I and recognition of a third factor of the anthrax toxin. J Gen Microbiol 1961; 26:49-63.
4. Smith H, Stoner HB. Anthrax toxic complex. Fed Proc 1967; 26(5):1554-1557.
5. Singh Y, Klimpel KR, Quinn CP et al. The carboxyl-terminal end of protective antigen is required for receptor binding and anthrax toxin activity. J Biol Chem 1991; 266(23):15493-15497.
6. Leppla SH. Anthrax toxin edema factor: A bacterial adenylate cyclase that increases cyclic AMP concentrations of eukaryotic cells. Proc Natl Acad Sci USA 1982; 79(10):3162-3166.
7. Klimpel KR, Arora N, Leppla SH. Anthrax toxin lethal factor contains a zinc metalloprotease consensus sequence which is required for lethal toxin activity. Mol Microbiol 1994; 13(6):1093-1100.
8. Duesbery NS, Webb CP, Leppla SH et al. Proteolytic inactivation of MAP-kinase-kinase by anthrax lethal factor. Science 1998; 280(5364):734-737.
9. Friedlander AM. Macrophages are sensitive to anthrax lethal toxin through an acid-dependent process. J Biol Chem 1986; 261(16):7123-7126.
10. Park JM, Greten FR, Li ZW et al. Macrophage apoptosis by anthrax lethal factor through p38 MAP kinase inhibition. Science 2002; 297(5589):2048-2051.
11. Popov SG, Villasmil R, Bernardi J et al. Lethal toxin of bacillus anthracis causes apoptosis of macrophages. Biochem Biophys Res Commun 2002; 293(1):349-355.
12. Popov SG, Villasmil R, Bernardi J et al. Effect of bacillus anthracis lethal toxin on human peripheral blood mononuclear cells. FEBS Lett 2002; 527(1-3):211-215.
13. Guidi-Rontani C, Duflot E, Mock M. Anthrax lethal toxin-induced mitogenic response of human T-cells. FEMS Microbiol Lett 1997; 157(2):285-289.
14. Pellizzari R, Guidi-Rontani C, Vitale G et al. Anthrax lethal factor cleaves MKK3 in macrophages and inhibits the LPS/IFNgamma-induced release of NO and TNFalpha. FEBS Lett 1999; 462(1-2):199-204.

15. Erwin JL, DaSilva LM, Bavari S et al. Macrophage-derived cell lines do not express proinflammatory cytokines after exposure to bacillus anthracis lethal toxin. Infect Immun 2001; 69(2):1175-1177.
16. Dang O, Navarro L, Anderson K et al. Cutting edge: anthrax lethal toxin inhibits activation of IFN-regulatory factor 3 by lipopolysaccharide. J Immunol 2004; 172(2):747-751.
17. Rossi Paccani S, Tonello F, Patrussi L et al. Anthrax toxins inhibit immune cell chemotaxis by perturbing chemokine receptor signalling. Cell Microbiol 2007; 9(4):924-929.
18. Kim J, Park H, Myung-Hyun J et al. The effects of anthrax lethal factor on the macrophage proteome: Potential activity on nitric oxide synthases. Arch Biochem Biophys 2008; 472(1):58-64
19. Alileche A, Serfass ER, Muehlbauer SM et al. Anthrax lethal toxin-mediated killing of human and murine dendritic cells impairs the adaptive immune response. PLoS Pathog 2005; 1(2):e19.
20. Agrawal A, Lingappa J, Leppla SH et al. Impairment of dendritic cells and adaptive immunity by anthrax lethal toxin. Nature 2003; 424(6946):329-334.
21. Cleret A, Quesnel Hellmann A, Mathieu J et al. Resident CD11c+ lung cells are impaired by anthrax toxins after spore infection. J Infect Dis 2006; 194(1):86-94.
22. Fang H, Xu L, Chen TY et al. Anthrax lethal toxin has direct and potent inhibitory effects on B-cell proliferation and immunoglobulin production. J Immunol 2006; 176(10):6155-6161.
23. Comer JE, Chopra AK, Peterson JW et al. Direct inhibition of T-lymphocyte activation by anthrax toxins in vivo. Infect Immun 2005; 73(12):8275-8281.
24. Fang H, Cordoba-Rodriguez R, Lankford CS et al. Anthrax lethal toxin blocks MAPK kinase-dependent IL-2 production in CD4+ T-cells. J Immunol 2005; 174(8):4966-4971.
25. Duverger A, Jackson RJ, van Ginkel FW et al. Bacillus anthracis edema toxin acts as an adjuvant for mucosal immune responses to nasally administered vaccine antigens. J Immunol 2006; 176(3):1776-1783.
26. Comer JE, Galindo CL, Zhang F et al. Murine macrophage transcriptional and functional responses to bacillus anthracis edema toxin. Microb Pathog 2006; 41(2-3):96-110.
27. Tournier JN, Quesnel Hellmann A, Mathieu J et al. Anthrax edema toxin cooperates with lethal toxin to impair cytokine secretion during infection of dendritic cells. J Immunol 2005; 174(8):4934-4941.
28. During RL, Li W, Hao B et al. Anthrax lethal toxin paralyzes neutrophil actin-based motility. J Infect Dis 2005; 192(5):837-845.
29. Wright GG, Mandell GL. Anthrax toxin blocks priming of neutrophils by lipopolysaccharide and by muramyl dipeptide. J Exp Med 1986; 164(5):1700-1709.
30. Crawford MA, Aylott CV, Bourdeau RW et al. Bacillus anthracis toxins inhibit human neutrophil NADPH oxidase activity. J Immunol 2006; 176(12):7557-7565.
31. Goletz TJ, Klimpel KR, Arora N et al. Targeting HIV proteins to the major histocompatibility complex class I processing pathway with a novel gp120-anthrax toxin fusion protein. Proc Natl Acad Sci USA 1997; 94(22):12059-12064.
32. Doling AM, Ballard JD, Shen H et al. Cytotoxic T-lymphocyte epitopes fused to anthrax toxin induce protective antiviral immunity. Infect Immun 1999; 67(7):3290-3296.
33. Lu Y, Friedman R, Kushner N et al. Genetically modified anthrax lethal toxin safely delivers whole HIV protein antigens into the cytosol to induce T-cell immunity. Proc Natl Acad Sci USA 2000; 97(14):8027-8032.
34. McEvers K, Elrefaei M, Norris P et al. Modified anthrax fusion proteins deliver HIV antigens through MHC Class I and II pathways. Vaccine 2005; 23(32):4128-4135.
35. Marshall BJ, Warren JR. Unidentified curved bacilli in the stomach of patients with gastritis and peptic ulceration. Lancet 1984; 1(8390):1311-1315.
36. Cover TL, Blaser MJ. Purification and characterization of the vacuolating toxin from helicobacter pylori. J Biol Chem 1992; 267(15):10570-10575.
37. Salama NR, Otto G, Tompkins L et al. Vacuolating cytotoxin of helicobacter pylori plays a role during colonization in a mouse model of infection. Infect Immun 2001; 69(2):730-736.
38. Telford JL, Ghiara P, Dell'Orco M et al. Gene structure of the helicobacter pylori cytotoxin and evidence of its key role in gastric disease. J Exp Med. 1994; 179(5):1653-1658.
39. Nakayama M, Kimura M, Wada A et al. helicobacter pylori VacA activates the p38/activating transcription factor 2-mediated signal pathway in AZ-521 cells. J Biol Chem 2004; 279(8):7024-7028.
40. Sewald X, Gebert-Vogl B, Prassl S et al. Integrin subunit CD18 Is the T-lymphocyte receptor for the helicobacter pylori vacuolating cytotoxin. Cell Host Microbe 2008; 3(1):20-29.
41. Gebert B, Fischer W, Weiss E et al. Helicobacter pylori vacuolating cytotoxin inhibits T-lymphocyte activation. Science 2003; 301(5636):1099-1102.
42. Boncristiano M, Paccani SR, Barone S et al. The Helicobacter pylori vacuolating toxin inhibits T-cell activation by two independent mechanisms. J Exp Med 2003; 198(12):1887-1897.

43. Sundrud MS, Torres VJ, Unutmaz D et al. Inhibition of primary human T-cell proliferation by helicobacter pylori vacuolating toxin (VacA) is independent of VacA effects on IL-2 secretion. Proc Natl Acad Sci USA 2004; 101(20):7727-7732.
44. Oswald-Richter K, Torres VJ, Sundrud MS et al. Helicobacter pylori VacA toxin inhibits human immunodeficiency virus infection of primary human T-cells. J Virol 2006; 80(23):11767-11775.
45. Molinari M, Salio M, Galli C et al. Selective inhibition of Ii-dependent antigen presentation by helicobacter pylori toxin VacA. J Exp Med 1998; 187(1):135-140.
46. Zheng PY, Jones NL. Helicobacter pylori strains expressing the vacuolating cytotoxin interrupt phagosome maturation in macrophages by recruiting and retaining TACO (coronin 1) protein. Cell Microbiol 2003; 5(1):25-40.
47. Supajatura V, Ushio H, Wada A et al. Cutting edge: VacA, a vacuolating cytotoxin of helicobacter pylori, directly activates mast cells for migration and production of proinflammatory cytokines. J Immunol 2002; 168(6):2603-2607.
48. de Bernard M, Cappon A, Pancotto L et al. The helicobacter pylori VacA cytotoxin activates RBL-2H3 cells by inducing cytosolic calcium oscillations. Cell Microbiol 2005; 7(2):191-198.
49. Bartlett JG. Antibiotic-associated diarrhea. Clin Infect Dis 1992; 15(4):573-581.
50. Banno Y, Kobayashi T, Kono H et al. Biochemical characterization and biologic actions of two toxins (D-1 and D-2) from clostridium difficile. Rev Infect Dis 1984; 6(Suppl 1):S11-20.
51. Just I, Selzer J, Wilm M et al. Glucosylation of rho proteins by clostridium difficile toxin B. Nature 1995; 375(6531):500-503.
52. Just I, Wilm M, Selzer J et al. The enterotoxin from clostridium difficile (ToxA) monoglucosylates the Rho proteins. J Biol Chem 1995; 270(23):13932-13936.
53. Lima AA, Lyerly DM, Wilkins TD et al. Effects of clostridium difficile toxins A and B in rabbit small and large intestine in vivo and on cultured cells in vitro. Infect Immun 1988; 56(3):582-588.
54. Flegel WA, Muller F, Daubener W et al. Cytokine response by human monocytes to clostridium difficile toxin A and toxin B. Infect Immun 1991; 59(10):3659-3666.
55. Linevsky JK, Pothoulakis C, Keates S et al. IL-8 release and neutrophil activation by clostridium difficile toxin-exposed human monocytes. Am J Physiol 1997; 273(6 Pt 1):G1333-1340.
56. Brito GA, Sullivan GW, Ciesla WP Jr et al. Clostridium difficile toxin A alters in vitro-adherent neutrophil morphology and function. J Infect Dis 2002; 185(9):1297-1306.
57. Mahida YR, Galvin A, Makh S et al. Effect of clostridium difficile toxin A on human colonic lamina propria cells: early loss of macrophages followed by T-cell apoptosis. Infect Immun 1998; 66(11):5462-5469.
58. Miller PD, Pothoulakis C, Baeker TR et al. Macrophage-dependent stimulation of T-cell-depleted spleen cells by clostridium difficile toxin A and calcium ionophore. Cell Immunol 1990; 126(1):155-163.
59. Calderon GM, Torres Lopez J, Lin TJ et al. Effects of toxin A from clostridium difficile on mast cell activation and survival. Infect Immun 1998; 66(6):2755-2761.
60. Prepens U, Just I, von Eichel-Streiber C et al. Inhibition of Fc epsilon-RI-mediated activation of rat basophilic leukemia cells by clostridium difficile toxin B (monoglucosyltransferase). J Biol Chem 1996; 271(13):7324-7329.
61. Wex CB, Koch G, Aktories K. Effects of clostridium difficile toxin B on activation of rat peritoneal mast cells. Naunyn Schmiedebergs Arch Pharmacol 1997; 355(3):328-334.
62. Meyer GK, Neetz A, Brandes G et al. Clostridium difficile toxins A and B directly stimulate human mast cells. Infect Immun 2007; 75(8):3868-3876.
63. Imamura K, Spriggs D, Ohno T et al. Effects of botulinum toxin Type D on secretion of tumor necrosis factor from human monocytes. Mol Cell Biol 1989; 9(5):2239-2243.
64. Jolivet-Reynaud C, Cavaillon JM, Alouf JE. Selective cytotoxicity of clostridium perfringens delta toxin on rabbit leukocytes. Infect Immun 1982; 38(3):860-864.
65. Mattoo S, Cherry JD. Molecular pathogenesis, epidemiology and clinical manifestations of respiratory infections due to bordetella pertussis and other bordetella subspecies. Clin Microbiol Rev 2005; 18(2):326-382.
66. Katada T, Tamura M, Ui M. The A protomer of islet-activating protein, pertussis toxin, as an active peptide catalyzing ADP-ribosylation of a membrane protein. Arch Biochem Biophys 1983; 224(1):290-298.
67. Moss J, Stanley SJ, Burns DL et al. Activation by thiol of the latent NAD glycohydrolase and ADP-ribosyltransferase activities of bordetella pertussis toxin (islet-activating protein). J Biol Chem 1983; 258(19):11879-11882.
68. Munoz JJ, Arai H, Bergman RK et al. Biological activities of crystalline pertussigen from bordetella pertussis. Infect Immun 1981; 33(3):820-826.
69. Carbonetti NH, Artamonova GV, Mays RM et al. Pertussis toxin plays an early role in respiratory tract colonization by bordetella pertussis. Infect Immun 2003; 71(11):6358-6366.

70. Hewlett E, Wolff J. Soluble adenylate cyclase from the culture medium of bordetella pertussis: purification and characterization. J Bacteriol 1976; 127(2):890-898.
71. Wolff J, Cook GH, Goldhammer AR et al. Calmodulin activates prokaryotic adenylate cyclase. Proc Natl Acad Sci USA 1980; 77(7):3841-3844.
72. Guermonprez P, Khelef N, Blouin E et al. The adenylate cyclase toxin of bordetella pertussis binds to target cells via the alpha(M)beta(2) integrin (CD11b/CD18). J Exp Med 2001; 193(9):1035-1044.
73. Goodwin MS, Weiss AA. Adenylate cyclase toxin is critical for colonization and pertussis toxin is critical for lethal infection by Bordetella pertussis in infant mice. Infect Immun 1990; 58(10):3445-3447.
74. Munoz JJ, Peacock MG. Action of pertussigen (pertussis toxin) on serum IgE and on Fc epsilon receptors on lymphocytes. Cell Immunol 1990; 127(2):327-336.
75. Mu HH, Sewell WA. Enhancement of interleukin-4 production by pertussis toxin. Infect Immun 1993; 61(7):2834-2840.
76. Roberts M, Bacon A, Rappuoli R et al. A mutant pertussis toxin molecule that lacks ADP-ribosyltransferase activity, PT-9K/129G, is an effective mucosal adjuvant for intranasally delivered proteins. Infect Immun 1995; 63(6):2100-2108.
77. Sewell WA, Munoz JJ, Vadas MA. Enhancement of the intensity, persistence and passive transfer of delayed-type hypersensitivity lesions by pertussigen in mice. J Exp Med 1983; 157(6):2087-2096.
78. Ryan M, McCarthy L, Rappuoli R et al. Pertussis toxin potentiates Th1 and Th2 responses to co-injected antigen: adjuvant action is associated with enhanced regulatory cytokine production and expression of the costimulatory molecules B7-1, B7-2 and CD28. Int Immunol 1998; 10(5):651-662.
79. Chen X, Howard OM, Oppenheim JJ. Pertussis toxin by inducing IL-6 promotes the generation of IL-17-producing CD4 cells. J Immunol 2007; 178(10):6123-6129.
80. Hofstetter HH, Grau C, Buttmann M et al. The PLPp-specific T-cell population promoted by pertussis toxin is characterized by high frequencies of IL-17-producing cells. Cytokine 2007; 40(1):35-43.
81. Munoz JJ, Bernard CC, Mackay IR. Elicitation of experimental allergic encephalomyelitis (EAE) in mice with the aid of pertussigen. Cell Immunol 1984; 83(1):92-100.
82. Linthicum DS, Munoz JJ, Blaskett A. Acute experimental autoimmune encephalomyelitis in mice. I adjuvant action of bordetella pertussis is due to vasoactive amine sensitization and increased vascular permeability of the central nervous system. Cell Immunol 1982; 73(2):299-310.
83. Caspi RR, Chan CC, Leake WC et al. Experimental autoimmune uveoretinitis in mice. Induction by a single eliciting event and dependence on quantitative parameters of immunization. J Autoimmun 1990; 3(3):237-246.
84. Amiel SA. The effects of bordetella pertussis vaccine on cerebral vascular permeability. Br J Exp Pathol 1976; 57(6):653-662.
85. Gonzalo JA, Gonzalez-Garcia A, Baixeras E et al. Pertussis toxin interferes with superantigen-induced deletion of peripheral T-cells without affecting T-cell activation in vivo. Inhibition of deletion and associated programmed cell death depends on ADP-ribosyltransferase activity. J Immunol 1994; 152(9):4291-4299.
86. Chen X, Winkler Pickett RT, Carbonetti NH et al. Pertussis toxin as an adjuvant suppresses the number and function of CD4+CD25+ T regulatory cells. Eur J Immunol 2006; 36(3):671-680.
87. Cassan C, Piaggio E, Zappulla JP et al. Pertussis toxin reduces the number of splenic Foxp3+ regulatory T-cells. J Immunol 2006; 177(3):1552-1560.
88. Hou W, Wu Y, Sun S et al. Pertussis toxin enhances Th1 responses by stimulation of dendritic cells. J Immunol 2003; 170(4):1728-1736.
89. Ben-Nun A, Mendel I, Kerlero de Rosbo N. Immunomodulation of murine experimental autoimmune encephalomyelitis by pertussis toxin: the protective activity, but not the disease-enhancing activity, can be attributed to the nontoxic B-oligomer. Proc Assoc Am Physicians 1997; 109(2):120-125.
90. Mielcarek N, Riveau G, Remoue F et al. Homologous and heterologous protection after single intranasal administration of live attenuated recombinant bordetella pertussis. Nat Biotechnol 1998; 16(5):454-457.
91. Ausiello CM, Fedele G, Urbani F et al. Native and genetically inactivated pertussis toxins induce human dendritic cell maturation and synergize with lipopolysaccharide in promoting T helper Type 1 responses. J Infect Dis 2002; 186(3):351-360.
92. Bagley KC, Abdelwahab SF, Tuskan RG et al. Pertussis toxin and the adenylate cyclase toxin from bordetella pertussis activate human monocyte-derived dendritic cells and dominantly inhibit cytokine production through a cAMP-dependent pathway. J Leukoc Biol 2002; 72(5):962-969.
93. He J, Gurunathan S, Iwasaki A et al. Primary role for Gi protein signaling in the regulation of interleukin 12 production and the induction of T helper cell Type 1 responses. J Exp Med 2000; 191(9):1605-1610.

94. Tonon S, Goriely S, Aksoy E et al. Bordetella pertussis toxin induces the release of inflammatory cytokines and dendritic cell activation in whole blood: Impaired responses in human newborns. Eur J Immunol 2002; 32(11):3118-3125.
95. Carbonetti NH, Artamonova GV, Van Rooijen N et al. Pertussis toxin targets airway macrophages to promote bordetella pertussis infection of the respiratory tract. Infect Immun 2007; 75(4):1713-1720.
96. Meade BD, Kind PD, Manclark CR. Lymphocytosis-promoting factor of bordetella pertussis alters mononuclear phagocyte circulation and response to inflammation. Infect Immun 1984; 46(3):733-739.
97. Zhang X, Morrison DC. Pertussis toxin-sensitive factor differentially regulates lipopolysaccharide-induced tumor necrosis factor-alpha and nitric oxide production in mouse peritoneal macrophages. J Immunol 1993; 150(3):1011-1018.
98. He YX, Hewlett E, Temeles D et al. Inhibition of interleukin 3 and colony-stimulating factor 1-stimulated marrow cell proliferation by pertussis toxin. Blood 1988; 71(5):1187-1195.
99. Wang SW, Parhar K, Chiu KJ et al. Pertussis toxin promotes macrophage survival through inhibition of acid sphingomyelinase and activation of the phosphoinositide 3-kinase/protein kinase B pathway. Cell Signal 2007; 19(8):1772-1783.
100. Mielcarek N, Hornquist EH, Johansson BR et al. Interaction of Bordetella pertussis with mast cells, modulation of cytokine secretion by pertussis toxin. Cell Microbiol 2001; 3(3):181-188.
101. Thomazzi SM, Souza MH, Melo Filho AA et al. Pertussis toxin from bordetella pertussis blocks neutrophil migration and neutrophil-dependent edema in response to inflammation. Braz J Med Biol Res 1995; 28(1):120-124.
102. Carbonetti NH, Artamonova GV, Andreasen C et al. Pertussis toxin and adenylate cyclase toxin provide a one-two punch for establishment of bordetella pertussis infection of the respiratory tract. Infect Immun 2005; 73(5):2698-2703.
103. Spangrude GJ, Sacchi F, Hill HR et al. Inhibition of lymphocyte and neutrophil chemotaxis by pertussis toxin. J Immunol 1985; 135(6):4135-4143.
104. Morse JH, Kong AS, Lindenbaum J et al. The mitogenic effect of the lymphocytosis promoting factor from Bordetella pertussis on human lymphocytes. J Clin Invest 1977; 60(3):683-692.
105. Tonon S, Badran B, Benghiat FS et al. Pertussis toxin activates adult and neonatal naive human CD4+ T-lymphocytes. Eur J Immunol 2006; 36(7):1794-1804.
106. Lyons AB. Pertussis toxin pretreatment alters the in vivo cell division behaviour and survival of B-lymphocytes after intravenous transfer. Immunol Cell Biol 1997; 75(1):7-12.
107. Cyster JG, Goodnow CC. Pertussis toxin inhibits migration of B and T-lymphocytes into splenic white pulp cords. J Exp Med 1995; 182(2):581-586.
108. Hormozi K, Parton R, Coote J. Adjuvant and protective properties of native and recombinant bordetella pertussis adenylate cyclase toxin preparations in mice. FEMS Immunol Med Microbiol 1999; 23(4):273-282.
109. Orr B, Douce G, Baillie S et al. Adjuvant effects of adenylate cyclase toxin of bordetella pertussis after intranasal immunisation of mice. Vaccine 2007; 25(1):64-71.
110. Ross PJ, Lavelle EC, Mills KH et al. Adenylate cyclase toxin from bordetella pertussis synergizes with lipopolysaccharide to promote innate interleukin-10 production and enhances the induction of Th2 and regulatory T-cells. Infect Immun 2004; 72(3):1568-1579.
111. Dadaglio G, Moukrim Z, Lo-Man R et al. Induction of a polarized Th1 response by insertion of multiple copies of a viral T-cell epitope into adenylate cyclase of bordetella pertussis. Infect Immun 2000; 68(7):3867-3872.
112. Friedman RL, Fiederlein RL, Glasser L et al. Bordetella pertussis adenylate cyclase: effects of affinity-purified adenylate cyclase on human polymorphonuclear leukocyte functions. Infect Immun 1987; 55(1):135-140.
113. Weingart CL, Weiss AA. Bordetella pertussis virulence factors affect phagocytosis by human neutrophils. Infect Immun 2000; 68(3):1735-1739.
114. Mobberley-Schuman PS, Weiss AA. Influence of CR3 (CD11b/CD18) expression on phagocytosis of bordetella pertussis by human neutrophils. Infect Immun 2005; 73(11):7317-7323.
115. Khelef N, Zychlinsky A, Guiso N. Bordetella pertussis induces apoptosis in macrophages: role of adenylate cyclase-hemolysin. Infect Immun 1993; 61(10):4064-4071.
116. Gueirard P, Druilhe A, Pretolani M et al. Role of adenylate cyclase-hemolysin in alveolar macrophage apoptosis during bordetella pertussis infection in vivo. Infect Immun 1998; 66(4):1718-1725.
117. Boyd AP, Ross PJ, Conroy H et al. Bordetella pertussis adenylate cyclase toxin modulates innate and adaptive immune responses: distinct roles for acylation and enzymatic activity in immunomodulation and cell death. J Immunol 2005; 175(2):730-738.
118. Pearson RD, Symes P, Conboy M et al. Inhibition of monocyte oxidative responses by bordetella pertussis adenylate cyclase toxin. J Immunol 1987; 139(8):2749-2754.

119. Njamkepo E, Pinot F, Francois D et al. Adaptive responses of human monocytes infected by bordetella pertussis: the role of adenylate cyclase hemolysin. J Cell Physiol 2000; 183(1):91-99.
120. Siciliano NA, Skinner JA, Yuk MH. Bordetella bronchiseptica modulates macrophage phenotype leading to the inhibition of CD4+ T-cell proliferation and the initiation of a Th17 immune response. J Immunol 2006; 177(10):7131-7138.
121. Skinner JA, Reissinger A, Shen H et al. Bordetella Type III secretion and adenylate cyclase toxin synergize to drive dendritic cells into a semimature state. J Immunol 2004; 173(3):1934-1940.
122. Spensieri F, Fedele G, Fazio C et al. Bordetella pertussis inhibition of interleukin-12 (IL-12) p70 in human monocyte-derived dendritic cells blocks IL-12 p35 through adenylate cyclase toxin-dependent cyclic AMP induction. Infect Immun 2006; 74(5):2831-2838.
123. Carbonetti NH, Tuskan RG, Lewis GK. Stimulation of HIV gp120-specific cytolytic T-lymphocyte responses in vitro and in vivo using a detoxified pertussis toxin vector. AIDS Res Hum Retroviruses 2001; 17(9):819-827.
124. Fayolle C, Sebo P, Ladant D et al. In vivo induction of CTL responses by recombinant adenylate cyclase of Bordetella pertussis carrying viral CD8+ T-cell epitopes. J Immunol 1996; 156(12):4697-4706.
125. Saron MF, Fayolle C, Sebo P et al. Anti-viral protection conferred by recombinant adenylate cyclase toxins from bordetella pertussis carrying a CD8+ T-cell epitope from lymphocytic choriomeningitis virus. Proc Natl Acad Sci USA 1997; 94(7):3314-3319.
126. Fayolle C, Ladant D, Karimova G et al. Therapy of murine tumors with recombinant bordetella pertussis adenylate cyclase carrying a cytotoxic T-cell epitope. J Immunol 1999; 162(7):4157-4162.
127. Sebo P, Moukrim Z, Kalhous M et al. In vivo induction of CTL responses by recombinant adenylate cyclase of bordetella pertussis carrying multiple copies of a viral CD8(+) T-cell epitope. FEMS Immunol Med Microbiol 1999; 26(2):167-173.
128. Fayolle C, Osickova A, Osicka R et al. Delivery of multiple epitopes by recombinant detoxified adenylate cyclase of bordetella pertussis induces protective antiviral immunity. J Virol 2001; 75(16):7330-7338.
129. Loucka J, Schlecht G, Vodolanova J et al. Delivery of a MalE CD4(+)-T-cell epitope into the major histocompatibility complex class II antigen presentation pathway by bordetella pertussis adenylate cyclase. Infect Immun 2002; 70(2):1002-1005.
130. Geoffroy C, Gaillard JL, Alouf JE et al. Purification, characterization and toxicity of the sulfhydryl-activated hemolysin listeriolysin O from listeria monocytogenes. Infect Immun 1987; 55(7):1641-1646.
131. Lety MA, Frehel C, Dubail I et al. Identification of a PEST-like motif in listeriolysin O required for phagosomal escape and for virulence in listeria monocytogenes. Mol Microbiol 2001; 39(5):1124-1139.
132. Guzman CA, Domann E, Rohde M et al. Apoptosis of mouse dendritic cells is triggered by listeriolysin, the major virulence determinant of listeria monocytogenes. Mol Microbiol 1996; 20(1):119-126.
133. Carrero JA, Calderon B, Unanue ER. Listeriolysin O from listeria monocytogenes is a lymphocyte apoptogenic molecule. J Immunol 2004; 172(8):4866-4874.
134. Nishibori T, Xiong H, Kawamura I et al. Induction of cytokine gene expression by listeriolysin O and roles of macrophages and NK cells. Infect Immun 1996; 64(8):3188-3195.
135. Kohda C, Kawamura I, Baba H et al. Dissociated linkage of cytokine-inducing activity and cytotoxicity to different domains of listeriolysin O from listeria monocytogenes. Infect Immun 2002; 70(3):1334-1341.
136. Nomura T, Kawamura I, Tsuchiya K et al. Essential role of interleukin-12 (IL-12) and IL-18 for gamma interferon production induced by listeriolysin O in mouse spleen cells. Infect Immun 2002; 70(3):1049-1055.
137. Sibelius U, Schulz EC, Rose F et al. Role of Listeria monocytogenes exotoxins listeriolysin and phosphatidylinositol-specific phospholipase C in activation of human neutrophils. Infect Immun 1999; 67(3):1125-1130.
138. Darji A, Stockinger B, Wehland J et al. T-cell anergy induced by antigen presenting cells treated with the hemolysin of listeria monocytogenes. Immunol Lett 1997; 57(1-3):33-37.
139. Yamamoto K, Kawamura I, Tominaga T et al. Listeriolysin O, a cytolysin derived from Listeria monocytogenes, inhibits generation of ovalbumin-specific Th2 immune response by skewing maturation of antigen-specific T-cells into Th1 cells. Clin Exp Immunol 2005; 142(2):268-274.
140. Yamamoto K, Kawamura I, Tominaga T et al. Listeriolysin O derived from listeria monocytogenes inhibits the effector phase of an experimental allergic rhinitis induced by ovalbumin in mice. Clin Exp Immunol 2006; 144(3):475-484.
141. Higgins DE, Shastri N, Portnoy DA. Delivery of protein to the cytosol of macrophages using Escherichia coli K-12. Mol Microbiol 1999; 31(6):1631-1641.
142. Radford KJ, Higgins DE, Pasquini S et al. A recombinant E. coli vaccine to promote MHC class I-dependent antigen presentation: Application to cancer immunotherapy. Gene Ther 2002; 9(21):1455-1463.

143. Hu PQ, Tuma-Warrino RJ, Bryan MA et al. Escherichia coli expressing recombinant antigen and listeriolysin O stimulate class I-restricted CD8+ T-cells following uptake by human APC. J Immunol 2004; 172(3):1595-1601.
144. Nitcheu-Tefit J, Dai MS, Critchley-Thorne RJ et al. Listeriolysin O expressed in a bacterial vaccine suppresses CD4+CD25 high regulatory T-cell function in vivo. J Immunol 2007; 179(3):1532-1541.
145. Mandal M, Lee KD. Listeriolysin O-liposome-mediated cytosolic delivery of macromolecule antigen in vivo: enhancement of antigen-specific cytotoxic T-lymphocyte frequency, activity and tumor protection. Biochim Biophys Acta 2002; 1563(1-2):7-17.
146. Sewell DA, Douven D, Pan ZK et al. Regression of HPV-positive tumors treated with a new listeria monocytogenes vaccine. Arch Otolaryngol Head Neck Surg 2004; 130(1):92-97.
147. Singh R, Dominiecki ME, Jaffee EM et al. Fusion to listeriolysin O and delivery by listeria monocytogenes enhances the immunogenicity of HER-2/neu and reveals subdominant epitopes in the FVB/N mouse. J Immunol 2005; 175(6):3663-3673.
148. Spangler BD. Structure and function of cholera toxin and the related Escherichia coli heat-labile enterotoxin. Microbiol Rev 1992; 56(4):622-647.
149. Sixma TK, Pronk SE, Kalk KH et al. Lactose binding to heat-labile enterotoxin revealed by X-ray crystallography. Nature 1992; 355(6360):561-564.
150. Zhang J, Van Meel JC, Pfaffendorf M et al. Cholera toxin but not pertussis toxin inhibits angiotensin II-enhanced contractions in the rat portal vein. Eur J Pharmacol 1993; 230(1):95-97.
151. Holmgren J, Lonnroth I, Svennerholm L. Tissue receptor for cholera exotoxin: Postulated structure from studies with GM1 ganglioside and related glycolipids. Infect Immun 1973; 8(2):208-214.
152. Holmgren J, Fredman P, Lindblad M et al. Rabbit intestinal glycoprotein receptor for Escherichia coli heat-labile enterotoxin lacking affinity for cholera toxin. Infect Immun 1982; 38(2):424-433.
153. Fukuta S, Magnani JL, Twiddy EM et al. Comparison of the carbohydrate-binding specificities of cholera toxin and Escherichia coli heat-labile enterotoxins LTh-I, LT-IIa and LT-IIb. Infect Immun 1988; 56(7):1748-1753.
154. Orlandi PA, Critchley DR, Fishman PH. The heat-labile enterotoxin of Escherichia coli binds to polylactosaminoglycan-containing receptors in CaCo-2 human intestinal epithelial cells. Biochemistry 1994; 33(43):12886-12895.
155. Karlsson KA, Teneberg S, Angstrom J et al. Unexpected carbohydrate cross-binding by Escherichia coli heat-labile enterotoxin. Recognition of human and rabbit target cell glycoconjugates in comparison with cholera toxin. Bioorg Med Chem 1996; 4(11):1919-1928.
156. Elson CO, Ealding W. Generalized systemic and mucosal immunity in mice after mucosal stimulation with cholera toxin. J Immunol 1984; 132(6):2736-2741.
157. Clements JD, Hartzog NM, Lyon FL. Adjuvant activity of Escherichia coli heat-labile enterotoxin and effect on the induction of oral tolerance in mice to unrelated protein antigens. Vaccine 1988; 6(3):269-277.
158. Vajdy M, Lycke NY. Cholera toxin adjuvant promotes long-term immunological memory in the gut mucosa to unrelated immunogens after oral immunization. Immunology 1992; 75(3):488-492.
159. Wilson AD, Bailey M, Williams NA et al. The in vitro production of cytokines by mucosal lymphocytes immunized by oral administration of keyhole limpet hemocyanin using cholera toxin as an adjuvant. Eur J Immunol 1991; 21(10):2333-2339.
160. Takahashi I, Marinaro M, Kiyono H et al. Mechanisms for mucosal immunogenicity and adjuvancy of Escherichia coli labile enterotoxin. J Infect Dis 1996; 173(3):627-635.
161. Yamamoto M, Kiyono H, Kweon MN et al. Enterotoxin adjuvants have direct effects on T-cells and antigen-presenting cells that result in either interleukin-4-dependent or -independent immune responses. J Infect Dis 2000; 182(1):180-190.
162. Boyaka PN, Ohmura M, Fujihashi K et al. Chimeras of labile toxin one and cholera toxin retain mucosal adjuvanticity and direct Th cell subsets via their B subunit. J Immunol 2003; 170(1):454-462.
163. Anastassiou ED, Yamada H, Francis ML et al. Effects of cholera toxin on human B-cells. cholera toxin induces B-cell surface DR expression while it inhibits anti-mu antibody-induced cell proliferation. J Immunol 1990; 145(8):2375-2380.
164. Arce S, Nawar HF, Muehlinghaus G et al. In vitro induction of immunoglobulin A (IgA)- and IgM-secreting plasma blasts by cholera toxin depends on T-cell help and is mediated by CD154 up-regulation and inhibition of gamma interferon synthesis. Infect Immun 2007; 75(3):1413-1423.
165. Cong Y, Weaver CT, Elson CO. The mucosal adjuvanticity of cholera toxin involves enhancement of costimulatory activity by selective up-regulation of B7.2 expression. J Immunol 1997; 159(11):5301-5308.
166. Martin M, Sharpe A, Clements JD et al. Role of B7 costimulatory molecules in the adjuvant activity of the heat-labile enterotoxin of Escherichia coli. J Immunol 2002; 169(4):1744-1752.

167. Agren LC, Ekman L, Lowenadler B et al. Genetically engineered nontoxic vaccine adjuvant that combines B-cell targeting with immunomodulation by cholera toxin A1 subunit. J Immunol 1997; 158(8):3936-3946.
168. Petrovska L, Lopes L, Simmons CP et al. Modulation of dendritic cell endocytosis and antigen processing pathways by Escherichia coli heat-labile enterotoxin and mutant derivatives. Vaccine 2003; 21(13-14):1445-1454.
169. Bagley KC, Abdelwahab SF, Tuskan RG et al. Cholera toxin indirectly activates human monocyte-derived dendritic cells in vitro through the production of soluble factors, including prostaglandin E(2) and nitric oxide. Clin Vaccine Immunol 2006; 13(1):106-115.
170. Gagliardi MC, Sallusto F, Marinaro M et al. Cholera toxin induces maturation of human dendritic cells and licences them for Th2 priming. Eur J Immunol 2000; 30(8):2394-2403.
171. de Jong EC, Vieira PL, Kalinski P et al. Microbial compounds selectively induce Th1 cell-promoting or Th2 cell-promoting dendritic cells in vitro with diverse th cell-polarizing signals. J Immunol 2002; 168(4):1704-1709.
172. Martin M, Metzger DJ, Michalek SM et al. Distinct cytokine regulation by cholera toxin and Type II heat-labile toxins involves differential regulation of CD40 ligand on CD4(+) T-cells. Infect Immun 2001; 69(7):4486-4492.
173. Lavelle EC, McNeela E, Armstrong ME et al. Cholera toxin promotes the induction of regulatory T-cells specific for bystander antigens by modulating dendritic cell activation. J Immunol 2003; 171(5):2384-2392.
174. Lycke N, Tsuji T, Holmgren J. The adjuvant effect of vibrio cholerae and Escherichia coli heat-labile enterotoxins is linked to their ADP-ribosyltransferase activity. Eur J Immunol 1992; 22(9):2277-2281.
175. Douce G, Turcotte C, Cropley I et al. Mutants of Escherichia coli heat-labile toxin lacking ADP-ribosyltransferase activity act as nontoxic, mucosal adjuvants. Proc Natl Acad Sci USA 1995; 92(5):1644-1648.
176. de Haan L, Verweij WR, Feil IK et al. Mutants of the Escherichia coli heat-labile enterotoxin with reduced ADP-ribosylation activity or no activity retain the immunogenic properties of the native holotoxin. Infect Immun 1996; 64(12):5413-5416.
177. Yamamoto S, Takeda Y, Yamamoto M et al. Mutants in the ADP-ribosyltransferase cleft of cholera toxin lack diarrheagenicity but retain adjuvanticity. J Exp Med 1997; 185(7):1203-1210.
178. Richards CM, Shimeld C, Williams NA et al. Induction of mucosal immunity against herpes simplex virus Type 1 in the mouse protects against ocular infection and establishment of latency. J Infect Dis 1998; 177(6):1451-1457.
179. Richards CM, Aman AT, Hirst TR et al. Protective mucosal immunity to ocular herpes simplex virus Type 1 infection in mice by using Escherichia coli heat-labile enterotoxin B subunit as an adjuvant. J Virol 2001; 75(4):1664-1671.
180. Douce G, Fontana M, Pizza M et al. Intranasal immunogenicity and adjuvanticity of site-directed mutant derivatives of cholera toxin. Infect Immun 1997; 65(7):2821-2828.
181. Verweij WR, de Haan L, Holtrop M et al. Mucosal immunoadjuvant activity of recombinant Escherichia coli heat-labile enterotoxin and its B subunit: induction of systemic IgG and secretory IgA responses in mice by intranasal immunization with influenza virus surface antigen. Vaccine 1998; 16(20):2069-2076.
182. Park SJ, Chun SK, Kim PH. Intraperitoneal delivery of cholera toxin B subunit enhances systemic and mucosal antibody responses. Mol Cells 2003; 16(1):106-112.
183. Millar DG, Hirst TR, Snider DP. Escherichia coli heat-labile enterotoxin B subunit is a more potent mucosal adjuvant than its vlosely related homologue, the B subunit of cholera toxin. Infect Immun 2001; 69(5):3476-3482.
184. Wu HY, Russell MW. Induction of mucosal immunity by intranasal application of a streptococcal surface protein antigen with the cholera toxin B subunit. Infect Immun 1993; 61(1):314-322.
185. Hirabayashi Y, Tamura SI, Shimada K et al. Involvement of antigen-presenting cells in the enhancement of the in vitro antibody responses by cholera toxin B subunit. Immunology 1992; 75(3):493-498.
186. Maeyama J, Isaka M, Yasuda Y et al. Cytokine responses to recombinant cholera toxin B subunit produced by bacillus brevis as a mucosal adjuvant. Microbiol Immunol 2001; 45(2):111-117.
187. Burkart V, Kim YE, Hartmann B et al. Cholera toxin B pretreatment of macrophages and monocytes diminishes their proinflammatory responsiveness to lipopolysaccharide. J Immunol 2002; 168(4):1730-1737.
188. Turcanu V, Hirst TR, Williams NA. Modulation of human monocytes by Escherichia coli heat-labile enterotoxin B-subunit; altered cytokine production and its functional consequences. Immunology 2002; 106(3):316-325.
189. Li TK, Fox BS. Cholera toxin B subunit binding to an antigen-presenting cell directly costimulates cytokine production from a T-cell clone. Int Immunol 1996; 8(12):1849-1856.

190. Millar DG, Hirst TR. Cholera toxin and Escherichia coli enterotoxin B-subunits inhibit macrophage-mediated antigen processing and presentation: Evidence for antigen persistence in nonacidic recycling endosomal compartments. Cell Microbiol 2001; 3(5):311-329.
191. Matousek MP, Nedrud JG, Harding CV. Distinct effects of recombinant cholera toxin B subunit and holotoxin on different stages of class II MHC antigen processing and presentation by macrophages. J Immunol 1996; 156(11):4137-4145.
192. Nashar TO, Betteridge ZE, Mitchell RN. Evidence for a role of ganglioside GM1 in antigen presentation: binding enhances presentation of Escherichia coli enterotoxin B subunit (EtxB) to CD4(+) T-cells. Int Immunol 2001; 13(4):541-551.
193. George-Chandy A, Eriksson K, Lebens M et al. Cholera toxin B subunit as a carrier molecule promotes antigen presentation and increases CD40 and CD86 expression on antigen-presenting cells. Infect Immun 2001; 69(9):5716-5725.
194. Isomura I, Yasuda Y, Tsujimura K et al. Recombinant cholera toxin B subunit activates dendritic cells and enhances antitumor immunity. Microbiol Immunol 2005; 49(1):79-87.
195. Francis ML, Ryan J, Jobling MG et al. Cyclic AMP-independent effects of cholera toxin on B-cell activation. II. Binding of ganglioside GM1 induces B-cell activation. J Immunol 1992; 148(7):1999-2005.
196. Nashar TO, Hirst TR, Williams NA. Modulation of B-cell activation by the B subunit of Escherichia coli enterotoxin: receptor interaction up-regulates MHC class II, B7, CD40, CD25 and ICAM-1. Immunology 1997; 91(4):572-578.
197. Schnitzler AC, Burke JM, Wetzler LM. Induction of cell signaling events by the cholera toxin B subunit in antigen-presenting cells. Infect Immun 2007; 75(6):3150-3159.
198. Bone H, Eckholdt S, Williams NA. Modulation of B-lymphocyte signalling by the B subunit of Escherichia coli heat-labile enterotoxin. Int Immunol 2002; 14(6):647-658.
199. Woogen SD, Ealding W, Elson CO. Inhibition of murine lymphocyte proliferation by the B subunit of cholera toxin. J Immunol 1987; 139(11):3764-3770.
200. Truitt RL, Hanke C, Radke J et al. Glycosphingolipids as novel targets for T-cell suppression by the B subunit of recombinant heat-labile enterotoxin. Infect Immun 1998; 66(4):1299-1308.
201. Lycke NY. Cholera toxin promotes B-cell isotype switching by two different mechanisms. cAMP induction augments germ-line Ig H-chain RNA transcripts whereas membrane ganglioside GM1-receptor binding enhances later events in differentiation. J Immunol 1993; 150(11):4810-4821.
202. Kim PH, Eckmann L, Lee WJ et al. Cholera toxin and cholera toxin B subunit induce IgA switching through the action of TGF-beta 1. J Immunol 1998; 160(3):1198-1203.
203. Nashar TO, Webb HM, Eaglestone S et al. Potent immunogenicity of the B subunits of Escherichia coli heat-labile enterotoxin: receptor binding is essential and induces differential modulation of lymphocyte subsets. Proc Natl Acad Sci USA 1996; 93(1):226-230.
204. Yankelevich B, Soldatenkov VA, Hodgson J et al. Differential induction of programmed cell death in CD8+ and CD4+ T-cells by the B subunit of cholera toxin. Cell Immunol 1996; 168(2):229-234.
205. Wang M, Bregenholt S, Petersen JS. The cholera toxin B subunit directly costimulates antigen-primed CD4+ T-cells ex vivo. Scand J Immunol 2003; 58(3):342-349.
206. Apostolaki M, Williams NA. Nasal delivery of antigen with the B subunit of Escherichia coli heat-labile enterotoxin augments antigen-specific T-cell clonal expansion and differentiation. Infect Immun 2004; 72(7):4072-4080.
207. Viola A, Schroeder S, Sakakibara Y et al. T-lymphocyte costimulation mediated by reorganization of membrane microdomains. Science 1999; 283(5402):680-682.
208. Sun JB, Rask C, Olsson T et al. Treatment of experimental autoimmune encephalomyelitis by feeding myelin basic protein conjugated to cholera toxin B subunit. Proc Natl Acad Sci USA 1996; 93(14):7196-7201.
209. Bergerot I, Ploix C, Petersen J et al. A cholera toxoid-insulin conjugate as an oral vaccine against spontaneous autoimmune diabetes. Proc Natl Acad Sci USA 1997; 94(9):4610-4614.
210. Ploix C, Bergerot I, Durand A et al. Oral administration of cholera toxin B-insulin conjugates protects NOD mice from autoimmune diabetes by inducing CD4+ regulatory T-cells. Diabetes 1999; 48(11):2150-2156.
211. Tarkowski A, Sun JB, Holmdahl R et al. Treatment of experimental autoimmune arthritis by nasal administration of a Type II collagen-cholera toxoid conjugate vaccine. Arthritis Rheum 1999; 42(8):1628-1634.
212. Sun JB, Xiao BG, Lindblad M et al. Oral administration of cholera toxin B subunit conjugated to myelin basic protein protects against experimental autoimmune encephalomyelitis by inducing transforming growth factor-beta-secreting cells and suppressing chemokine expression. Int Immunol 2000; 12(10):1449-1457.
213. Phipps PA, Stanford MR, Sun JB et al. Prevention of mucosally induced uveitis with a HSP60-derived peptide linked to cholera toxin B subunit. Eur J Immunol 2003; 33(1):224-232.

214. Rask C, Holmgren J, Fredriksson M et al. Prolonged oral treatment with low doses of allergen conjugated to cholera toxin B subunit suppresses immunoglobulin E antibody responses in sensitized mice. Clin Exp Allergy 2000; 30(7):1024-1032.
215. Luross JA, Heaton T, Hirst TR et al. Escherichia coli heat-labile enterotoxin B subunit prevents autoimmune arthritis through induction of regulatory CD4+ T-cells. Arthritis Rheum 2002; 46(6):1671-1682.
216. Ola TO, Williams NA. Protection of nonobese diabetic mice from autoimmune diabetes by Escherichia coli heat-labile enterotoxin B subunit. Immunology 2006; 117(2):262-270.
217. Tamura S, Hatori E, Tsuruhara T et al. Suppression of delayed-type hypersensitivity and IgE antibody responses to ovalbumin by intranasal administration of Escherichia coli heat-labile enterotoxin B subunit-conjugated ovalbumin. Vaccine 1997; 15(2):225-229.
218. George Chandy A, Hultkrantz S, Raghavan S et al. Oral tolerance induction by mucosal administration of cholera toxin B-coupled antigen involves T-cell proliferation in vivo and is not affected by depletion of CD25+ T-cells. Immunology 2006; 118(3):311-320.
219. Sun JB, Cuburu N, Blomquist M et al. Sublingual tolerance induction with antigen conjugated to cholera toxin B subunit induces Foxp3+CD25+CD4+ regulatory T-cells and suppresses delayed-type hypersensitivity reactions. Scand J Immunol 2006; 64(3):251-259.
220. Sun JB, Raghavan S, Sjoling A et al. Oral tolerance induction with antigen conjugated to cholera toxin B subunit generates both Foxp3+CD25+ and Foxp3-CD25- CD4+ regulatory T-cells. J Immunol 2006; 177(11):7634-7644.
221. Ong KW, Wilson AD, Hirst TR et al. The B subunit of Escherichia coli heat-labile enterotoxin enhances CD8+ cytotoxic-T-lymphocyte killing of Epstein-Barr virus-infected cell lines. J Virol 2003; 77(7):4298-4305.
222. Loregian A, Papini E, Satin B et al. Intranuclear delivery of an antiviral peptide mediated by the B subunit of Escherichia coli heat-labile enterotoxin. Proc Natl Acad Sci USA 1999; 96(9):5221-5226.
223. De Haan L, Hearn AR, Rivett AJ et al. Enhanced delivery of exogenous peptides into the class I antigen processing and presentation pathway. Infect Immun 2002; 70(6):3249-3258.
224. Facciabene A, Aurisicchio L, Elia L et al. Vectors encoding carcinoembryonic antigen fused to the B subunit of heat-labile enterotoxin elicit antigen-specific immune responses and antitumor effects. Vaccine 2007; 26(1):47-58.
225. Hajishengallis G, Hollingshead SK, Koga T et al. Mucosal immunization with a bacterial protein antigen genetically coupled to cholera toxin A2/B subunits. J Immunol 1995; 154(9):4322-4332.
226. Stanford M, Whittall T, Bergmeier LA et al. Oral tolerization with peptide 336-351 linked to cholera toxin B subunit in preventing relapses of uveitis in behcet's disease. Clin Exp Immunol 2004; 137(1):201-208.
227. Paananen A, Mikkola R, Sareneva T et al. Inhibition of human natural killer cell activity by cereulide, an emetic toxin from bacillus cereus. Clin Exp Immunol 2002; 129(3):420-428.
228. Stenmark H, Olsnes S, Madshus IH. Elimination of the disulphide bridge in fragment B of diphtheria toxin: effect on membrane insertion, channel formation and ATP binding. Mol Microbiol 1991; 5(3):595-606.
229. Shaw CA, Starnbach MN. Stimulation of CD8+ T-cells following diphtheria toxin-mediated antigen delivery into dendritic cells. Infect Immun 2006; 74(2):1001-1008.
230. Nawar HF, Arce S, Russell MW et al. Mutants of Type II heat-labile enterotoxin LT-IIa with altered ganglioside-binding activities and diminished toxicity are potent mucosal adjuvants. Infect Immun 2007; 75(2):621-633.
231. Arce S, Nawar HF, Russell MW et al. Differential binding of Escherichia coli enterotoxins LT-IIa and LT-IIb and of cholera toxin elicits differences in apoptosis, proliferation and activation of lymphoid cells. Infect Immun 2005; 73(5):2718-2727.
232. Hajishengallis G, Nawar H, Tapping RI et al. The Type II heat-labile enterotoxins LT-IIa and LT-IIb and their respective B pentamers differentially induce and regulate cytokine production in human monocytic cells. Infect Immun 2004; 72(11):6351-6358.
233. Hajishengallis G, Tapping RI, Martin MH et al. Toll-like receptor 2 mediates cellular activation by the B subunits of Type II heat-labile enterotoxins. Infect Immun 2005; 73(3):1343-1349.
234. Svensson LA, Tarkowski A, Thelestam M et al. The impact of haemophilus ducreyi cytolethal distending toxin on cells involved in immune response. Microb Pathog 2001; 30(3):157-166.
235. Xu T, Lundqvist A, Ahmed HJ et al. Interactions of haemophilus ducreyi and purified cytolethal distending toxin with human monocyte-derived dendritic cells, macrophages and CD4+ T-cells. Microbes Infect 2004; 6(13):1171-1181.
236. Coutanceau E, Decalf J, Martino A et al. Selective suppression of dendritic cell functions by mycobacterium ulcerans toxin mycolactone. J Exp Med 2007; 204(6):1395-1403.

237. Torrado E, Adusumilli S, Fraga AG et al. Mycolactone-mediated inhibition of tumor necrosis factor production by macrophages infected with mycobacterium ulcerans has implications for the control of infection. Infect Immun 2007; 75(8):3979-3988.
238. Blocker D, Berod L, Fluhr JW et al. Pasteurella multocida toxin (PMT) activates RhoGTPases, induces actin polymerization and inhibits migration of human dendritic cells, but does not influence macropinocytosis. Int Immunol 2006; 18(3):459-464.
239. Bagley KC, Abdelwahab SF, Tuskan RG et al. Pasteurella multocida toxin activates human monocyte-derived and murine bone marrow-derived dendritic cells in vitro but suppresses antibody production in vivo. Infect Immun 2005; 73(1):413-421.
240. Donnelly JJ, Ulmer JB, Hawe LA et al. Targeted delivery of peptide epitopes to class I major histocompatibility molecules by a modified pseudomonas exotoxin. Proc Natl Acad Sci USA 1993; 90(8):3530-3534.
241. Lippolis JD, Denis-Mize KS, Brinckerhoff LH et al. Pseudomonas exotoxin-mediated delivery of exogenous antigens to MHC class I and class II processing pathways. Cell Immunol 2000; 203(2):75-83.
242. Holt PS, Misfeldt ML. Biological effects of pseudomonas aeruginosa exotoxin A: lymphoproliferation of T-lymphocytes in athymic mice. Eur J Epidemiol 1988; 4(1):25-32.
243. Zehavi-Willner T. Induction of murine cytolytic T-lymphocytes by pseudomonas aeruginosa exotoxin A. Infect Immun 1988; 56(1):213-218.
244. Michalkiewicz J, Stachowski J, Barth C et al. Effect of pseudomonas aeruginosa exotoxin A on IFN-gamma synthesis: expression of costimulatory molecules on monocytes and activity of NK cells. Immunol Lett 1999; 69(3):359-366.
245. Holt PS, Misfeldt ML. Alteration of murine immune response by pseudomonas aeruginosa exotoxin A. Infect Immun 1984; 45(1):227-233.
246. Jenkins CE, Swiatoniowski A, Issekutz AC et al. Pseudomonas aeruginosa exotoxin A induces human mast cell apoptosis by a caspase-8 and -3-dependent mechanism. J Biol Chem 2004; 279(35):37201-37207.
247. Tesh VL, Ramegowda B, Samuel JE. Purified Shiga-like toxins induce expression of proinflammatory cytokines from murine peritoneal macrophages. Infect Immun 1994; 62(11):5085-5094.
248. Ramegowda B, Tesh VL. Differentiation-associated toxin receptor modulation, cytokine production and sensitivity to Shiga-like toxins in human monocytes and monocytic cell lines. Infect Immun 1996; 64(4):1173-1180.
249. van Setten PA, Monnens LA, Verstraten RG et al. Effects of verocytotoxin-1 on nonadherent human monocytes: binding characteristics, protein synthesis and induction of cytokine release. Blood 1996; 88(1):174-183.
250. Liu J, Akahoshi T, Sasahana T et al. Inhibition of neutrophil apoptosis by verotoxin 2 derived from Escherichia coli O157:H7. Infect Immun 1999; 67(11):6203-6205.
251. Lee RS, Tartour E, van der Bruggen P et al. Major histocompatibility complex class I presentation of exogenous soluble tumor antigen fused to the B-fragment of shiga toxin. Eur J Immunol 1998; 28(9):2726-2737.
252. Noakes KL, Teisserenc HT, Lord JM et al. Exploiting retrograde transport of Shiga-like toxin 1 for the delivery of exogenous antigens into the MHC class I presentation pathway. FEBS Lett 1999; 453(1-2):95-99.
253. Haicheur N, Bismuth E, Bosset S et al. The B subunit of Shiga toxin fused to a tumor antigen elicits CTL and targets dendritic cells to allow MHC class I-restricted presentation of peptides derived from exogenous antigens. J Immunol 2000; 165(6):3301-3308.
254. King AJ, Sundaram S, Cendoroglo M et al. Shiga toxin induces superoxide production in polymorphonuclear cells with subsequent impairment of phagocytosis and responsiveness to phorbol esters. J Infect Dis 1999; 179(2):503-507.
255. LaCasse EC, Saleh MT, Patterson B et al. Shiga-like toxin purges human lymphoma from bone marrow of severe combined immunodeficient mice. Blood 1996; 88(5):1561-1567.

Chapter 2

Innate Immune Evasion by Staphylococci

Karin Ellen Veldkamp and Jos A.G. van Strijp*

Abstract

When bacteria invade the human host, they are directly confronted with a serious threat, the human innate immune system. This chapter describes the challenge that a staphylococci face and recent findings on how this bacterium counteracts the massive attack of this innate immune system. In order to survive within the human host, staphylococci have evolved a wide variety of small, excreted proteins that interfere with subsequent steps of the human innate immune system cascade.

Introduction

Especially in the last years the molecular details of the innate immune system and its mechanism of action have been revealed by numerous groups worldwide. We first have to distinguish different important feature of that innate immune system with respect to this particular bacterium.

Staphylococci are nonmotile, nonspore forming, Gram-positive, facultative anaerobic cocci, belonging to the Firmicutes. Colonies on blood agar are round convex, classically golden, but may vary from white to yellow, often surrounded by a clear zone of hemolysis. *Staphylococcus aureus* (*S. aureus*) is a normal commensal of the skin and mucous membranes. Within a few days after birth, the skin, perineal area and sometimes the gastrointestinal tract of infants are colonized by their immediate human surroundings.[1] Older children and up to 20-40 percent of the adults can become carriers but then *S. aureus* is most commonly found in the anterior nares. Organisms resident in the nares are probably the principle cause of endogenous infections.[2,3]

Many *S. aureus* strains are coated with an external polysaccharide layer, which has been called a microcapsule because it can only be visualized by electron microscopy after antibody labeling. The major cell wall component of *S. aureus* is peptidoglycan (PG). It is a polysaccharide polymer composed of alternating subunits of N-acetyl glucosamine and N-acetyl muramic acid. Pentapeptide side chains of alternating L- and D-amino acids are linked to the muramic acid residue and are cross-linked by a pentaglycin bridge attached to L-lysine of one chain and D-alanine on the other chain. This tight three-dimensional structure makes up for 40 to 60% of the dry weight of the cell wall.[4] Teichoic acids contribute 30 to 40% of the cell wall weight.[5] Some of these phosphate-containing polymers are covalently bound to PG and are called cell wall teichoic acids. In contrast to Eukarya and Archaea, bacteria start protein translation with a formyl-methionine instead of a methionine. After protein processing, formylated peptides are formed, which are partially degraded by deformylases, but a vast amount is released in the bacterial surroundings.

On its cell surface *S. aureus* express proteins that promote attachment to host proteins, such as collagen and fibronectin which are part of the extracellular matrix.[6] Most strains of *S. aureus*

*Corresponding Author: Jos A. G. van Strijp—Medical Microbiology, UMC Utrecht G04-614, Heidelberglaan 100, 3584 CX Utrecht, The Netherlands. Email: j.vanstrijp@umcutrecht.nl

Pathogen-Derived Immunomodulatory Molecules, edited by Padraic G. Fallon.

express fibronectin and fibrinogen-binding proteins. Fibrinogen/fibrin binding protein is a glycoprotein, which is also called clumping factor as the binding of fibrinogen results in clumping of whole bacteria.

During growth *S. aureus* produces a wide variety of exoproteins, especially in the postexponential phase of the growth cycle.[7] Nearly all strains secrete cytotoxins, which includes four hemolysins (alpha, beta, gamma and delta) and a group of enzymes like nucleases, proteases, lipases, hyaluronidase, catalase and collagenase. Hemolysins attack erythrocytes from various species, as well as other cells. Alpha-hemolysin is produced by most clinical isolates and acts on a wide variety of eukaryotic cell membranes by forming pores.[8,9] In a mouse model, Alpha-hemolysin expression by independent *S. aureus* strains directly correlated with their virulence. Passive and active immunization in this model showed protection against *S. aureus* challenge.[10] Beta-hemolysin, a sphingomyelinase is particularly produced by isolates from animals with for example mastitis.[11] Virtually every strain of *S. aureus* makes gamma-hemolysin which consist of two nonassociated secreted proteins, S and F component acting synergistically at membrane level by forming pores.[12,13] Delta-hemolysin is produced by 97% of *S. aureus* strains and is capable of causing damage in a variety of mammalian cells by forming channels in the cell membrane[11] and enhances priming of neutrophils and stimulates TNF-α production in monocytes.[14] Delta toxin now belongs to a class of secreted staphylococcal peptides that have a remarkable ability to recruit, activate and subsequently lyse human neutrophils, thus eliminating the main cellular defense against *S. aureus* infection. These peptides (phenol soluble modulins) are produced at high concentrations by and contribute significantly to the strains' ability to cause disease in animal models of infection. A recent study revealed a contribution of these factors to enhanced virulence of community acquired methicillin resistant Staphylococci.[15]

The main function of most staphylococcal excreted enzymes is to convert macromolecular substrates from the host tissue, like proteins, lipids, polysaccharides and nucleic acids, into low molecular weight nutrients necessary for bacterial growth and to contribute to invasiveness of *S. aureus* by breaking down extracellular matrix. Some strains produce one or more additional exoproteins, which include toxic shock syndrome toxin-1 (TSST-1), the staphylococcal enterotoxins (SEA, SEB, SEC, SED, SEE, SEG, SHE and SEI), the exfoliative toxins (ETA and ETB) and PV-leukocidin. Kuroda et al sequenced the genome of 2 methicillin-resistant *S. aureus* (MRSA) strains and this revealed several other open reading frames encoding for exoproteins including a cluster composed with enterotoxin genes encoding for enterotoxin homologues.[16] TSST-1 causes menstruation- and wound-associated toxic shock syndrome. Two former names for TSST-1 were staphylococcal pyrogenic exotoxin C and staphylococcal enterotoxin F. Staphylococcal enterotoxins (SE's) are heat-stable toxins causing staphylococcal food-poisoning by increasing dramatically intestinal peristalsis. TSST-1 and the staphylococcal enterotoxins are also known as pyrogenic toxin superantigens. Superantigens are able to activate subsets of T-lymphocytes without antigenic specificity by interacting directly with MHC class II molecules on macrophages and with the Vb chain of T-cell receptors. This causes a cytokine release leading to major systemic effects.[17,18] Molecular studies of the SEs have shown that superantigenicity and the capacity for causing food poisoning are determined by separate parts of the protein.[19,20] Exfoliative toxins (ET-A and ET-B) cause separation of the dermis at the granular cell layer resulting in extensive scalding, which is called staphylococcal scalded syndrome (SSSS).[21,22]

Panton-Valentine leukocidin (PV-leukocidin) damages membranes of leukocytes. Like gamma-hemolysin it consists of two components, which act synergistically, but leukocidin is leukocidic and not haemolytic.[12,13] The epidemiologic association of PVL bacteriophage with *S. aureus* strains from patients with necrotizing pneumonia led to research that illustrated that PVL is a key virulence factor in pulmonary infections.[23] Others however found opposite results in different mouse models and illustrated a major role for alpha-hemolysin.[24]

Host Defense and the Battle against *S. aureus* and Its Products

S. aureus has a complex array of toxins and enzymes for attacking the host, so that it becomes a highly pathogenic invader able to cause serious local and systemic infections. Once the outer barriers have been breached, staphylococci encounter additional host defenses mediated by either humoral or cellular components (Figure 1). Phagocytosis and intracellular killing by phagocytes is the major principle of clearing. *S. aureus* and is most efficiently phagocytosed after opsonization by both complement and antibodies.

Complement

Complement was first identified as a heat-labile principle in serum that "complemented" antibodies in the killing of bacteria. Now is known that complement is a system of more than 30 proteins in plasma and on cell surfaces that interact with each other in an orderly fashion that is referred to as the 'complement cascade'. Complement and complement evasion by bacteria is addressed in a different chapter ("Bacterial complement escape" by Jongerius, Ram and Rooijakkers) and in recent reviews.[25,26]

Phagocytes

Phagocytosis was discovered by Elie Metchnikoff in 1882.[27] He noted the power of certain blood cells to move toward bacteria and foreign substances and ingest them and alerted the scientific community to the importance of phagocytosis in immunity.[28] In general there are two types of phagocytic cells: granulocytes, most importantly polymorphonuclear neutrophils and mononuclear phagocytes like blood monocytes and tissue macrophages. neutrophils and monocytes, which circulate in blood and migrate through tissues, are attracted to sites of inflammation. Phagocytosis and complement activation trigger the production of cytokines. Mononuclear phagocytes are an important source of pro-inflammatory cytokines like IL-1, TNF-α, IL-6 and IFNα. Especially IL-1 and TNF-α increase expression of adhesion molecules on endothelial cells, neutrophils and monocytes.[29] Production of IL-8 by endothelial cells, monocytes and neutrophils in response to cytokines or microbial products, further modulates expression of adhesion molecules and provides a chemotactic signal that promotes emigration of neutrophils from the blood to the inflammatory site.

Distinct steps can be recognized in the emigration cascade of leukocytes.[30] Rolling of leukocytes along the vessel wall depending on a loosely adhesive contact between leukocytes and endothelial cells during which selectins (CD62) are the main adhesion molecules: P-selectin and E-selectin on the endothelial cells of the blood vessel and L-selectin on the leukocytes. Firm adhesion of leukocytes to the endothelium and activation of these leukocytes leads to flattening of the leukocyte. The integrins predominate this adhesion step. The most important integrins expressed on leukocytes are LFA-1 (CD11a/CD18), Mac-1 (CD11b/CD18), P150,95 (CD11c/CD18) and VLA-4 CD49d/CD29). Their counterstuctures on the endothelium are intracellular adhesion molecule-1 (ICAM-1) and -2 for LFA-1, ICAM-1 for Mac-1 and VCAM-1 for VLA-4.[31] Diapedesis, the actual penetration of the leukocytes between endothelial cells, is mediated by PECAM-1 on the endothelial cells binding to PECAM-1 on the leukocyte.[32]

Chemotaxis

For migration towards the actual site of infection, directed locomotion of phagocytes is triggered by specific substances, termed chemotactic factors. The phagocytes carry several receptors on their cell surface by which they are able to recognize these chemotactic factors, which are secreted by either activated host cells, activated complement or by the invading micro-organism itself to form a gradient by which phagocytes are able to determine the right direction of migration. Well known chemoattractants for phagocytes are platelet activating factor (PAF), arachidonate metabolite leukotriene B4 (LTB4), anaphylatoxin complement fragment C5a, bacterial derived formylated peptides like N-formyl-methionyl-leucyl-phenylalanine (fMLP) and chemokines like interleukine-8 (IL-8). LTB4 is formed by neutrophils, C5a is released during activation of complement when C5 is cleaved by C5 convertase into C5a and C5b to form the membrane-attack complex. PAF, originally isolated from stimulated basophils, can be produced by a variety of cells, such as neutrophils, platelets, monocytes, macrophages and endothelial cells and has been suggested as a mediator of inflammation, platelet and neutrophil activation, plasma extravasation and anaphylactic shock. N-formylated peptides are derived from bacterial protein degradation or to arise from mitochondrial proteins upon tissue damage.[33-35] IL-8 is produced by macrophages, neutrophils, endothelial cells, mesothelium. It belongs to the family of the chemokines, which are small proteins that are structurally and functionally related with chemotactic as well as cytokine-like activation properties.[36,37] The most prominent shared feature of the chemokines is a conserved motif of two cysteine pairs linked together with sulfide bonds. There are at least four families of chemokines. The two major families are subdivided on the basis of the N terminal region where two cysteine residues are separated by one amino acid , CXC or alpha chemokines, or adjacent to each other, CC or beta chemokines.[36] IL-8, the first recognized chemokine, belongs to the CXC chemokines. Whether alpha chemokines are chemotactic for neutrophils depends on whether the sequence preceding the CXC sequence contains glutamic acid-leucine-arginine. If not then the chemokine will act on lymphocytes.

The chemoattractants as well as the chemokines exert their function by binding to their receptors. These receptors belong to the superfamily of G-protein coupled seven transmembrane receptors or 'serpentine' receptors.[38] Receptor-ligand binding leads to a cascade of cellular activation, including the generation of inositol triphosphate, the release of intracellular calcium and the activation of protein kinase C (PKC), which eventually results in actin polymerization.[39,40] Chemokine-receptor signaling also activates small guanosine triphosphate-binding proteins of the Ras and Rho families.[41] Rho proteins are involved in cell motility through regulation of actin-dependent processes such as membrane ruffling, pseudopod formation and assembly of focal adhesion complexes. Thus, binding of chemoattractants and chemokines to their receptors activate multiple intracellular signaling pathways that regulate the intracellular machinery necessary to propel the cell in its chosen direction.

Bacterial products can induce chemotaxis such as formylated peptides and phenol soluble modulins. Next to that, the micro-organism or its products can induce the release of chemoattractants from host cells.[42] Endothelial cells infected with *S. aureus* express both IL-8 mRNA and the protein[43] and after stimulation with purified LTA, blood monocytes also show IL-8 mRNA expression.[44] Espersen et al studied the chemotactic activity of peptidoglycan for peripheral blood neutrophils in human serum.[45] They showed that PG was able to induce chemotactic activity in normal human serum, but not in heat-treated or C5-deficient human serum, indicating that PG treatment of serum resulted in generation of chemoattractants by activation of complement. PV-leukocidin could directly induce the release of LTB4 and IL-8 in neutrophils, as was shown by Konig et al.[46] Immunomodulatory effects of TSST-1 were seen in neutrophils. After 10 minutes incubation with TSST-1 neutrophils showed an enhanced expression for the FPR and LTB4 receptor. This enhancement was not seen after 30 minutes incubation, which according to the authors, could be due to negative feedback mechanisms.[47]

Phagocytosis and Killing

Many micro-organisms are not readily ingested by phagocytes because antiphagocytic components are present on the microbial surface. Ingestion of such organisms requires the effective neutralization of these antiphagocytic structures. This is accomplished by the deposition of specific IgG antibody or of the complement fragment C3b on the microbial surface. Optimal opsonization for phagocytosis usually requires the participation of both these substances. C3b and specific immunoglobulin function as ligands connecting the micro-organism to the phagocyte at the specific C3b receptor and Fc receptor on the cell membrane of the neutrophils. Opsonin-independent phagocytosis by neutrophils is less efficient because it depends on the contact angle between micro-organism and the surface on which it rests. So opsonin-independent phagocytosis is more efficient in small tissue spaces than on smooth open surfaces. The mechanism by which neutrophils ingest opsonized micro-organisms and enclose them in phagosomes depends on the interaction between opsonic ligands distributed homogeneously over the surface of the particle and receptors on the phagocyte membrane. This interaction results in the circumferential flow of the phagocyte membrane around the particle and its subsequent enclosure within the phagosome.[48,49] *S. aureus* eradication from the host depends mainly on opsonization, phagocytosis and subsequent killing by neutrophils.

In addition to initiating microbial ingestion, the interaction of opsonic ligands with receptors in the cell membrane results in stimulation of the membrane. This stimulation triggers degranulation and a burst of oxidative metabolism as well as a release of oxidative products and granule contents into the phagosome and extracellular milieu. The metabolic burst is characterized by markedly increased oxygen consumption by neutrophils, production of a number of active metabolites of oxygen (including superoxide, singlet oxygen, hydroxyl radical and hydrogen peroxide) and increased activity of the hexose monophosphate shunt.[50,51] Although the active metabolites of oxygen may themselves possess microbicidal activity, this activity is markedly enhanced by the granule contents, particularly myeloperoxide, present in the phagosome. Other granule-bound antibacterial molecules are defensins, azurocidin, cathepsin G and bactericidal/permeability-increasing (BPI) protein, which kill the microbes by permeabilizing the membrane. Myeloperoxide, hydrogen peroxide and a halide ion acting in conjunction form one of the neutrophils most potent microbicidal mechanisms.[52]

The importance of the active metabolites of oxygen is shown in diseases with defects in staphylococcal killing like in chronic granulomatous disease (CGD). CGD is an inherited disease characterized by severe and recurrent staphylococcal and other bacterial and fungal infections, manifested in most cases in early childhood. Phagocytic cells of CGD patients are unable to produce superoxide anions and their efficiency in bacterial killing is significantly impaired. Chronic granulomatous disease (CGD) results from a defect in any of the 4 subunits of the NADPH oxidase complex. This NADPH oxidase complex allows phagocytes to rapidly convert O_2 to superoxide anion which generates other antimicrobial reactive oxygen intermediates, such as H_2O_2, hydroxyl anion and peroxynitrite anion.[53,54]

Innate Immune Evasion

In 2000 we first described the chemotaxis inhibitory activity of staphylococci.[55] Later this activity was pinpointed to a molecule called CHIPS for chemotaxis inhibitory protein of Staphylococci.[56] Since then a lot of small excreted molecules have been identified by us and by others that act in various ways to counteract the inflammatory capacity of the innate immune system.[25,26,57-62] Here we provide a concise overview of the molecules that have recently been identified and act on specific parts of the human innate immune system.

Complement Evasion

A vast number of staphylococcal complement evading strategies have been described. Staphylococcal complement inhibitor (SCIN) is a highly efficient complement inhibitor.[63-66] It is a 10 kD, excreted protein that blocks all three complement pathways: the lectin, classical and alternative pathway. SCIN efficiently prevent phagocytosis and killing of staphylococci and C5a production. SCIN specifically acts on both surface-bound C3 convertases (C3bBb and C4b2a). SCIN has two homologues in *S. aureus* SCIN-B and SCIN-C.[67] They have the same mechanism of action as SCIN has. Most S. *aureus* strains have next to SCIN, either the SCIN-B or the SCIN-C gene. The genes for SCIN-B or the SCIN-C are located on a genomic island that also harbors the genes for Efb (extracellular fibrinogen binding protein) and Ecb (Extracellular complement binding protein) The C-terminal part of Efb as well as the Ecb molecule are potent inhibitors of C5 convertases.[25,67,68,69] Protein A is well known for its capacity to bind the Fc part of IgG. Through this binding, protein A blocks Fc-receptor mediated phagocytosis on one hand and complement activation via C1q on the other.[70] Recently Protein A also recognizes TNF-Receptor 1, a receptor for tumor-necrosis factor-alpha (TNF-α).[71] The staphylococcal superantigen-like proteins (SSLs) are

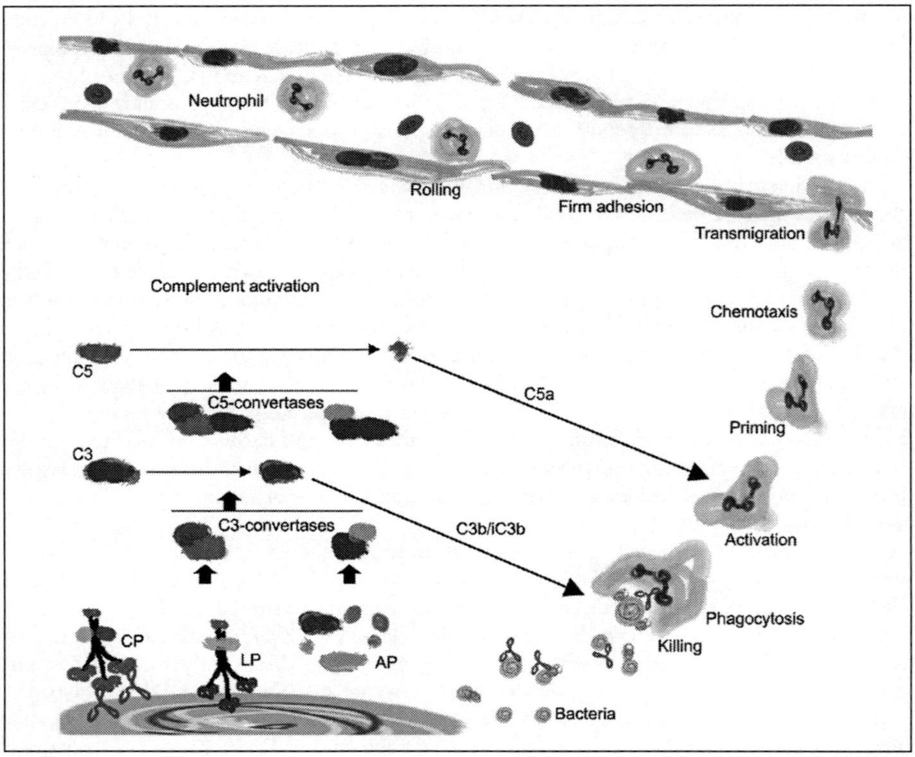

Figure 1. Acute inflammatory response. Subsequent steps in neutrophil extravasation, movement toward the site of infection and removal of invading bacteria.

close relatives of the superantigens but are located on a separate gene cluster within a 19-kb region of the pathogenicity island SaPIn2. SSL-7 (23.2 kD) binds human IgA1 and IgA2 resulting in an inhibited IgA binding to cell surface FcαRI (CD89). In addition to that, SSL7 binds complement factor C5 from human, primate, sheep, pig and rabbit serum and inhibited complement-mediated hemolysis and serum killing of a Gram-negative organism Escherichia coli.[72]

Rolling, Adhesion and Transmigration Inhibition

Yet another member of the SSL family, SSL-5 was described to inhibit the first step in Neutrophil mediated inflammatory events. SSL-5 inhibits the interaction between P-selectin on endothelial cells and P-selectin glycoprotein ligand-1 (PSGL-1) on neutrophils, thereby inhibiting neutrophil rolling on the endothelium suggesting that *S. aureus* uses SSL5 to prevent neutrophil extravasation towards the site of infection, even at the earliest possible step.[73] The interaction with sialyl-lewis-X (sLe(X)) is crucial and this is now illustrated by cocrystal studies. sLe(X) bound to a specific site on the surface of the C-terminal domain of SSL5 in a conformation identical with that bound by P-selectin.[74,75]

The next step in neutrophil extravasation is inhibited by the Extracellular adherence protein (Eap) of *S. aureus*. Eap inhibits direct interactions with the host adhesive proteins intercellular adhesion molecule 1 (ICAM-1), fibrinogen or vitronectin, Eap disrupted beta(2)-integrin and urokinase receptor mediated leukocyte adhesion in vitro. In a bacterial peritonitis mouse model, Eap-expressing *S. aureus* (or isolated Eap) elicited less neutrophil influx as compared to an Eap-negative strain.[76,77] Further Eap protects in an ICAM-1 dependent experimental autoimmune encephalomyelitis (EAE) in mice, indicating that Eap could possibly represent a treatment for autoimmune neuroinflammatory disorders such as multiple sclerosis.[78] In recent reviews many other possible interactions of Eap with numerous other ligands were described.[60]

Evading Neutrophil Chemotaxis and Activation

Leukocyte migration is a key event both in host defense against invading pathogens as well as in inflammation. Here we discuss three proteins that specifically interfere with this aspect of the innate immune system; CHIPS and FLIPr.

The very early signs of bacterial invasion, C5a and formylated peptides are recognized by the innate immune system through two related receptors on neutrophils, the C5aR and FPR. The excreted Chemotaxis inhibitory protein of *S. aureus* (CHIPS) is a 14.1 kD protein that specifically impairs the response of neutrophils and monocytes to formylated peptides and C5a.[55,56,79] CHIPS inhibits neutrophil recruitment in vitro and in vivo at high CHIPS concentrations. Also phagocyte activation by C5a and fortmylated peptides is completely blocked by CHIPS. The CHIPS gene is present in 62% of all *S. aureus* isolates. CHIPS binds specifically and directly to the C5aR and FPR and thereby totally blocks ligand binding and signal transduction. The apparent KD values of CHIPS for the C5aR and FPR were 1.1 nM and 35.4 nM, respectively.[80,81]

The anaphylatoxin C5a plays an important role in host immunity and pathological inflammatory processes. For C5a a two-site binding model was proposed, in which initial binding of C5a to the C5aR N-terminus is followed by interaction of its C-terminal tail with an effector domain on the receptor. We found that CHIPS exclusively binds the C5aR N-terminus. Deletion and mutation experiments of the C5aR N-terminus revealed that the binding site of CHIPS is contained in a short stretch of 9 amino acids, of which the aspartic acid residues at position 10, 15 and 18 plus the glycine at position 12 are crucial. Binding studies with C5aR/C3aR and C5aR/IL8RA chimeras confirmed that CHIPS binds only to the C5aR N terminus without involvement of its extracellular loops. CHIPS may provide new strategies to block the C5aR, which may lead to the development of new C5aR antagonists.[74,80,82] Next to its action on the C5aR, CHIPS also blocks the action of the formylated peptide receptor (FPR).[80,81] N-formyl peptides, such as fMet-Leu-Phe, are some of the first identified and most potent chemoattractants for phagocytic leukocytes are spill-over products of the bacterial protein production machinery. Both the C5aR and FPR belong to the superfamily of seven-transmembrane, heterotrimeric G protein-coupled receptors (GPCR). Although their ligands are highly diverse and receptor-specific, all GPCRs have a comparable architecture and share 20 to 30% amino acid homology, mainly in the transmembrane regions. Small molecules, such as fMet-Leu-Phe, bind to their GPCR in the interhelical region, while larger molecules, such as C5a, use a two-site motif for binding their GPCR.

Using different monoclonal antibodies, a set of CHIPS-spanning peptides and truncated rCHIPS molecules, the two activities of CHIPS could be separated on the molecule. The C5aR modulating capacity was assigned to a not yet completely defined region in the C-terminus of CHIPS. The FPR modulating capacity of CHIPS is located at the far N-terminal region of the molecule.[82] The two phenylalanines at position one and three are crucial for FPR blocking activity.

The receptor FPR-like-1 (FPRL1) belongs also to the family of formyl peptide receptors. Another protein from *S. aureus* impairs the neutrophil responses to FPRL1 agonists. FPRL1 inhibitory protein (FLIPr) inhibited the calcium mobilization in neutrophils stimulated with MMK-1, WKYMVM, prion-protein fragment PrP(106-126) and amyloid beta (1-42). All these ligands are known FRPL1 agonists. Stimulation with low concentrations of fMLP was partly inhibited. Directed migration was also completely prevented toward MMK-1 and partly toward fMLP. Fluorescence-labeled FLIPr efficiently bound to neutrophils, monocytes, B cells and NK cells. HEK293 cells transfected with human C5aR, FPR, FPRL1 and FPRL2 clearly showed that FLIPr directly bound to FPRL1 and, at higher concentrations, also to FPR but not to C5aR and FPRL2. FLIPr can reveal unknown inflammatory ligands crucial during *S. aureus* infections. As a novel described FPRL1 antagonist, it might lead to the development of therapeutic agents in FPRL1-mediated inflammatory components of diseases such as systemic amyloidosis, Alzheimer's and prion disease.[83]

Evading Phagocytosis and Killing

Protein A inhibits phagocytosis by binding the Fc part of IgG and blocking the access to neutrophil Fc receptors.[70] Alternatively staphylococci can inhibit IgG mediated effector functions by the excretion of staphylokinase (SAK). *S. aureus* expresses several plasminogen (PLG)-binding receptors at its surface. SA, after excretion, activates surface-bound plasminogen (PLG) into plasmin (PL). Surface-bound plasmin cleaves both IgG and C3b. We showed (226) that PL, formed by the conversion of PLG by SAK at physiological concentrations, leads to removal of these important opsonic molecules necessary for recognition by immune cells. PL attacks human IgG as well as human (i)C3b at the bacterial cell wall, causing impaired phagocytosis by human neutrophils. PL cleaves IgG at position Lys 222 and thus removes the entire Fc fragment, including the glycosylation site (Asn 297) necessary for recognition by C1q thereby inhibiting the activation of the classical pathway of complement. Also PL-activity created by PLG and SAK led to decrease of (i)C3b molecules at the staphylococcal surface. PLG+rSAK cleaved C3b in both the α-chain as well as in the β-chain. The decrease of C3b molecules will indirectly diminish C3 convertases as well as C5 convertases.

SSL-7 (23.2 kD) binds human IgA1 and IgA2 resulting in an inhibited IgA binding to cell surface FcαRI (CD89).[72] Through this mechanism *S. aureus* interferes with IgA-mediated effector functions. The three-dimensional structure of the complex of SSL7 with Fc of human IgA1 shows that SSL7 and Fc αRI bind many of the same key residues in the Fc region of human IgA, thereby explaining how SSL7 potently inhibits IgA-dependent cellular effector functions mediated by FcαRI, such as phagocytosis, degranulation and respiratory burst. Two SSL7 molecules interact with the Fc (one per heavy chain) primarily at the junction between the Cα2 and Cα3 domains. The binding site on each IgA chain is extensive, with SSL7 shielding most of the lateral surface of the Cα3 domain.[85,86]

Alfa-Defensins are peptides that are present in and are secreted by neutrophils and provide antimicrobial protection mediated by disruption of the integrity of bacterial cell walls. Staphylokinase (SAK) (16 kD) is excreted by *S. aureus* and is well-known to activate human plasminogen. Tarkowski et al showed that binding of defensins by SAK inhibits the bactericidal effect of α-defensins. Notably, the site within staphylokinase that binds α-defensins is different from its plasminogen-binding site. In vivo, staphylococcal strains producing SAK were protected against the bactericidal effect of α-defensins. Therefore SAK mediated inhibition of α-defensins is considered to be an important defense mechanism of *S. aureus*.[87]

Resistance to the metabolic burst of neutrophils can help to gain advantage within the host. All *S. aureus* species are catalase positive. Catalase inactivates toxic hydrogen peroxide and free radicals produced within phagocytic cells following ingestion of microbes[70] The production of catalase by *S. aureus* can be a defense mechanisms against killing in the phagosome. Catalase is able to convert hydrogen peroxide into nontoxic H_2O and O_2. By this convertion catalase production can counteract staphylococcal phagocytic killing, which is mediated by toxic oxygen radicals produced by neutrophils.

Staphylococcal Innate Immune Evasion

Here we have summarized a new group of immune evasion molecules in *S. aureus* (Figure 2). Besides being important in Staphylococcal pathogenicity these factors can teach us to recognize crucial steps in innate immunity pathways. Evolution seems to have tuned these factors to act on essential steps in our innate immune system and that specific element is useful in the understanding of fundamentally crucial steps of the innate immune system. From a therapeutic point of view, these proteins have a high potential in anti-inflammatory therapy. Per definition, molecules that help to evade acute innate immune mechanism are anti-inflammatory compounds. Their high specificity partially explains their nontoxic nature. Being soluble molecules, it is tempting to speculate that they could be used as injectables in severe acute inflammatory disorders. Pre-existing antibodies against all these proteins will complicate this approach but the targets that are attacked by all of these factors might prove to be the essential targets to tackle in anti-inflammatory therapy.

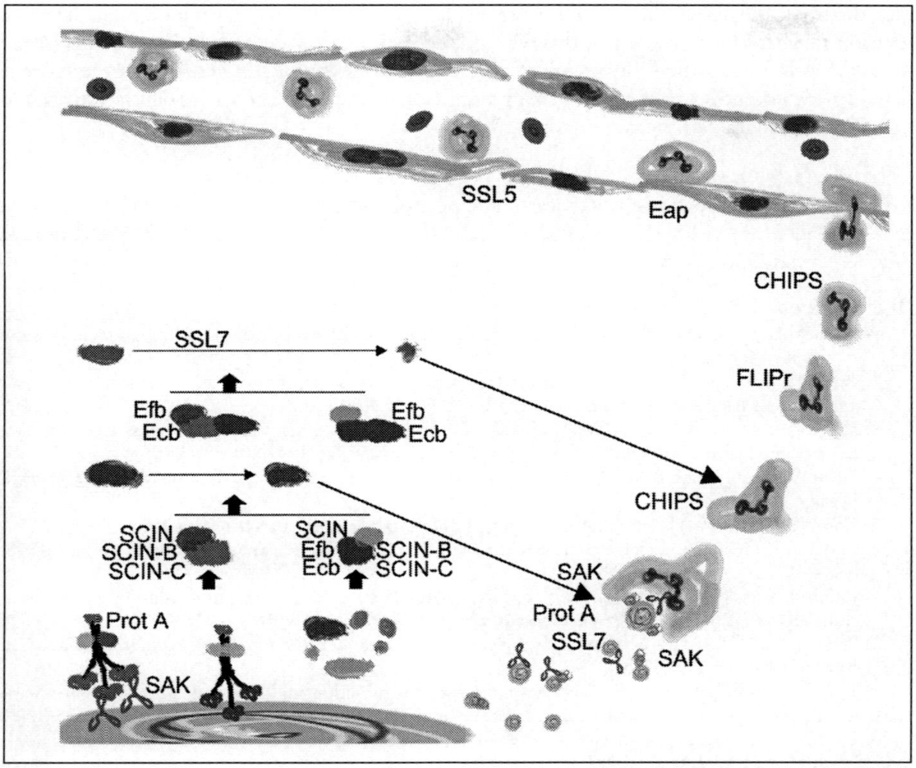

Figure 2. Staphylococcal innate immune evasion. The molecules produced by *S. aureus* that counteract the different steps of our innate immune system. SSL-5 (Staphylococcal Superantigen-like 5), Eap (Extracellular adherence protein), CHIPS (Chemotaxis inhibitory protein of Staphylococci), FLIPr (FPRL-1 inhibitory protein), SSL-7 (Staphylococcal Superantigen-like 7), Efb (Extracellular fibrinogen binding protein), Ecb (Extracellular complement binding protein), prot A (protein A), SCIN (Staphylococcal Complement inhibitor), SCIN B (Staphylococcal Complement inhibitor-B), SCIN C (Staphylococcal Complement inhibitor-C), SAK (Staphylokinase).

Bacteriophages serve as a driving force in microbial evolution, adaptation to new environments and the pathogenesis of human bacterial infections. Two newly discovered immune modulators, chemotaxis inhibitory protein of *S. aureus* (CHIPS) and staphylococcal complement inhibitor (SCIN), cluster on the conserved 3′ end of beta-hemolysin (hlb)-converting bacteriophages (betaC-phis). Since these betaC-phis also carry the genes for the immune evasion molecules staphylokinase and enterotoxin A, this 8-kb region at the 3′ end of betaC-phi represents an innate immune evasion cluster (IEC).[65]

Extensive phage dynamics (intragenomic translocation, duplication, transfer between hosts, recombination events) during infection was shown by analyzing cocolonizing and consecutive isolates of patients. This activity leads to the splitting of the strain population into various subfractions exhibiting different virulence potentials (Hlb-production and/or production of immune evasion molecules). Thus, phage-inducing conditions and strong selection for survival of the bacterial host after phage movement are typical for the infectious situation.[88] Also other immune evasion molecules are located on specific clusters, such as SaPIn2 with all the SSL's[72] and immune evasion cluster with Efb, Ecb, SCIN B/C, FLIPr.[67]

In a relatively short period of time we and others have identified an impressive amount of highly specific innate immune evasion molecules in a single micro-organism. Probably this is just the tip of the iceberg. If we can document the complete armamentarium of innate immune evasion mechanisms in *S. aureus*, this will indirectly provide an increasing insight into innate immune evasion and the fundaments of bacterial pathophysiology in general. Furthermore it will provide insight into our own innate immune system and open the way to develop smart and specific anti-inflammatory compounds.

Acknowledgements

Work described in this chapter was made possible by grants from STW (#UKG-06609), EU FP6 (STREP #512093) (TOK #517128), NGI(#050-71-028), NWO (#91206020) and the Vanderes foundation (#156).

References

1. Wu S, Shen L. Plasmid analysis and phage typing in the study of staphylococcal colonization and disease in newborn infants. Chin Med Sci J 1993; 8:157-161.
2. Perl TM, Golub JE. New approaches to reduce Staphylococcus aureus nosocomial infection rates: treating S. aureus nasal carriage. Ann Pharmacother 1998; 32:S7-16.
3. Kluytmans J, van Belkum A, Verbrugh H. Nasal carriage of Staphylococcus aureus: epidemiology, underlying mechanisms and associated risks. Clin Microbiol Rev 1997; 10:505-520.
4. Norse SI. Staphylococci. In: Davis BD, Dulbecco R, Eisen HN, et al, eds. Microbiology. Hagertown: Harper and Row, 1980:624.
5. Knox KW, Wicken AJ. Immunological properties of teichoic acids. Bacteriol Rev 1973; 37:215-257.
6. Foster TJ, Hook M. Surface protein adhesins of Staphylococcus aureus. Trends Microbiol 1998; 6:484-488.
7. Coleman G, Jakeman CM, Martin N. Patterns of total extracellular protein secretion by a number of clinically isolated strains of Staphylococcus aureus. J Gen Microbiol 1978; 107:189-192.
8. Sperber WH. The identification of staphylococci in clinical and food microbiology laboratories. Crit Rev Clin Lab Sci 1976; 7:121-184.
9. Bhakdi S, Muhly M, Fussle R. Correlation between toxin binding and hemolytic activity in membrane damage by staphylococcal alpha-toxin. Infect Immun 1984; 46:318-323.
10. Bubeck Wardenburg J, Schneewind O. Vaccine protection against Staphylococcus aureus pneumonia J Exp Med 2008; 205:287-294.
11. Möllby R. Isolation and properties of membrane damaging toxins. In: Easmon CSF, Adlam C, eds. Staphylococci and staphylococcal infections. 2 ed. London: Academic Press, 1983:619-669.
12. Prevost G, Cribier B, Couppie P et al. Panton-Valentine leucocidin and gamma-hemolysin from Staphylococcus aureus ATCC 49775 are encoded by distinct genetic loci and have different biological activities. Infect Immun 1995; 63:4121-4129.
13. Prevost G, Couppie P, Prevost P et al. Epidemiological data on Staphylococcus aureus strains producing synergohymenotropic toxins. J Med Microbiol 1995; 42:237-245.
14. Schmitz FJ, Veldkamp KE, Van Kessel KP et al. Delta-toxin from Staphylococcus aureus as a costimulator of human neutrophil oxidative burst. J Infect Dis 1997; 176:1531-1537.
15. Wang R, Braughton KR, Kretschmer D et al. Identification of novel cytolytic peptides as key virulence determinants for community-associated MRSA. Nat Med 2007; 13:1510-1514.
16. Kuroda M, Ohta T, Uchiyama I et al. Whole genome sequencing of meticillin-resistant Staphylococcus aureus. Lancet 2001; 357:1225-1240.
17. Marrack P, Kappler J. The staphylococcal enterotoxins and their relatives. Science 1990; 248:705-711.
18. Murray DL, Ohlendorf DH, Schlievert PM. Staphylococcal and streptococcal superantigens: Their role in human diseases. ASM News 1995; 61:229-235.
19. Alber G, Hammer DK, Fleischer B. Relationship between enterotoxic- and T-lymphocyte-stimulating activity of staphylococcal enterotoxin B. J Immunol 1990; 144:4501-4506.
20. Harris TO, Grossman D, Kappler JW et al. Lack of complete correlation between emetic and T-cell-stimulatory activities of staphylococcal enterotoxins. Infect Immun 1993; 61:3175-3183.
21. Bailey CJ, Lockhart BP, Redpath MB et al. The epidermolytic (exfoliative) toxins of Staphylococcus aureus. Med Microbiol Immunol 1995; 184:53-61.
22. Elias PM, Fritsch P, Dahl MV et al. Staphylococcal toxic epidermal necrolysis: pathogenesis and studies on the subcellular site of action of exfoliatin. J Invest Dermatol 1975; 65:501-512.
23. Labandeira-Rey M, Couzon F, Boisset S et al. Staphylococcus aureus Panton-Valentine Leukocidin Causes Necrotizing Pneumonia Science 2007; 315:1130-1133.
24. Bubeck Wardenburg J, Bae T, Otto M et al. Poring over pores: alpha-hemolysin and Panton-Valentine leukocidin in Staphylococcus aureus pneumonia. Nat Med 2007; 13:1405-1406.

25. Lambris JD, Ricklin D, Geisbrecht BV. Complement evasion by human pathogens. Nat Rev Microbiol 2008; 6:132-142.
26. Rooijakkers SH, van Strijp JA. Bacterial complement evasion. Mol Immunol 2007; 44:23-32.
27. Metschnikoff E. Sur la lutte des cellules de l'organisme contre l'invasion des microbes. Annales de l'institut Pasteur 2001; 1:321.
28. Heifets L. Centennial of Metchnikoff's discovery. J Reticuloendothel Soc 1982; 31:381-391.
29. Meager A. Cytokine regulation of cellular adhesion molecule expression in inflammation. Cytokine Growth Factor Rev 1999; 10:27-39.
30. Dunon D, Piali L, Imhof BA. To stick or not to stick: the new leukocyte homing paradigm. Curr Opin Cell Biol 1996; 8:714-723.
31. Carlos TM, Harlan JM. Leukocyte-endothelial adhesion molecules. Blood 1994; 84:2068-2101.
32. Muller WA, Weigl SA, Deng X et al. PECAM-1 is required for transendothelial migration of leukocytes. J Exp Med 1993; 178:449-460.
33. Marasco WA, Phan SH, Krutzsch H et al. Purification and identification of formyl-methionyl-leucyl-phenylalanine as the major peptide neutrophil chemotactic factor produced by Escherichia coli. J Biol Chem 1984; 259:5430-5439.
34. Carp H. Mitochondrial N-formylmethionyl proteins as chemoattractants for neutrophils. J Exp Med 1982; 155:264-275.
35. Schiffmann E, Corcoran BA, Wahl SM. N-formyl-methionyl peptides and chemoattractants for leukocytes. Proc Natl Acad Sci USA 1975; 72:1059.
36. Baggiolini M, Dewald B, Moser B. Interleukin-8 and related chemotactic cytokines—CXC and CC chemokines. Adv Immunol 1994; 55:97-179.
37. Graves DT, Jiang Y. Chemokines, a family of chemotactic cytokines. Crit Rev Oral Biol Med 1995; 6:109-118.
38. Kelvin DJ, Michiel DF, Johnston JA et al. Chemokines and serpentines: the molecular biology of chemokine receptors. J Leukoc Biol 1993; 54:604-612.
39. Bokoch GM. Chemoattractant signaling and leukocyte activation. Blood 1995; 86:1649-1660.
40. Premack BA, Schall TJ. Chemokine receptors: gateways to inflammation and infection. Nat Med 1996; 2:1174-1178.
41. Laudanna C, Campbell JJ, Butcher EC. Role of Rho in chemoattractant-activated leukocyte adhesion through integrins. Science 1996; 271:981-983.
42. Rot A, Henderson LE, Sowder R et al. Staphylococcus aureus tetrapeptide with high chemotactic potency and efficacy for human leukocytes. J Leukoc Biol 1989; 45:114-120.
43. Yao L, Lowy FD, Berman JW. Interleukin-8 gene expression in Staphylococcus aureus-infected endothelial cells. Infect Immun 1996; 64:3407-3409.
44. Standiford TJ, Arenberg DA, Danforth JM et al. Lipoteichoic acid induces secretion of interleukin-8 from human blood monocytes: a cellular and molecular analysis. Infect Immun 1994; 62:119-125.
45. Riber U, Espersen F, Wilkinson BJ et al. Neutrophil chemotactic activity of peptidoglycan. A comparison between Staphylococcus aureus and Staphylococcus epidermidis. APMIS 1990; 98:881-886.
46. Konig B, Prevost G, Piemont Y et al. Effects of Staphylococcus aureus leukocidins on inflammatory mediator release from human granulocytes. J Infect Dis 1995; 171:607-613.
47. Hensler T, Koller M, Geoffroy C et al. Staphylococcus aureus toxic shock syndrome toxin 1 and Streptococcus pyogenes erythrogenic toxin A modulate inflammatory mediator release from human neutrophils. Infect Immun 1993; 61:1055-1061.
48. Griffin FM Jr, Griffin JA, Silverstein SC. Studies on the mechanism of phagocytosis II. The interaction of macrophages with anti-immunoglobulin IgG-coated bone marrow-derived lymphocytes. J Exp Med 1976; 144:788-809.
49. Griffin FM Jr, Griffin JA, Leider JE et al. Studies on the mechanism of phagocytosis I. Requirements for circumferential attachment of particle-bound ligands to specific receptors on the macrophage plasma membrane. J Exp Med 1975; 142:1263-1282.
50. Babior BM. Oxygen-dependent microbial killing by phagocytes (second of two parts). N Engl J Med 1978; 298:721-725.
51. Babior BM. Oxygen-dependent microbial killing by phagocytes (first of two parts). N Engl J Med 1978; 298:659-668.
52. Klebanoff SJ. Antimicrobial mechanisms in neutrophilic polymorphonuclear leukocytes. Semin Hematol 1975; 12:117-142.
53. Segal BH, Leto TL, Gallin JI et al. Genetic, biochemical and clinical features of chronic granulomatous disease. Medicine 2000; 79:170-200.
54. Johnston RB Jr. Clinical aspects of chronic granulomatous disease. Curr Opin Hematol 2001; 8:17-22.

55. Veldkamp KE, Heezius HC, Verhoef J et al. Modulation of Neutrophil Chemokine Receptors by Staphylococcus aureus Supernate. Infect Immun 2000; 68:5908-5913.
56. de Haas CJ, Veldkamp KE, Peschel A et al. Chemotaxis inhibitory protein of Staphylococcus aureus, a bacterial anti-inflammatory agent. J Exp Med 2004; 199:687-695.
57. Fedtke I, Gotz F, Peschel A. Bacterial evasion of innate host defenses—the Staphylococcus aureus lesson. Int J Med Microbiol 2004; 294:189-194.
58. Foster TJ. Immune evasion by staphylococci. Nat Rev Microbiol 2005; 3:948-958.
59. van Belkum A. Staphylococcal colonization and infection: homeostasis versus disbalance of human(innate) immunity and bacterial virulence. Curr Opin Infect Dis 2006; 19:339-344.
60. Chavakis T, Preissner KT, Herrmann M. The anti-inflammatory activities of Staphylococcus aureus. Trends Immunol 2007; 28:408-418.
61. Nizet V. Understanding how leading bacterial pathogens subvert innate immunity to reveal novel therapeutic targets. J Allergy Clin Immunol 2007; 120:13-22
62. Rooijakkers SH, van Kessel KP, van Strijp JA. Staphylococcal innate immune Evasion. Trends Microbiol 2005; 13:596-601.
63. Rooijakkers SH, Ruyken M, Roos A et al. Immune evasion by a staphylococcal complement inhibitor that acts on C3 convertases. Nat Immunol 2005; 6:920-927.
64. Rooijakkers SH, Ruyken M, van Roon J et al. Early expression of SCIN and CHIPS drives instant immune evasion by Staphylococcus aureus. Cell Microbiol 2006; 8:1282-1293.
65. van Wamel WJ, Rooijakkers SH, Ruyken M et al. The innate immune modulators staphylococcal complement inhibitor and chemotaxis inhibitory protein of Staphylococcus aureus are located on beta-hemolysin-converting bacteriophages. J Bacteriol 2006; 188:1310-1315.
66. Rooijakkers SH, Milder FJ, Bardoel BW et al. Staphylococcal complement inhibitor: structure and active sites. J Immunol 2007; 179:2989-2998.
67. Jongerius I, Kohl J, Pandey MK et al. Staphylococcal complement evasion by various convertase-blocking molecules. J Exp Med 2007; 204:2461-2471.
68. Lee LY, Liang X, Höök M et al. Identification and characterization of the C3 binding domain of the Staphylococcus aureus extracellular fibrinogen-binding protein (Efb). J Biol Chem 2004; 279:50710-50716.
69. Lee LY, Höök M, Haviland D et al. Inhibition of complement activation by a secreted Staphylococcus aureus protein. J Infect Dis 2004; 190:571-579.
70. Verhoef J et al. Staphylococci and other micrococci. In: Cohen J, Powderly WG, eds. Infectious Disease. New York: Elsevier (Mosby), 2004:2119-2132.
71. Gómez MI, Lee A, Reddy B et al. Staphylococcus aureus protein A induces airway epithelial inflammatory responses by activating TNFR1. Nat Med 2004; 10:842-848.
72. Langley R, Wines B, Willoughby N et al. The staphylococcal superantigen-like protein 7 binds IgA and complement C5 and inhibits IgA-Fc alpha RI binding and serum killing of bacteria. J Immunol 2005; 174:2926-2933.
73. Bestebroer J, Poppelier MJ, Ulfman LH et al. Staphylococcal superantigen-like 5 binds PSGL-1 and inhibits P-selectin-mediated neutrophil rolling. Blood 2007; 109:2936-2943.
74. Haas PJ, de Haas CJ, Poppelier MJ et al. The structure of the C5a receptor-blocking domain of chemotaxis inhibitory protein of Staphylococcus aureus is related to a group of immune evasive molecules. J Mol Biol 2005; 353:859-872.
75. Baker HM, Basu I, Chung MC et al. Crystal structures of the staphylococcal toxin SSL5 in complex with sialyl Lewis X reveal a conserved binding site that shares common features with viral and bacterial sialic acid binding proteins. J Mol Biol 2007; 374:1298-1308.
76. Chavakis T, Hussain M, Kanse SM et al. Staphylococcus aureus extracellular adherence protein serves as anti-inflammatory factor by inhibiting the recruitment of host leukocytes. Nat Med 2002; 8:687-693.
77. Haggar A, Ehrnfelt C, Holgersson J et al. The extracellular adherence protein from Staphylococcus aureus inhibits neutrophil binding to endothelial cells. Infect Immun 2004; 72:6164-6167.
78. Xie C, Alcaide P, Geisbrecht BV et al. Suppression of experimental autoimmune encephalomyelitis by extracellular adherence protein of Staphylococcus aureus. J Exp Med 2006; 203:985-994.
79. Wright AJ, Higginbottom A, Philippe D et al. Characterisation of receptor binding by the chemotaxis inhibitory protein of Staphylococcus aureus and the effects of the host immune response. Mol Immunol 2007; 44:2507-2517.
80. Postma B, Kleibeuker W, Poppelier MJ et al. Residues 10-18 within the C5a receptor N terminus compose a binding domain for chemotaxis inhibitory protein of Staphylococcus aureus. J Biol Chem 2005; 280:2020-2027.
81. Postma B, Poppelier MJ, van Galen JC et al. Chemotaxis inhibitory protein of Staphylococcus aureus binds specifically to the C5a and formylated peptide receptor. J Immunol 2004; 172:6994-7001.

82. Haas PJ, de Haas CJ, Kleibeuker W et al. N-terminal residues of the chemotaxis inhibitory protein of Staphylococcus aureus are essential for blocking formylated peptide receptor but not C5a receptor. J Immunol 2004; 173:5704-5711.
83. Prat C, Bestebroer J, de Haas CJ et al. A new staphylococcal anti-inflammatory protein that antagonizes the formyl peptide receptor-like 1. J Immunol 2006; 177:8017-8026.
84. Rooijakkers SH, van Wamel WJ, Ruyken M et al. Anti-opsonic properties of staphylokinase. Microbes Infect 2005; 7:476-484.
85. Wines BD, Willoughby N, Fraser JD et al. A competitive mechanism for staphylococcal toxin SSL7 inhibiting the leukocyte IgA receptor, Fc alphaRI, is revealed by SSL7 binding at the C α2/C α3 interface of IgA. J Biol Chem 2006; 281:1389-1393.
86. Ramsland PA, Willoughby N, Trist HM et al. Structural basis for evasion of IgA immunity by Staphylococcus aureus revealed in the complex of SSL7 with Fc of human IgA1. Proc Natl Acad Sci USA 2007; 104:15051-15056.
87. Jin T, Bokarewa M, Foster T et al. Staphylococcus aureus resists human defensins by production of staphylokinase, a novel bacterial evasion mechanism. J Immunol 2004; 172:1169-1176.
88. Goerke C, Wirtz C, Flückiger U et al. Extensive phage dynamics in Staphylococcus aureus contributes to adaptation to the human host during infection. Mol Microbiol 2006; 61:1673-1685.

CHAPTER 3

Bacterial Complement Escape

Ilse Jongerius, Sanjay Ram and Suzan Rooijakkers*

Abstract

Complement activation is a crucial step in our innate immune defense against invading bacteria. Complement proteins can quickly recognize invading bacteria and subsequently label them for phagocytosis or kill them by direct lysis. In order to survive in the human host, bacterial pathogens have evolved a number of excreted and membrane-bound proteins that interfere with several steps of the complement cascade. In this chapter we summarize the most successful complement-modulating strategies by human bacterial pathogens.

Introduction

The complement system is a major mediator of the innate immune system, our first line of defense against invading micro-organisms.[1,2] Complement consists of more than thirty proteins in plasma and on cell surfaces and its activation results in a quick and effective defense against invading microbes. An important eradication strategy is the opsonization of foreign substances with C3b and iC3b which marks them for uptake by neutrophils via complement receptors (CR).[3,4] Furthermore, Gram-negative bacteria can be directly killed via the formation of C5b-9, the membrane attack complex (MAC).[5] The complement system also forms a bridge between the innate and adaptive immune system since C3d, the degradation product of C3b and iC3b, facilitates antigen presentation to B-cells.[6,7]

The complement system comprises three different activation routes to recognize extrinsic substances (Fig. 1): the classical (CP), the lectin (LP) and the alternative (AP) pathway. The CP is activated by binding of C1q to IgG or IgM molecules bound to the microbial surface.[8] The C1q-attached serine protease C1s in turn cleaves C4 resulting in formation of the anaphylatoxic peptide C4a and C4b.[9,10] The C4b molecule covalently binds to the bacterial surface due to exposure of its internal thioester that reacts with hydroxyl (creating an ester bond) or amino groups (creating an amide bond).[11] Then C2 binds to surface-bound C4b whereupon it is cleaved by activated C1s to form the CP C3 convertase C4b2a.[10]

The LP is highly similar to the CP and its activation also results in the formation of C4b2a.[12] The recognition molecules of the LP are mannan-binding lectin (MBL) and ficolins (L-, H- or M-ficolin). These lectins are structurally similar to C1q although they recognize microbial sugar patterns instead of immune complexes.[13] MBL and ficolins recognize neutral sugars (preferentially mannose, N-acetylglucosamine (GlcNAc) and fucose) when presented in a repetitive manner such as on the surface of a range of microbes.[12,14] MBL and ficolins are associated with MBL-associated serine protease (MASP)-1, MASP-2, MASP-3 and a nonprotease small MBL-associated protein (sMAP or MAP19).[13] MASP-2 is the only protease known to be responsible for cleavage of C4 and C2 to generate the C3 convertase C4b2a.[14] Other less well-defined functions of MASPs are activation of the AP by direct cleavage of C3 by MASP-1[15] and activation of the coagulation system by MASP-2.[16]

*Corresponding Author: Suzan Rooijakkers—Medical Microbiology, HP G04.614, University Medical Center Utrecht, The Netherlands. Email: s.h.m.rooijakkers@umcutrecht.nl

Pathogen-Derived Immunomodulatory Molecules, edited by Padraic G. Fallon.

The AP is either activated by spontaneous hydrolysis of the internal thioester bond in C3 (forming C3(H_2O)) or by covalent attachment of C3b to bacterial surfaces via the CP and LP.[5] Factor B binds to both surface-bound C3b and fluid-phase C3(H_2O) and is in turn cleaved by factor D to form a fluid-phase C3 convertase C3(H_2O)Bb or a surface-bound C3bBb complex.[17,18]

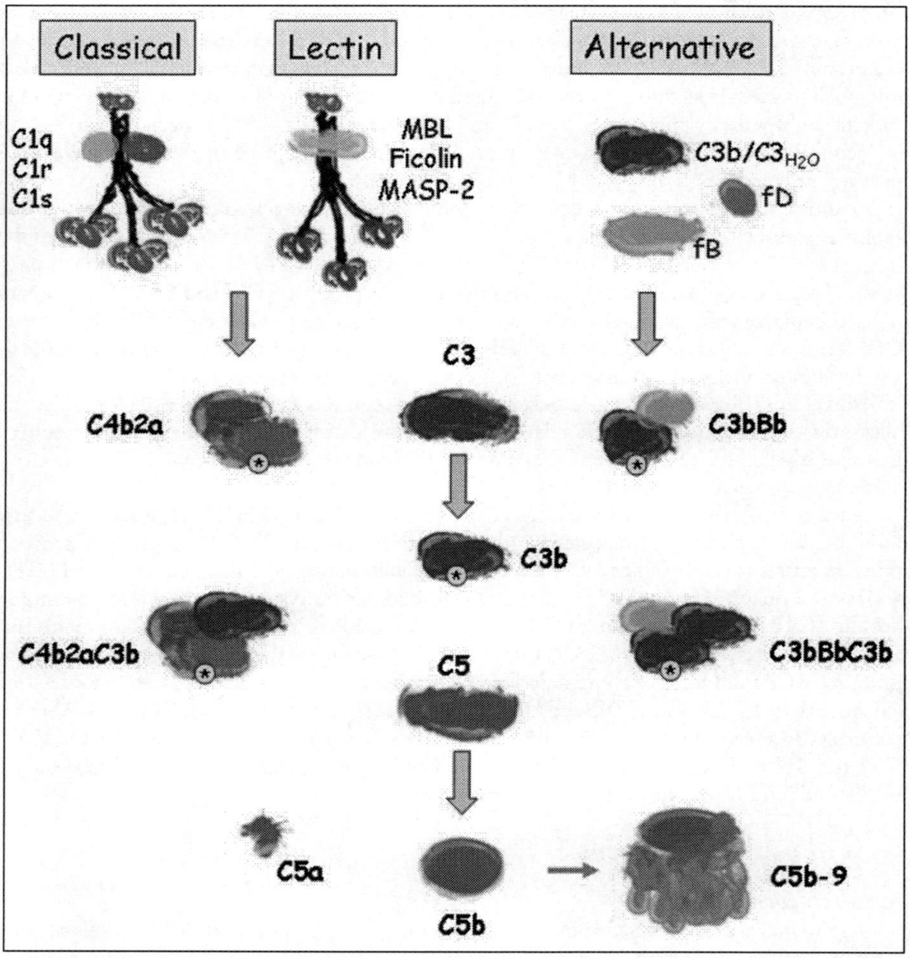

Figure 1. Schematic overview of the complement system. Complement activation can occur via three different pathways. The antibody-dependent Classical Pathway starts when C1q in the C1q-C1r_2-C1s_2 complex recognizes antibodies that are bound to the microbial surface. In the Lectin Pathway, Mannose Binding Lectin (MBL) and Ficolins recognize microbial sugar patterns and activate the MBL-associated serine protease 2 (MASP-2). Both C1s and MASP-2 can cleave complement proteins C4 and C2 to generate the CP/LP C3 convertase: C4b2a. Within this complex, C4b is covalently (*) attached to the microbial surface. The Alternative Pathway C3 convertase (C3bBb) is generated after binding of factor B (fB) to surface-bound C3b or fluid-phase C3(H_2O). Factor B is subsequently cleaved by factor D (fD) to generate C3bBb. Both C3 convertases C4b2a and C3bBb cleave C3 into covalently bound C3b (*) and an anaphylatoxin C3a. C3b contributes to phagocytosis, antigen presentation and formation of C5 convertases, C4b2a3b and C3bBb3b. C5 convertases cleave C5 into an anaphylatoxin C5a and C5b, which forms a complex with complement proteins C6, C7, C8 and C9 to generate the membrane attack complex (MAC) and mediate microbial lysis.

The AP C3 convertase is positively regulated by properdin, a multimeric protein that increases the stability of the complex.[19]

The C3 convertase enzymes fulfill a central role in the complement cascade, since all initiation pathways result in their formation (C4b2a for the CP/LP and C3bBb for the AP). C3 convertases are bimolecular complexes that only have a short lifespan of 1-2 minutes because C2a and Bb dissociate from the complex in an irreversible way.[18] C3 is the most abundant complement protein in serum (1 mg/ml) and its cleavage to C3b and C3a is an important step in complement activation.[20,21] C3a is a small fluid-phase chemo-attractant while C3b covalently attaches to the microbial surface via its thioester. Both C3b and its degradation product iC3b mark the microbe for efficient uptake by phagocytes via CR1 and CR3, respectively.[1] Further degradation of iC3b by factor H and I results in the formation of C3d which is recognized by CR2 on B-cells, supporting antigen presentation.[21]

In addition to its role in phagocytosis, C3b molecules are also needed for initiation of the terminal pathway. The binding of an additional C3b molecule to C3 convertases results in the formation of C5 convertases: C4b2aC3b (CP/LP) and $C3b_2Bb$ (AP).[22,23] These convertases preferentially cleave C5 instead of C3 and form two C5 split products (C5a and C5b). C5a is a very potent chemotactic factor which recruits and activates phagocytes by binding to the C5a receptor. C5b forms a soluble complex with C6 and subsequent binding of C7 results in hydrophilic to amphiphilic transition and induces expression of a metastable site through which C5b-7 is inserted into target lipid bilayer membranes. Subsequent incorporation of C8 and multiple C9 molecules allows the complex to traverse lipid bilayers forming complete transmembrane channels resulting in osmotic lysis of the cell. The MAC can directly kill Gram-negative bacteria while Gram-positive bacteria are protected against the MAC by their thick cell wall.[24]

Since complement activation can also damage host tissues it is tightly regulated by several fluid-phase and cell bound molecules. Convertase activity is regulated by a large group of proteins called regulators of complement activation (RCA). Fluid-phase molecules include Factor H (fH) and factor H-like protein 1 (FHL-1) that accelerate decay of the AP convertase, while C4-binding protein (C4BP) dissociates the CP/LP convertase. Host cells express membrane-bound CR1 and decay-accelerating factor (DAF) that dissociate C4b2a. Furthermore, membrane cofactor protein (MCP) and most of above described convertase regulators are also cofactors for factor I (FI)-mediated degradation of C4b and C3b.[6] Finally, the CP/LP enzymes C1r, C1s and MASPs are controlled by the serine protease inhibitor C1 Inhibitor (C1-INH).[10] The formation of the MAC is regulated by molecules that prevent insertion of C5b-7 into lipid membranes: membrane-bound CD59 and fluid-phase vitronectin and clusterin.[25]

Gram-Positive Pathogens

Staphylococcus aureus

Staphylococcus aureus causes a number of community- and hospital derived infections ranging from uncomplicated wound infections to severe diseases as septicemia or endocarditis. *S. aureus* expresses a large number of proteins that specifically impair the effectiveness of the innate immune system.[26-28] These include factors that modulate antimicrobial peptides and phagocytic cells but also complement modulators (Table 1). The first anti-opsonic molecule identified in *S. aureus* was Staphylococcal protein A (SpA) (42 kDa), a surface protein that can also be released in the extracellular milieu.[29] By binding the Fc part of IgG molecules, SpA covers the bacterial surface with IgG molecules in a manner that prevents recognition by Fc-receptors on phagocytes. This way, SpA blocks Fc-receptor mediated phagocytosis. Furthermore, since SpA interferes with C1q binding it prevents classical pathway complement fixation as well.[30,31]

The staphylococcal cell wall-associated protein Sbi (45 kDa) is also involved in immunoglobulin binding.[32] The Sbi extracellular region (Sbi-E) consists of four major domains including two IgG-binding regions. Recent data have indicated that the other two domains have

Table 1. Pathogens that directly inactivate complement proteins

Pathogen	Reference
IgG	
Staphylococcus aureus (SpA, Sbi, SAK)^	30-32,34
Streptococcus pyogenes (M-protein, IdeS, SpeB)	52,62,63
Porphyromonas gingivalis (Arg- and Lys- gingipains)	106
C1q	
Pseudomonas aeruginosa (elastase, alkaline protease)	98
C3/C3b	
Staphylococcus aureus (Sbi, SAK)	33,34
Streptococcus pyogenes (SpeB)	65,66
Streptococcus pneumoniae (PspC)	78
Pseudomonas aeruginosa (elastase, alkaline protease)	98
Serratia liquefaciens (metalloprotease)	99
Yersinia pestis (Pla)	101
Porphyromonas gingivalis (Arg- and Lys-gingipains)	105
C4	
Serratia liquefaciens (metalloprotease)	99
AP C3 convertases	
Staphylococcus aureus (SCIN, SCIN-B, SCIN-C, Efb, Ecb)	40,43
CP/LP C3 convertases	
Staphylococcus aureus (SCIN, SCIN-B, SCIN-C)	40
C5 convertases	
Staphylococcus aureus (Efb, Ecb)	43
C5a	
Streptococcus pyogenes (scpA)	67
Streptococcus agalactiae (scpB)	71
C5	
Staphylococcus aureus (SSL7)	38
Porphyromonas gingivalis (Arg- and Lys- gingipains)	103
C5b-9	
Streptococcus pyogenes (SIC, CRS)	59,60
Serratia liquefaciens (metalloprotease)	99
Salmonella typhimurium (Rck)	120
Yersinia enterocolitica (Ail)	121
Borrelia burgdorferi (CD59-like)	137
Properdin	
Streptococcus pyogenes (SpeB)	64

^bacterial complement inhibitor is indicated in parenthesis.

complement-inhibitory functions since they bind C3 via the thioester-containing C3dg fragment and the C3a anaphylatoxin domain.[33]

Another way for *S. aureus* to inhibit phagocytosis is via the production of Staphylokinase (SAK) (16 kDa).[34] SAK targets plasminogen (PLG) on the bacterial surface and activates it into plasmin (PL). PL can cleave two major opsonic molecules, IgG (at the hinge region) and C3b.[35,36] It was shown that the conversion of PLG to PL by SAK leads to removal of IgG and C3b on the bacterial surface. Thus, SAK impairs both IgG and complement dependent phagocytosis by human neutrophils.[34]

Staphylococcal superantigen-like (SSL) protein 7 is a 23 kDa protein that belongs to a family of 11 different SSL's that are clustered on staphylococcal pathogenicity island 2 (SapIn2).[37,38] SSL7 binds to IgA but also prevents formation of the MAC via specific binding to C5.[38] SSL7 binding to C5 possibly also prevents formation of C5a which seems more important for clearance of *S. aureus* because its thick cell wall renders it resistant to MAC-mediated lysis.

The importance for inhibiting C5a-mediated responses was already indicated by the Chemotaxis inhibitory protein of *S. aureus* (CHIPS) (14 kDa).[39] This extracellular protein binds two major chemotactic receptors on neutrophils, the C5a receptor (C5aR) and the formyl peptide receptor (FPR). By binding these receptors, CHIPS effectively blocks neutrophil chemotaxis towards C5a and formylated peptides (e.g., fMet-Leu-Phe) (fMLP).

Finally, *S. aureus* excretes a number of proteins that specifically block convertase activity. The first described convertase inhibitor is Staphylococcal complement inhibitor (SCIN) (10 kDa), an excreted protein that binds C3 convertases and interferes with all complement pathways.[40] SCIN binding to surface-bound C3 convertases has two major consequences. First, the binding of SCIN to C4b2a and C3bBb impairs the enzymatic activity of the convertases. Second, SCIN stabilizes C3 convertases which impairs the formation of new convertases. These actions result in very potent inhibition of phagocytosis and C5a production.[40-42] Next to production of SCIN, *S. aureus* also excretes the proteins SCIN-B and SCIN-C. Both proteins are homologues of SCIN that share the exact same mechanism of action.[43]

Extracellular fibrinogen binding protein (Efb) (16 kDa) and Extracellular complement binding protein (Ecb) (10 kDa) (also known as Efb homologous protein Ehp[44]) also target surface-bound convertases.[43] Efb was first shown to bind fibrinogen but subsequently was shown to bind to the C3d region of C3 as well.[45-47] Ecb is a homologue of Efb that also binds C3, but does not harbor the fibrinogen-binding domain.[43,44] Efb and Ecb were found to specifically block all C3b containing convertases (C3 convertases of the AP and C5 convertases of the CP, LP and the AP), thereby inhibiting C3b deposition via the AP and more potently block C5a formation via all pathways. Inhibition of C5a formation was demonstrated in vivo using a mouse model for immune complex diseases. Addition of Ecb prior to induction of the Arthus reaction resulted in complete inhibition of neutrophil influx via C5a production.[43] Next to targeting C3b containing convertases it was also shown that fluid phase interactions of Efb and Ecb with C3 can inhibit proteolytic activation of C3.[44,46]

Streptococci

Group A *streptococcus* (GAS, *Streptococcus pyogenes*) is a highly virulent human pathogen responsible for a large number of severe (septicemia, necrotizing fasciitis) and mild (pharyngitis) infections. The best known virulence factor in GAS is the surface-bound M-protein, which exists in more than 100 antigenically distinguishable variants that all have different functions.[48-50] Several M proteins are known to bind FH, FHL-1 and Fc regions of IgG. IgG binding results in inhibition of the CP while binding of FH and FHL-1 interferes with activation of the AP. Recently it was shown that several GAS also use M-proteins to attract C4BP to the surface and thereby interfere with activation of the classical pathway.[51,52] Furthermore, cell-bound C4BP still functions as cofactor for FI and thus contributes to degradation of C4b.

The highly virulent GAS serotype M1, does not use M proteins to bind FH or FHL-1 but expresses the surface-bound Fba protein to bind these complement regulators.[53] Fba was shown to contribute to M1 GAS survival in human whole blood. Furthermore, it was demonstrated that binding of Fba to FHL-1 promotes entry of GAS into human epithelial cells.[53,54]

Recently, streptococcal collagen-like protein Scl1.6 of M6-type GAS was characterized as a FH/FHL-1 binding protein. Scl1.6-bound FH retained cofactor activity for factor I-mediated cleavage of C3b suggesting a role in GAS evasion of complement-mediated opsonization.[55]

Mrp4 (M-related protein 4) was shown to bind to fibrinogen.[56] Inactivation of Mrp4 results in loss of fibrinogen binding and a twofold increase in C3b deposition on the bacterial surface. It was show that fibrinogen binding was responsible for inhibition of the CP resulting in a novel mechanism to resist phagocytosis. This might be a common mechanism among GAS since half of the clinical isolates express Mrp.

GAS also express molecules that specifically bind to CR3 and FcγRIIIB. M5 protein interferes with CR3-dependent association between GAS and neutrophils, inhibiting the ingestion of bacteria. Isolated human neutrophils killed an M5-negative GAS strain, but not the wild-type parent strain.[57] The GAS Mac-1-like protein Mac shares homology with the alpha-subunit of CR3. Despite the homology with CR3 its function is not related to complement modulation. Mac binds to CD16 (FcγRIIIB) and thereby blocks phagocytosis and production of reactive oxygen species, resulting in decreased pathogen killing.[58]

Streptococcal Inhibitor of Complement (SIC) (31 kDa) is the only bacterial protein that specifically binds components of the terminal complement pathway. SIC is excreted by M1 type GAS and binds to soluble C5b-7 to prevent its insertion into cell membranes.[59,60] The protein Closely Related to SIC (CRS) was found in serotype M57 and shares many characteristics of SIC, including its ability to bind C6 and C7. Since streptococci are naturally resistant to complement-mediated cytolysis, MAC inhibition was suggested not to be the sole function of SIC. In serotypes M12 and M55 the protein Distantly Related to SIC (DRS) was found. DRS has a limited sequence similarity with the C-proximal half of SIC and its biological function is unkown.[61]

Finally, GAS produce several proteases that modulate complement functions. The cysteine proteinases IdeS and SpeB cleave microbe-bound IgG molecules in the hinge region to remove the Fc region and prevent Fc-mediated phagocytosis and killing.[62,63] Next to cleavage of IgG, SpeB can also cleave properdin and C3.[64-66] This way, SpeB blocks opsonization and bactericidal activity by neutrophils to GAS. Terao et al showed that a SpeB-deficient *S. pyogenes* had a lower survival rate in mice than its isogenic parent.[66] The cell wall anchored C5a peptidase (scpA) cleaves the anaphylatoxin C5a fragment but not its precursor C5 and thereby inhibits chemotaxis of neutrophils towards the site of infection.[67] ScpA is a well-established virulence factor as determined in animal models where mutant strains were better cleared than wild-type strains.[68]

Group B *streptococcus* (GBS, *S. agalactiae*) is a major cause of severe neonatal infections including pneumonia, meningitis and septicemia. Relatively few complement modulators have been described in GBS thus far. Like GAS, GBS expresses a surface-bound protein (β protein or Bac) that binds to FH via its C-terminal domain. By capturing FH to the bacterial surface, β protein mediates FI-mediated degradation of C3b degradation.[69] β protein also binds to the Fc-part of human IgA but the biological relevance is still unknown.[70] GBS also produce the cell wall anchored C5a peptidase (called scpB in GBS).[71] Interestingly, ScpB is highly specific for human C5a.[72]

S. pneumoniae causes a variety of infections including upper respiratory infections, meningitis and pneumonia. A possible role for the surface-bound pneumolysin in complement evasion by *S. pneumoniae* remains uncertain since several studies demonstrated that it specifically activates the CP in an IgG-independent manner.[73,74] Pneumococcal surface protein A (PspA) was also proposed as a complement inhibitor although the exact mechanism of action remains unknown.[75] Nevertheless, PspA is required for full virulence in mouse models of pneumococcal infections. Pneumococcal surface protein C (PspC, also known as SpsA, CbpA and Hic) binds to FH via its N-terminal part. This binding results in inhibition of the AP.[76,77] Furthermore, PspC can also directly bind to C3, IgA and the human polymeric Ig receptor.[78] Recent studies have demonstrated that PspA and PcpC have a synergistic effect on complement inhibition.[79]

Gram-Negative Pathogens

Unlike their Gram-positive counterparts, Gram-negative bacteria do not possess a cell wall and are susceptible to direct killing by complement via insertion of C5b-9. Therefore most Gram-negative pathogens have evolved mechanisms to block complement activation at various stages along the cascade and in some instances a single pathogen may possess multiple such mechanisms. Because of the extensive number of studies describing complement evasion strategies used by numerous Gram-negative pathogens, this section discusses the broad mechanisms of complement evasion with select illustrative examples.

Complement Regulation by Smooth Lipopolysaccharide and Capsule

Several Gram-negative bacteria express smooth lipopolysaccharide (LPS) which consists of repeating saccharide units beyond the core saccharide; each repeat is called an O-unit that usually comprises two to six sugars and the entire structure made up of repeating units is called the O-antigen. In addition, several of these pathogens express capsular polysaccharide which also plays a key role in virulence.

Smooth LPS from organisms such as *Klebsiella pneumoniae* can directly activate the alternative pathway.[80] Direct binding of C1q to outer membrane components, for example, OmpK36 of *K. pneumoniae*,[81] porin II of *Aeromonas hydrophila*[82] or to the surface of *Salmonella minnesota*[83] can activate the classical pathway in an antibody independent fashion. Bacterial resistance to complement-mediated killing may be the result of either an inability to activate complement or a failure of activated complement to effect killing. *K. pneumoniae* strains bearing certain capsular polysaccharide (called the K antigen) types, for example, K1, K10 or K16, impede complement activation by masking the O-antigen.[84] Similarly, the S-layer protein of certain Aeromonas species prevents complement activation by masking LPS.[82] Although the O-antigen activates the alternative pathway of complement, C3b activated in this manner binds distant from the bacterial membrane and fails to form C5b-9.[80] Isogenic rough mutants (LPS that lack O-antigenic repeats) bind C3b close to the bacterial membrane with formation of C5b-9 and are serum sensitive. The O-antigen may regulate the classical pathway by preventing C1q from binding to somatic antigens.[85,86]

Many medically important Gram-negative pathogens such as *Neisseria meningitidis, N. gonorrhoeae, Haemophilus influenzae* Type b, nontypeable *H. influenzae, H. ducreyi, Campylobacter jejuni, Pasteurella multocida* and *Yersinia pestis* do not express an O-antigen. Instead, their LPS consist of up to six glycan residues that extend beyond the core heptose and therefore are often referred to as lipooligosaccharide (LOS). Some (for example, *N. meningitidis, H. influenzae* Type b, *C. jejuni, P. multocida* and *Y. pestis*) express a capsule, which enhances virulence probably by preventing opsonophagocytosis and/or enhancing serum resistance. The molecular basis for capsule-mediated serum resistance has not been fully elucidated.

Phase Variation and Lipooligosaccharide Sialylation

Pathogens such as *N. gonorrhoeae*, nontypeable *Haemophilus influenzae* and *Moraxella catarrhalis* that lack a capsule and also do not express smooth LPS must rely on alternative means to evade complement. As discussed below, binding of complement inhibitory proteins is an important mechanism of complement evasion for some of these bacteria. In addition, these species (and others, such as Campylobacter and Helicobacter) have the ability to alter their surface structures through a mechanism called phase variation, that is, stochastic, high-frequency switches in expression of surface structures which enables a bacterium to generate diversity.[87-89] This diversity allows a population to "preemptively" escape an immune response to the surface antigen. A mechanism of phase variation involves changes in the length of simple sequence repeat regions within genes. Several LOS biosynthesis genes are phase variable which allows a single strain to vary LOS glycans which facilitates immune escape. The LOS structures of Neisseria, Haemophilus and Campylobacter can mimic host molecules such as glycosphingolipids and GM1 gangliosides;[90,91] while this may provide a mechanism of immune escape in some instances, the generation of autoantibodies can result in neuronal damage, as seen with Guillain-Barre syndrome following *C. jejuni* infections.[92] In addition, certain LOS structures can be substituted with sialic acid which may serve to decrease

C3 and C4 binding to bacteria because LOS has been identified as a target for these complement components on Neisseriae and Haemophilus.[93-96] Increased factor H binding upon sialylation of the lacto-*N*-neotetraose LOS structure is unique to *N. gonorrhoeae* and requires the concomitant presence of gonococcal porin.[97]

Proteolytic Cleavage of Complement Components

Several bacteria secrete proteases that degrade complement components. These include *Pseudomonas aeruginosa* elastase and alkaline protease that cleave C1q and C3,[98] *Serratia liquefaciens* metalloprotease that degrades C4, C3, C5-9,[99] *S. marcescens* protease that inactivates complement,[100] *Y. pestis* plasminogen (Pla) protease which also possesses C3 cleaving activity[101] and *Porphyromonas gingivalis* Arg- and Lys-gingipains that degrade Ig, C3, C5 and the C5a receptor.[102-106] Activation of C1 and cleavage of C5 by *P. gingivalis* results in complement activation, C5a generation and neutrophil recruitment with resulting inflammatory response that characterizes periodontal disease.[107,108]

Rather interestingly, protease-deficient mutants of *P. gingivalis* bind greater amounts of both C3b and C5b-9 and are phagocytosed more readily,[105] yet remain serum resistant. Recent data have shown that complement resistance in *P. gingivalis* does not depend on protease production but requires expression of cell surface polysaccharide, which is an anionic branched mannan structure.[109,110] Similarly, loss of *Y. pestis* Pla expression does not affect resistance to human complement, although Pla-negative mutants show decreased virulence murine infection models.[100,111] Complement cleaving proteases may serve to limit anaphylatoxin generation and neutrophil influx rather than serving as the primary mechanism of serum resistance.

Binding of Complement Inhibitory Molecules to the Bacterial Surface

Over the past decade numerous pathogens have been shown to bind to complement inhibitory molecules which results in complement regulation on the bacterial surface. The main fluid-phase complement inhibitory molecules are C1-INH, C4BP, factor H and its alternatively spliced variant, FHL-1. Table 2 lists the pathogens and the molecule(s) on these bacteria that bind to complement inhibitory molecules. The bacteria bind these inhibitors in a manner that permits bound proteins to remain functionally active and perform their complement regulation functions.

In certain instances, binding of a complement regulator may not necessarily enhance serum resistance. As an example, C4BP binds to *N. gonorrhoeae* Type IV pilin protein through a relatively weak interaction that is best appreciated in hypoosmolar buffers.[112] It was postulated, although not formally proven, that this interaction could contribute to serum resistance. In contrast, binding of C4BP to gonococcal porin protein directly contributes to complement evasion.[113] Another example is the C4BP-*Moraxella catarrhalis* interaction, where purified C4BP binds to bacteria and preincubation of bacteria with C4BP followed by addition of C4BP-depleted serum results in cleavage of C4b.[114] However, minimal C4BP binding occurs when normal human serum is used as a source of C4BP and cofactor activity is not seen in this setting.[115] The same *M. catarrhalis* protein (UspA2) that binds pure C4BP also binds to vitronectin; the latter prevents C9 polymerization within the MAC complex and enhances serum resistance.[115]

Binding of complement inhibitors to some pathogens is restricted to certain mammalian hosts. Host-specific serum resistance in turn may play a role in determining species specificity of infections. As an example, most strains of *N. gonorrhoeae* causes natural infection only in humans, bind only human C4BP and resist only human serum. Select gonococcal strains that express the porin 1B molecule bind chimpanzee C4BP and are resistant to chimpanzee serum.[116] Interestingly, experimental urethritis in chimpanzees can be caused by porin 1B, but not porin 1A, strains;[117] the latter do not bind chimpanzee C4BP and are sensitive to killing by chimpanzee complement. In contrast, *B. burgdorferi* can infect a wide range of animals binds to factor H from different species.[118] Similarly, *Y. pestis* that infects a wide array of animals and is resistant to complement from several sources[119] exhibits broad specificity of C4BP binding.[116]

Table 2. Pathogens that capture human complement regulators

Pathogen	Reference
C1 inhibitor	
Bordetella pertussis (protein coded by *bvg* locus)[A]	144
Factor H and/or factor H-like protein 1 (FHL-1)	
Streptococcus pyogenes (M-protein, Fba, Scl1.6)	49,53,55
Streptococcus agalactiae (β protein)	69
Streptococcus pneumoniae (PspC)	76,77
Yersinia enterocolitica (YadA protein)	145
Neisseria gonorrhoeae (porin 1A; porin 1B when lipooligosaccharide is sialylated)	146
Neisseria meningitidis (lipoprotein GNA1870)	147
Pseudomonas aeruginosa (elongation factor Tuf)	148
Borrelia burgdorferi (CRASP, Erp)	135
Borrelia afzelii (CRASP, Erp)	135
Borrelia hermsii (FHBP19, FHBP28)	138,139
Treponema denticola (FhbB)	141
C4b-binding protein	
Streptococcus pyogenes (M-protein)	52
Escherichia coli K1	149
Neisseria gonorrhoeae (porin 1A, select porin 1B and Type IV pilin)	112,113
Bordetella pertussis (filamentous hemagglutinin)	150
Moraxella catarrhalis (ubiquitous surface protein (Usp)A2)	114
Neisseria meningitidis (porin A)	151
Nontypeable *Haemophilus influenzae*, select strains	152
Vitronectin	
Moraxella catarrhalis (UspA2)	115
Haemophilus influenzae (Haemophilus surface fibrils (Hsf))	153

[A]bacterial ligand for complement regulator is indicated in parenthesis.

Prevention of C9 Polymerization and Membrane Attack Complex Insertion

Killing of Gram-negative bacteria requires assembly of terminal complement components (MAC) in the bacterial membrane. Polymerization of C9 within this complex is important for effective killing. One such mechanism of preventing effective C5b-9 formation is binding of vitronectin to bacteria such as *M. catarrhalis* as described above, which restricts C9 polymerization to three molecules of C9 per MAC complex. Expression of the Rck (Resistance to complement killing) protein of *S. typhimurium* restricts C9 polymerization.[120] Proteins homologous to Rck in *Y. enterocolitica* and *Y. pestis* called Ail (Adhesion and invasion locus) also mediate resistance to complement.[119,121] Although the molecular basis for serum resistance mediated by Rck and Ail proteins have not been fully elucidated, it is possible that these proteins may also serve as ligands for vitronectin, which in turn prevents C9 polymerization.

Another mechanism of complement regulation at the level of the terminal complement pathway involves extrusion of the membrane attack complex as occurs with *Salmonella minnesota*. C5b-7 complexes that are assembled on the bacterium are released upon the incorporation of C8 into the complex.[122,123] The molecular configuration of C5b-9 in the bacterial membrane also appears to be an important factor in determining the fate of the bacterium; Joiner et al showed that C5b-9 complexes (bactericidal and nonbactericidal MAC) were associated with different outer membrane proteins in serum sensitive and serum resistant *N. gonorrhoeae*, respectively.[124]

Blocking Antibodies

As early as 1894, Pfeiffer described a phenomenon where animals given an excess of immune serum were sometimes more susceptible to the challenge organism.[125] In 1942, Thomas et al demonstrated that convalescent sera obtained from patients with meningococcal disease sometimes had less bactericidal activity than acute (nonimmune) serum.[126] Subsequently it was demonstrated that certain immune IgA1 directed against serogroup C meningococcal capsule,[127] or immune IgG against serogroup W-135 meningococcal capsule[128] could block killing by otherwise bactericidal antibodies. The Fab fragment of anti-capsular IgA1 was sufficient to block IgG-mediated killing. Other pathogens against which blocking Ig has been demonstrated include *Brucella abortus*,[129] *B. melitensis*[130] and *P. aeruginosa*.[131,132]

The role of blocking antibodies in enhancing serum resistance and in virulence of *N. gonorrhoeae* has been best characterized. Antibodies against Rmp (Reduction modifiable protein, or Protein III) antibodies were shown to block killing by otherwise bactericidal antibodies. Rather interestingly, the presence of blocking antibodies (either in the presence or absence of bactericidal antibodies) resulted in increased C3 and C9 binding to gonococci.[133] It has been postulated that in the presence of blocking antibodies, C3 is diverted to different targets and results in the formation of nonbactericidal C5b-9.[133] Plummer et al subsequently showed that women with antibodies against *N. gonorrhoeae* Rmp were at increased risk of acquiring gonorrhea.[134] While the exact mechanism of blocking still remains enigmatic, this under-appreciated complement evasion mechanism may be used by diverse bacterial species.

Spirochetes

Borreliae

Lyme disease is a vector-borne inflammatory disorder that can be caused by several Borreliae of which *Borrelia burgdorferi, B. afzelii* and *B. garinii* are the best known. These spirochetes are well-known to escape the complement system by acquisition of host regulators (Table 2).[49] *B. burgdorferi* and *B. afzelii* express five Complement regulator-acquiring surface proteins (CRASP) that bind FH and/or FHL-1.[135] Furthermore, the Osp E/F related proteins (Erp) were also described to bind factor H.[136] *B. burgdorferi* can also resist formation of the MAC by the CD59-like (80 kDa) surface protein that shares functional similarities with CD59, the human cell-bound inhibitor of MAC.[137] CD59-like prevents polymerisation of C9 and thus formation of MAC.

Relapsing fever is a rapidly progressive and severe septic disease caused by Borreliae that are either transmitted by soft bodied ticks (e.g., *B. hermsii* and *B. parkeri*) or by lice (*B. recurrentis*). *B. hermsii* and *B. parkerii* and *B. recurrentis* have all been shown to bind FH.[138,139] In *B. hermsii*, two proteins were found to be responsible for FH binding: FHBP19/FhbA and FHBP28. *B. recurrentis* was shown to also capture C4BP in a way that retained functional activity of the CP regulator.[140] A more detailed overview of complement modulation by Borreliae is provided by Kraiczy and Wurzner.[49]

Treponema

Treponema denticola belongs to the normal flora of the human oral cavity and is an important cause of acute and chronic to periodontal disease. *T. denticola* encodes a 14 kDa surface protein that specifically binds FHL-1 but not fH.[141,142] This FHL-1 binding protein B (FhbB) captures FHL-1 to the surface of *T. denticola* although it was not shown to serve as a cofactor for factor I-mediated cleavage of C3b. Therefore it was hypothesized that FhbB might facilitate adherence to FHL-1 expressing cells.

Discussion

Human bacterial pathogens have evolved a wide variety of strategies to prevent elimination via the complement system. Since complement activation occurs via multiple recognition molecules and effector mechanisms, bacterial complement modulators are found to target several levels of the cascade. Some bacteria modulate the initial steps of complement activation by blocking recognition molecules or masking their surface. Other bacteria produce surface-exposed or secreted proteins that specifically block complement proteolysis. A large number of these proteins are aimed at the central enzymatic steps (convertases) either directly or by binding human regulators of these convertases. All these individual strategies are effective ways to circumvent complement attack and in some cases the in vivo efficacy has been proven. Nevertheless, most bacteria use more than one strategy to withstand the complement attack, preferably with different molecules. Both the redundancy and multiplicity of complement evasion strategies are critical determinants for the survival of a successful pathogen in a population.

Because Gram-negative bacteria and spirochetes do not posses a cell wall, they have evolved ways to block incorporation of the lytic C5b-9 complex. However, regardless of their thick cell wall, terminal pathway evasion has also been observed in both *S. aureus* (SSL7) and Group A *streptococcus* (SIC). Although the importance of SSL7 probably lies in blocking the formation of the C5a anaphylatoxin, SIC seems to be specific for the MAC. Gram-negatives seem to preferentially use surface-bound molecules to evade complement, rather than secreted proteins that appear to abound among Gram-positive pathogens. It remains to be determined whether this difference truly exists or is caused by a bias on part of investigators. Nevertheless, both types of molecules are highly important subjects of research since they have therapeutic applications. Being soluble, secreted anti-inflammatory proteins are likely candidates for therapeutic treatment of complement-associated diseases. Surface-bound proteins could be exploited to design molecules and/or vaccines that interfere with their action and thus help to treat infectious diseases. In order to achieve this, more insight into the molecular interactions is required. The rapidly increasing knowledge of the structural organization of complement proteins will certainly help to understand the mechanisms of the inhibitory actions of these bacterial complement inhibitors and facilitate their use in therapeutics.[143]

Because the complement system is an evolutionary conserved protein cascade, the observed host specificity of bacterial complement inhibitors is very striking. The Gram-positive proteins SAK, SCIN and ScpB and the Neisserial porin molecules display strict human specificity. In case of Neisserial porin which exclusively binds human C4BP, host-specificity of the complement inhibitor binding may contribute to host specificity of the infection. Understanding the basis for species specificity of complement evasion will hopefully lead to development of better animal models for infectious diseases, for example, creating transgenic mice for human complement inhibitory proteins to study pathogens such as *N. gonorrhoeae*.

Understanding how bacteria block complement activation is essential to further our insight into bacterial pathogenesis. The ongoing discovery of bacterial complement modulatory proteins and their exact mechanism of action will greatly enhance our understanding of bacteria-host interactions.

Acknowledgements

The authors thank Jos van Strijp for his artistic representation of the complement system (Fig. 1). Sanjay Ram is supported by National Institutes of Health grants AI054544 and AI32725. Suzan Rooijakkers is supported by the Netherlands Organization for Scientific Research (NWO)-VENI (916-76-037).

References

1. Walport MJ. Complement. First of two parts. N Engl J Med 2001; 344(14):1058-1066.
2. Walport MJ. Complement. Second of two parts. N Engl J Med 2001; 344(15):1140-1144.
3. Gasque P. Complement: a unique innate immune sensor for danger signals. Mol Immunol 2004; 41(11):1089-1098.
4. Song WC, Sarrias MR, Lambris JD. Complement and innate immunity. Immunopharmacology 2000; 49(1-2):187-198.
5. Ramm LE, Whitlow MB, Mayer MM. Transmembrane channel formation by complement: functional analysis of the number of C5b6, C7, C8 and C9 molecules required for a single channel. Proc Natl Acad Sci USA 1982; 79(15):4751-4755.
6. Kirkitadze MD, Barlow PN. Structure and flexibility of the multiple domain proteins that regulate complement activation. Immunol Rev 2001; 180:146-161.
7. Sahu A, Lambris JD. Structure and biology of complement protein C3, a connecting link between innate and acquired immunity. Immunol Rev 2001; 180:35-48.
8. Duncan AR, Winter G. The binding site for C1q on IgG. Nature 1988; 332(6166):738-740.
9. Dodds AW, Sim RB, Porter RR et al. Activation of the first component of human complement (C1) by antibody-antigen aggregates. Biochem J 1978; 175(2):383-390.
10. Sim RB, Laich A. Serine proteases of the complement system. Biochem Soc Trans 2000; 28(5):545-550.
11. Law SK, Dodds AW. The internal thioester and the covalent binding properties of the complement proteins C3 and C4. Protein Sci 1997; 6(2):263-274.
12. Matsushita M, Fujita T. Ficolins and the lectin complement pathway. Immunol Rev 2001; 180:78-85.
13. Fujita T, Matsushita M, Endo Y. The lectin-complement pathway—its role in innate immunity and evolution. Immunol Rev 2004; 198:185-202.
14. Holmskov U, Thiel S, Jensenius JC. Collections and ficolins: humoral lectins of the innate immune defense. Annu Rev Immunol 2003; 21:547-578.
15. Matsushita M, Fujita T. Cleavage of the third component of complement (C3) by mannose-binding protein-associated serine protease (MASP) with subsequent complement activation. Immunobiology 1995; 194(4-5):443-448.
16. Krarup A, Wallis R, Presanis JS et al. Simultaneous activation of complement and coagulation by MBL-associated serine protease 2. PLoS ONE 2007; 2(7):e623.
17. Ponnuraj K, Xu Y, Macon K et al. Structural analysis of engineered Bb fragment of complement factor B: insights into the activation mechanism of the alternative pathway C3-convertase. Mol Cell 2004; 14(1):17-28.
18. Xu Y, Narayana SV, Volanakis JE. Structural biology of the alternative pathway convertase. Immunol Rev 2001; 180:123-135.
19. Hourcade DE. The role of properdin in the assembly of the alternative pathway C3 convertases of complement. J Biol Chem 2006; 281(4):2128-2132.
20. Janssen BJ, Huizinga EG, Raaijmakers HC et al. Structures of complement component C3 provide insights into the function and evolution of immunity. Nature 2005; 437(7058):505-511.
21. Lambris JD. The multifunctional role of C3, the third component of complement. Immunol Today 1988; 9(12):387-393.
22. Pangburn MK, Rawal N. Structure and function of complement C5 convertase enzymes. Biochem Soc Trans 2002; 30(Pt 6):1006-1010.
23. Rawal N, Pangburn MK. Structure/function of C5 convertases of complement. Int Immunopharmacol 2001; 1(3):415-422.
24. Joiner K, Brown E, Hammer C et al. Studies on the mechanism of bacterial resistance to complement-mediated killing. III. C5b-9 deposits stably on rough and Type 7S. pneumoniae without causing bacterial killing. J Immunol 1983; 130(2):845-849.
25. Davies A, Lachmann PJ. Membrane defence against complement lysis: the structure and biological properties of CD59. Immunol Res 1993; 12(3):258-275.
26. Foster TJ. Immune evasion by staphylococci. Nat Rev Microbiol 2005; 3(12):948-958.
27. Rooijakkers SH, van Kessel KP, van Strijp JA. Staphylococcal innate immune evasion. Trends Microbiol 2005; 13(12):596-601.
28. Rooijakkers SH, van Strijp JA. Bacterial complement evasion. Mol Immunol 2007; 44(1-3):23-32.
29. Silverman GJ, Goodyear CS, Siegel DL. On the mechanism of staphylococcal protein A immunomodulation. Transfusion 2005; 45(2):274-280.
30. Forsgren A, Sjoquist J. "Protein A" from S. aureus. I. Pseudo-immune reaction with human gamma-globulin. J Immunol 1966; 97(6):822-827.
31. Goward CR, Scawen MD, Murphy JP et al. Molecular evolution of bacterial cell-surface proteins. Trends Biochem Sci 1993; 18(4):136-140.

32. Zhang L, Jacobsson K, Vasi J et al. A second IgG-binding protein in Staphylococcus aureus. Microbiology 1998; 144(Pt 4):985-991.
33. Burman J, Leung E, Isenman DE et al. Interaction of human complement with Sbi, a staphylococcal immunoglobulin-binding protein. Mol Immunol 2007; 44(16):3982
34. Rooijakkers SH, van Wamel WJ, Ruyken M et al. Anti-opsonic properties of staphylokinase. Microbes Infect 2005; 7(3):476-484.
35. Harpel PC, Sullivan R, Chang TS. Binding and activation of plasminogen on immobilized immunoglobulin G. Identification of the plasmin-derived Fab as the plasminogen-binding fragment. J Biol Chem 1989; 264(1):616-624.
36. Seya T, Nagasawa S, Matsukura M et al. Generation of C3d,g and C3d by urokinase-treated plasma in association with fibrinolysis. Complement 1985; 2(2-3):165-174.
37. Fitzgerald JR, Reid SD, Ruotsalainen E et al. Genome diversification in Staphylococcus aureus: Molecular evolution of a highly variable chromosomal region encoding the Staphylococcal exotoxin-like family of proteins. Infect Immun 2003; 71(5):2827-2838.
38. Langley R, Wines B, Willoughby N et al. The staphylococcal superantigen-like protein 7 binds IgA and complement C5 and inhibits IgA-Fc alpha RI binding and serum killing of bacteria. J Immunol 2005; 174(5):2926-2933.
39. de Haas CJ, Veldkamp KE, Peschel A et al. Chemotaxis inhibitory protein of Staphylococcus aureus, a bacterial antiinflammatory agent. J Exp Med 2004; 199(5):687-695.
40. Rooijakkers SH, Ruyken M, Roos A et al. Immune evasion by a staphylococcal complement inhibitor that acts on C3 convertases. Nat Immunol 2005; 6(9):920-927.
41. Rooijakkers SH, Ruyken M, van RJ et al. Early expression of SCIN and CHIPS drives instant immune evasion by Staphylococcus aureus. Cell Microbiol 2006; 8(8):1282-1293.
42. Rooijakkers SH, Milder FJ, Bardoel BW et al. Staphylococcal complement inhibitor: structure and active sites. J Immunol 2007; 179(5):2989-2998.
43. Jongerius I, Kohl J, Pandey MK et al. Staphylococcal complement evasion by various convertase-blocking molecules. J Exp Med 2007; 204(10):2461-2471.
44. Hammel M, Sfyroera G, Pyrpassopoulos S et al. Characterization of Ehp, a secreted complement inhibitory protein from Staphylococcus aureus. J Biol Chem 2007; 282(41):30051-30061.
45. Boden MK, Flock JI. Cloning and characterization of a gene for a 19 kDa fibrinogen-binding protein from Staphylococcus aureus. Mol Microbiol 1994; 12(4):599-606.
46. Hammel M, Sfyroera G, Ricklin D et al. A structural basis for complement inhibition by Staphylococcus aureus. Nat Immunol 2007; 8(4):430-437.
47. Lee LY, Liang X, Hook M et al. Identification and characterization of the C3 binding domain of the Staphylococcus aureus extracellular fibrinogen-binding protein (Efb). J Biol Chem 2004; 279(49):50710-50716.
48. Jarva H, Jokiranta TS, Wurzner R et al. Complement resistance mechanisms of streptococci. Mol Immunol 2003; 40(2-4):95-107.
49. Kraiczy P, Wurzner R. Complement escape of human pathogenic bacteria by acquisition of complement regulators. Mol Immunol 2006; 43:31-44.
50. Nizet V. Understanding how leading bacterial pathogens subvert innate immunity to reveal novel therapeutic targets. J Allergy Clin Immunol 2007; 120(1):13-22.
51. Pérez-Caballero D, García-Laorden I, Cortés G et al. Interaction between complement regulators and Streptococcus pyogenes: binding of C4b-binding protein and factor H/factor H-like protein 1 to M18 strains involves two different cell surface molecules. J Immunol 2004; 173(11):6899-6904.
52. Thern A, Stenberg L, Dahlback B et al. Ig-binding surface proteins of Streptococcus pyogenes also bind human C4b-binding protein (C4BP), a regulatory component of the complement system. J Immunol 1995; 154(1):375-386.
53. Pandiripally V, Gregory E, Cue D. Acquisition of regulators of complement activation by Streptococcus pyogenes serotype M1. Infect Immun 2002; 70(11):6206-6214.
54. Pandiripally V, Wei L, Skerka C et al. Recruitment of complement factor H-like protein 1 promotes intracellular invasion by group A streptococci. Infect Immun 2003; 71(12):7119-7128.
55. Caswell CC, Han R, Hovis KM et al. The Scl1 protein of M6-type group A Streptococcus binds the human complement regulatory protein, factor H and inhibits the alternative pathway of complement. Mol Microbiol 2008; 67(3):584-596.
56. Courtney HS, Hasty DL, Dale JB. Anti-phagocytic mechanisms of Streptococcus pyogenes: binding of fibrinogen to M-related protein. Mol Microbiol 2006; 59(3):936-947.
57. Weineisen M, Sjobring U, Fallman M et al. Streptococcal M5 protein prevents neutrophil phagocytosis by interfering with CD11b/CD18 receptor-mediated association and signaling. J Immunol 2004; 172(6):3798-3807.

58. Lei B, DeLeo FR, Hoe NP et al. Evasion of human innate and acquired immunity by a bacterial homolog of CD11b that inhibits opsonophagocytosis. Nat Med 2001; 7(12):1298-1305.
59. Akesson P, Sjoholm AG, Bjorck L. Protein SIC, a novel extracellular protein of Streptococcus pyogenes interfering with complement function. J Biol Chem 1996; 271(2):1081-1088.
60. Fernie-King BA, Seilly DJ, Willers C et al. Streptococcal inhibitor of complement (SIC) inhibits the membrane attack complex by preventing uptake of C567 onto cell membranes. Immunology 2001; 103(3):390-398.
61. Hartas J, Sriprakash KS. Streptococcus pyogenes strains containing emm12 and emm55 possess a novel gene coding for distantly related SIC protein. Microb Pathog 1999; 26(1):25-33.
62. von Pawel-Rammingen U, Johansson BP, Bjorck L. IdeS, a novel streptococcal cysteine proteinase with unique specificity for immunoglobulin G. EMBO J 2002; 21(7):1607-1615.
63. von Pawel-Rammingen U, Björck L. IdeS and SpeB: immunoglobulin-degrading cysteine proteinases of Streptococcus pyogenes. Curr Opin Microbiol 2003; 6(1):50-55.
64. Tsao N, Tsai WH, Lin YS et al. Streptococcal pyrogenic exotoxin B cleaves properdin and inhibits complement-mediated opsonophagocytosis. Biochem Biophys Res Commun 2006; 339(3):779-784.
65. Kuo CF, Lin YS, Chuang WJ et al. Degrading Complement 3 by Streptococcal Pyrogenic Exotoxin B Inhibits Complement Activation and Neutrophil Opsonophagocytosis. Infect Immun 2008; 76(3):1116-1169.
66. Terao Y, Mori Y, Yamaguchi M et al. Group A streptococcal cysteine protease degrades C3 (C3b) and contributes to evasion of innate immunity. J Biol Chem 2008; 283(10):6253-6260.
67. Wexler DE, Chenoweth DE, Cleary PP. Mechanism of action of the group A streptococcal C5a inactivator. Proc Natl Acad Sci USA 1985; 82(23):8144-8148.
68. Ji Y, McLandsborough L, Kondagunta A et al. C5a peptidase alters clearance and trafficking of group A streptococci by infected mice. Infect Immun 1996; 64(2):503-510.
69. Areschoug T, Stalhammar-Carlemalm M, Karlsson I et al. Streptococcal beta protein has separate binding sites for human factor H and IgA-Fc. J Biol Chem 2002; 277(15):12642-12648.
70. Heden LO, Frithz E, Lindahl G. Molecular characterization of an IgA receptor from group B streptococci: sequence of the gene, identification of a proline-rich region with unique structure and isolation of N-terminal fragments with IgA-binding capacity. Eur J Immunol 1991; 21(6):1481-1490.
71. Chmouryguina I, Suvorov A, Ferrieri P et al. Conservation of the C5a peptidase genes in group A and B streptococci. Infect Immun 1996; 64(7):2387-2390.
72. Bohnsack JF, Chang JK, Hill HR. Restricted ability of group B streptococcal C5a-ase to inactivate C5a prepared from different animal species. Infect Immun 1993; 61(4):1421-1426.
73. Mitchell TJ, Andrew PW, Saunders FK et al. Complement activation and antibody binding by pneumolysin via a region of the toxin homologous to a human acute-phase protein. Mol Microbiol 1991; 5(8):1883-1888.
74. Paton JC, Rowan-Kelly B, Ferrante A. Activation of human complement by the pneumococcal toxin pneumolysin. Infect Immun 1984; 43(3):1085-1087.
75. McDaniel LS, Yother J, Vijayakumar M et al. Use of insertional inactivation to facilitate studies of biological properties of pneumococcal surface protein A (PspA). J Exp Med 1987; 165(2):381-394.
76. Dave S, Brooks-Walter A, Pangburn MK et al. PspC, a pneumococcal surface protein, binds human factor H. Infect Immun 2001; 69(5):3435-3437.
77. Jarva H, Janulczyk R, Hellwage J et al. Streptococcus pneumoniae evades complement attack and opsonophagocytosis by expressing the pspC locus-encoded Hic protein that binds to short consensus repeats 8-11 of factor H. J Immunol 2002; 168(4):1886-1894.
78. Kerr AR, Paterson GK, McCluskey J et al. The contribution of PspC to pneumococcal virulence varies between strains and is accomplished by both complement evasion and complement-independent mechanisms. Infect Immun 2006; 74(9):5319-5324.
79. Li J, Glover DT, Szalai AJ et al. PspA and PspC minimize immune adherence and transfer of pneumococci from erythrocytes to macrophages through their effects on complement activation. Infect Immun 2007; 75(12):5877-5885.
80. Merino S, Camprubi S, Alberti S et al. Mechanisms of Klebsiella pneumoniae resistance to complement-mediated killing. Infect Immun 1992; 60:2529-2535.
81. Alberti S, Marques G, Hernandez-Alles S et al. Interaction between complement subcomponent C1q and the Klebsiella pneumoniae porin OmpK36. Infect Immun 1996; 64:4719-4725.
82. Merino S, Nogueras MM, Aguilar A et al. Activation of the complement classical pathway (C1q binding) by mesophilic Aeromonas hydrophila outer membrane protein. Infect Immun 1998; 66:3825-3831.
83. Clas F, Loos M. Antibody-independent binding of the first component of complement (C1) and its subcomponent C1q to the S and R forms of Salmonella minnesota. Infect Immun 1981; 31:1138-1144.

84. Tomas JM, Camprubi S, Merino S et al. Surface exposure of O1 serotype lipopolysaccharide in Klebsiella pneumoniae strains expressing different K antigens. Infect Immun 1991; 59:2006-2011.
85. Alberti S, Marques G, Camprubi S et al. C1q binding and activation of the complement classical pathway by Klebsiella pneumoniae outer membrane proteins. Infect Immun 1993; 61:852-860.
86. Eisenschenk FC, Houle JJ, Hoffmann EM. Mechanism of serum resistance among Brucella abortus isolates. Vet Microbiol 1999; 68:235-244.
87. Peak IR, Jennings MP, Hood DW et al. Tetrameric repeat units associated with virulence factor phase variation in Haemophilus also occur in Neisseria spp. and Moraxella catarrhalis. FEMS Microbiol Lett 1996; 137:109-114.
88. Peak IR, Jennings MP, Hood DW et al. Tetranucleotide repeats identify novel virulence determinant homologues in Neisseria meningitidis. Microb Pathog 1999; 26:13-23.
89. Saunders NJ, Peden JF, Hood DW et al. Simple sequence repeats in the Helicobacter pylori genome. Mol Microbiol 1998; 27:1091-1098.
90. Mandrell RE, Apicella MA. Lipo-oligosaccharides (LOS) of mucosal pathogens: molecular mimicry and host-modification of LOS. Immunobiology 1993; 187:382-402.
91. Mandrell RE, Griffiss JM, Macher BA. Lipooligosaccharides (LOS) of Neisseria gonorrhoeae and Neisseria meningitidis have components that are immunochemically similar to precursors of human blood group antigens. Carbohydrate sequence specificity of the mouse monoclonal antibodies that recognize cross-reacting antigens on LOS and human erythrocytes. J Exp Med 1988; 168:107-126.
92. Yuki NK, Susuki M. Koga Y et al. Carbohydrate mimicry between human ganglioside GM1 and Campylobacter jejuni lipooligosaccharide causes Guillain-Barre syndrome. Proc Natl Acad Sci USA 2004; 101:11404-11409.
93. Edwards JL, Apicella MA. The role of lipooligosaccharide in Neisseria gonorrhoeae pathogenesis of cervical epithelia: lipid A serves as a C3 acceptor molecule. Cell Microbiol 2002; 4:585-598.
94. Ho DK, Ram S, Nelson KL et al. lgtC expression modulates resistance to C4b deposition on an invasive nontypeable Haemophilus influenzae. J Immunol 2007; 178:1002-1012.
95. Lewis LA, Ram S, Prasad A et al. Defining targets for complement components C4b and C3b on the pathogenic neisseriae. Infect Immun 2008; 76:339-350.
96. Ram S, Cox AD, Wright JC et al. Neisserial lipooligosaccharide is a target for complement component C4b: Inner core phosphoethanolamine residues define C4b linkage specificity. J Biol Chem 2003; 278:50853-50862.
97. Madico G, Ngampasutadol J, Gulati S et al. Factor H Binding and Function in Sialylated Pathogenic Neisseriae is Influenced by Gonococcal, but Not Meningococcal, Porin. J Immunol 2007; 178:4489-4497.
98. Hong YQ, Ghebrehiwet B. Effect of Pseudomonas aeruginosa elastase and alkaline protease on serum complement and isolated components C1q and C3. Clin Immunol Immunopathol 1992; 62:133-138.
99. Wolf U, Bauer D, Traub WH. Metalloproteases of Serratia liquefaciens: degradation of purified human serum proteins. Zentralbl Bakteriol 1991; 276:16-26.
100. Molla A, Akaike T, Maeda H. Inactivation of various proteinase inhibitors and the complement system in human plasma by the 56-kilodalton proteinase from Serratia marcescens. Infect Immun 1989; 57:1868-1871.
101. Sodeinde OA, Subrahmanyam YV, Stark K et al. A surface protease and the invasive character of plague. Science 1992; 258:1004-1007.
102. Jagels MA, Travis J, Potempa J et al. Proteolytic inactivation of the leukocyte C5a receptor by proteinases derived from Porphyromonas gingivalis. Infect Immun 1996; 64:1984-1991.
103. Discipio RG, Daffern PJ, Kawahara M et al. Cleavage of human complement component C5 by cysteine proteinases from Porphyromonas (Bacteroides) gingivalis. Prior oxidation of C5 augments proteinase digestion of C5. Immunology 1996; 87:660-667.
104. Jagels MA, Ember JA, Travis J et al. Cleavage of the human C5A receptor by proteinases derived from Porphyromonas gingivalis: cleavage of leukocyte C5a receptor. Adv Exp Med Biol 1996; 389:155-164.
105. Schenkein HA, Fletcher HM, Bodnar M et al. Increased opsonization of a prtH-defective mutant of Porphyromonas gingivalis W83 is caused by reduced degradation of complement-derived opsonins. J Immunol 1995; 154:5331-5337.
106. Grenier D. Inactivation of human serum bactericidal activity by a trypsinlike protease isolated from Porphyromonas gingivalis. Infect Immun 1992; 60:1854-1857.
107. Popadiak K, Potempa J, Riesbeck K et al. Biphasic effect of gingipains from Porphyromonas gingivalis on the human complement system. J Immunol 2007; 178:7242-7250.
108. Wingrove JA, DiScipio RG, Chen Z et al. Activation of complement components C3 and C5 by a cysteine proteinase (gingipain-1) from Porphyromonas (Bacteroides) gingivalis. J Biol Chem 1992; 267:18902-18907.

109. Slaney JM, Curtis MA. Mechanisms of evasion of complement by Porphyromonas ging\valis. Front Biosci 2008; 13:188-196.
110. Slaney JM, Gallagher A, Aduse-Opoku J et al. Mechanisms of resistance of Porphyromonas gingivalis to killing by serum complement. Infect Immun 2006; 74:5352-5361.
111. Lathem WW, Price PA, Miller VL et al. A plasminogen-activating protease specifically controls the development of primary pneumonic plague. Science 2007; 315:509-513.
112. Blom AM, Rytkonen A, Vasquez P et al. A novel interaction between Type IV pili of Neisseria gonorrhoeae and the human complement regulator C4B-binding protein. J Immunol 2001; 166:6764-6770.
113. Ram S, Cullinane M, Blom A et al. Binding of C4b-binding Protein to Porin: A molecular mechanism of serum resistance of Neisseria gonorrhoeae. J Exp Med 2001; 193:281-296.
114. Nordstrom T, Blom AM, Forsgren A et al. The emerging pathogen Moraxella catarrhalis interacts with complement inhibitor C4b binding protein through ubiquitous surface proteins A1 and A2. J Immunol 2004; 173:4598-4606.
115. Attia AS, Ram S, Rice PA et al. Binding of vitronectin by the Moraxella catarrhalis UspA2 protein interferes with late stages of the complement cascade. Infect Immun 2006; 74:1597-1611.
116. Ngampasutadol J, Ram S, Blom AM et al. Human C4b-binding protein selectively interacts with Neisseria gonorrhoeae and results in species-specific infection. Proc Natl Acad Sci USA 2005; 102:17142-17147.
117. Arko RJ, Kraus SJ, Brown WJ et al. Neisseria gonorrhoeae: effects of systemic immunization on resistance of chimpanzees to urethral infection. J Infect Dis 1974; 130:160-164.
118. Stevenson B, El-Hage N, Hines MA et al. Differential binding of host complement inhibitor factor H by Borrelia burgdorferi Erp surface proteins: a possible mechanism underlying the expansive host range of Lyme disease spirochetes. Infect Immun 2002; 70:491-497.
119. Bartra SS, Styer KL, O'Bryant DM et al. Resistance of Yersinia pestis to Complement-Dependent Killing is Mediated by the Ail Outer Membrane Protein. Infect Immun 2008; 76 (2): 612-622.
120. Heffernan EJ, Reed S, Hackett J et al. Mechanism of resistance to complement-mediated killing of bacteria encoded by the Salmonella typhimurium virulence plasmid gene rck. J Clin Invest 1992; 90:953-964.
121. Bliska JB, Falkow S. Bacterial resistance to complement killing mediated by the Ail protein of Yersinia enterocolitica. Proc Natl Acad Sci USA 1992; 89:3561-3565.
122. Joiner KA, Hammer CH, Brown EJ et al. Studies on the mechanism of bacterial resistance to complement-mediated killing. I. Terminal complement components are deposited and released from Salmonella minnesota S218 without causing bacterial death. J Exp Med 1982; 155:797-808.
123. Joiner KA, Hammer CH, Brown EJ et al. Studies on the mechanism of bacterial resistance to complement-mediated killing. II. C8 and C9 release C5b67 from the surface of Salmonella minnesota S218 because the terminal complex does not insert into the bacterial outer membrane. J Exp Med 1982; 155:809-919.
124. Joiner KA, Warren KA, Hammer C et al. Bactericidal but not nonbactericidal C5b-9 is associated with distinctive outer membrane proteins in Neisseria gonorrhoeae. J Immunol 1985; 134:1920-1925.
125. Pfeiffer R. Z Hyg Infektionskr 1895; 20:215.
126. Thomas L, Dingle JH. Investigations of meningococcal infection. III. Bactericidal action of normal and immune sera for the meningococcus. J Clin Invest 1943; 22:375-385.
127. Jarvis GA, Griffiss JM. Human IgA1 blockade of IgG-initiated lysis of Neisseria meningitidis is a function of antigen-binding fragment binding to the polysaccharide capsule. J Immunol 1991; 147:1962-1967.
128. Selander B, Kayhty H, Wedege E et al. Vaccination responses to capsular polysaccharides of Neisseria meningitidis and Haemophilus influenzae Type b in two C2-deficient sisters: alternative pathway-mediated bacterial killing and evidence for a novel type of blocking IgG. J Clin Immunol 2000; 20:138-149.
129. Hall WH, Manion RE, Zinneman HH. Blocking serum lysis of Brucella abortus by hyperimmune rabbit immunoglubulin A. J Immunol 1971; 107:41-46.
130. Glenchur H, Hall WH, Zinneman HH. Blocking antibodies in rabbits infected with Brucella melitensis. Proc Soc Exp Biol Med 1959; 101:422-425.
131. Guttman RM, Waisbren BA. Bacterial blocking activity of specific IgG in chronic Pseudomonas aeruginosa infection. Clin Exp Immunol 1975; 19:121-130.
132. Waisbren BA, Brown I. A factor in the serum of patients with persisting infection that inhibits the bactericidal activity of normal serum against the organism that is causing the infection. J Immunol 1966; 97:431-437.
133. Joiner KA, Scales R, Warren KA et al. Mechanism of action of blocking immunoglobulin G for Neisseria gonorrhoeae. J Clin Invest 1985; 76:1765-1772.
134. Plummer FA, Chubb H, Simonsen JN et al. Antibody to Rmp (outer membrane protein 3) increases susceptibility to gonococcal infection. J Clin Invest 1993; 91:339-343.
135. Kraiczy P, Skerka C, Brade V et al. Further characterization of complement regulator-acquiring surface proteins of Borrelia burgdorferi. Infect Immun 2001; 69(12):7800-7809.

136. Alitalo A, Meri T, Lankinen H et al. Complement inhibitor factor H binding to Lyme disease spirochetes is mediated by inducible expression of multiple plasmid-encoded outer surface protein E paralogs. J Immunol 2002; 169(7):3847-3853.
137. Pausa M, Pellis V, Cinco M et al. Serum-resistant strains of Borrelia burgdorferi evade complement-mediated killing by expressing a CD59-like complement inhibitory molecule. J Immunol 2003; 170(6):3214-3222.
138. McDowell JV, Wolfgang J, Tran E et al. Comprehensive analysis of the factor h binding capabilities of borrelia species associated with lyme disease: delineation of two distinct classes of factor H binding proteins. Infect Immun 2003; 71(6):3597-3602.
139. Hovis KM, McDowell JV, Griffin L et al. Identification and characterization of a linear-plasmid-encoded factor H-binding protein (FhbA) of the relapsing fever spirochete Borrelia hermsii. J Bacteriol 2004; 186(9):2612-2618.
140. Meri T, Cutler SJ, Blom AM et al. Relapsing fever spirochetes Borrelia recurrentis and B. Duttonii acquire complement regulators C4b-binding protein and factor H. Infect Immun 2006; 74(7):4157-4163.
141. McDowell JV, Frederick J, Stamm L et al. Identification of the gene encoding the FhbB protein of Treponema denticola, a highly unique factor H-like protein 1 binding protein. Infect Immun 2007; 75(2):1050-1054.
142. McDowell JV, Lankford J, Stamm L et al. Demonstration of factor H-like protein 1 binding to Treponema denticola, a pathogen associated with periodontal disease in humans. Infect Immun 2005; 73(11):7126-7132.
143. Gros P, Milder FJ, Janssen BJ. Complement driven by conformational changes. Nat Rev Immunol 2008; 8(1):48-58.
144. Marr N, Luu RA, Fernandez RC. Bordetella pertussis binds human C1 esterase inhibitor during the virulent phase, to evade complement-mediated killing. J Infect Dis 2007; 195:585-588.
145. China B, Sory MP, N'Guyen BT et al. Role of the YadA protein in prevention of opsonization of Yersinia enterocolitica by C3b molecules. Infect Immun 1993; 61:3129-3136.
146. Ram S, McQuillen DP, Gulati S et al. Binding of complement factor H to loop 5 of porin protein 1A: a molecular mechanism of serum resistance of nonsialylated Neisseria gonorrhoeae. J Exp Med 1998; 188:671-680.
147. Madico G, Welsch JA, Lewis LA et al. The meningococcal vaccine candidate GNA1870 binds the complement regulatory protein factor H and enhances serum resistance. J Immunol 2006; 177:501-510.
148. Kunert A, Losse J, Gruszin C et al. Immune evasion of the human pathogen Pseudomonas aeruginosa: elongation factor Tuf is a factor H and plasminogen binding protein. J Immunol 2007; 179:2979-2988.
149. Prasadarao NV, Blom AM, Villoutreix BO et al. A novel interaction of outer membrane protein A with C4b binding protein mediates serum resistance of Escherichia coli K1. J Immunol 2002; 169:6352-6360.
150. Berggard K, Johnsson E, Mooi FR et al. Bordetella pertussis binds the human complement regulator C4BP: role of filamentous hemagglutinin. Infect Immun 1997; 65:3638-3643.
151. Jarva H, Ram S, Vogel U et al. Binding of the complement inhibitor C4bp to serogroup B Neisseria meningitidis. J Immunol 2005; 174:6299-6307.
152. Hallstrom T, Jarva H, Riesbeck K et al. Interaction with C4b-binding protein contributes to nontypeable Haemophilus influenzae serum resistance. J Immunol 2007; 178:6359-6366.
153. Hallstrom T, Trajkovska E, Forsgren A et al. Haemophilus influenzae surface fibrils contribute to serum resistance by interacting with vitronectin. J Immunol 2006; 177:430-436.

Chapter 4

Modulation of Innate Immune Signalling Pathways by Viral Proteins

Orla Mulhern, Barry Harrington and Andrew G. Bowie*

Abstract

In recent years an explosion of information on the various strategies viruses employ to penetrate and hijack the host cell has led to an increased understanding of both viruses themselves and the host immune response. Despite their simplicity viruses have evolved a number of strategies to not only evade the host immune response but also modulate immune signalling to favour their replication and survival within the cell. The innate immune response provides the host with an early reaction against viruses. This response relies heavily upon the recognition of pathogen-associated molecular patterns (PAMPs) by a number of host pattern recognition receptors (PRRs), leading to activation of innate signalling pathways and altered gene expression. In this chapter we outline the signalling pathways that respond to viral infection and the various methods that viruses utilize to evade detection and modulate the innate immune response to favour their survival.

Introduction

The host immune response relies upon the detection of PAMPs to initiate the innate immune response and ensuing adaptive response. These PAMPs are usually highly conserved molecular patterns that are produced by the invading organism and are essential to the life cycle of the organism. Viral PAMPs include DNA, dsRNA, ssRNA and various viral proteins. Upon recognition of these PAMPs by host PRRs, signalling pathways are initiated which result in the activation of transcription factors such as the interferon (IFN) regulatory factors (IRFs) and nuclear factor κB (NFκB). Altered gene expression in response to this, especially the upregulation of Type I IFNs, induces an antiviral state in the cell. Given the importance that host PRRs play in challenging viral infection, it is critical that viruses adopt mechanism to avoid them. Indeed viruses not only evade, but also subvert PRRs, by turning particular signalling pathways on for the benefit of the life cycle of the virus. Interestingly, by using viruses to study the interaction between host and pathogen many lessons can be learnt, not only related to viral evasion, but also to the host immune response. Thus viruses can serve as immunological tools to define important host responses. The viruses referred to in this chapter and their classification and common abbreviation are shown in Table 1.

PKR

Double-stranded (ds) RNA-dependent protein kinase (PKR) was the first PRR shown to recognise a product of viral replication, namely dsRNA. PKR is induced by IFN and in a critical host antiviral defence mechanism and as such it is both a PRR and an IFN effector protein.[1] The discovery of PKR explained the observation that in cells pretreated with IFN, the translation of viral mRNAs were blocked.[1,2] PKR is activated in response to viral, cellular or synthetic dsRNA of 30

*Corresponding Author: Andrew G Bowie—School of Biochemistry and Immunology, Trinity College Dublin, Dublin 2, Ireland. Email: agbowie@tcd.ie

Pathogen-Derived Immunomodulatory Molecules, edited by Padraic G. Fallon.

Table 1. Virus nomenclature

Abbreviation	Virus	Family	Genome
AdV	Adenovirus	*Adenoviridae*	dsDNA
CMV	Cytomegalovirus	*Herpesviridae*	dsDNA
EBV	Epstein-barr virus	*Herpesviridae*	dsDNA
EMCV	Encephalomyocarditis virus	*Picornaviridae*	ssRNA (+)
HCV	Hepatitis C virus	*Flaviviridae*	ssRNA (+)
HIV	Human immunodeficiency virus	*Retroviridae*	ssRNA (RT)
HPV	Human papilloma virus	*Papvaviridae*	dsDNA
HSV	Herpes simplex virus	*Herpesviridae*	dsDNA
IAV	Influenza A virus	*Orthomyxoviridae*	ssRNA (–)
JEV	Japanese encephalitis virus	*Flaviviridae*	ssRNA (+)
MMTV	Mouse mammary tumor virus	*Retroviridae*	ssRNA (RT)
PV	Polio virus	*Picornaviridae*	ssRNA (+)
RSV	Respiratory syncytical virus	*Paramyxoviridae*	ssRNA (–)
RV	Rabies virus	*Rhabdoviridae*	ssRNA (–)
VACV	Vaccinia virus	*Poxviridae*	dsDNA
VSV	Vesicular stomatitis virus	*Rhabdoviridae*	ssRNA (–)
WNV	West nile virus	*Flaviviridae*	ssRNA (+)

+: Positive-stranded RNA; –: Negative-stranded RNA; RT: Reverse transcribed RNA.

nucleotides or longer, such as poly(I:C). In addition PKR can be activated by single-stranded (ss) RNA molecules with internal dsRNA structure, as in the case of adenovirus (AdV) virus-associated Type I (VAI) RNA and human immunodeficiency virus (HIV) transactivation-responsive element (TAR) RNA.[3,4] PKR can also be activated by cell stresses and through binding to PKR-associated activator (PACT).[5,6] More recently it has been demonstrated that single stranded RNA (ssRNA) with a 5′-triphosphate (ppp), which may have limited secondary structure, can activate PKR.[7]

PKR inhibits translation through the phosphorylation of the alpha subunit of eukaryotic initiation factor 2 (eIF-2α). PKR not only signals to eIF-2α but also to the signal transducer and activator of transcription (STAT), NFκB, IRF1, p53, c-Jun N-terminal MAP kinase (JNK) and p38 MAP kinase pathways.[8-10] PKR may also function in Toll-like receptor 3 (TLR3) signalling.[11] As shown in Figure 1, unactivated PKR is in a monomeric form in unstimulated cells. Recognition and binding of dsRNA molecules by PKR occurs through the two N-terminal dsRNA binding motifs in PKR.[12] Conformational changes then result in PKR homodimerisation and autophosphorylation. This autophosphorylation occurs in a stretch of amino acids termed the activation segment. Within this region phosphorylation of residues thr-446 and thr-451 are of particular importance in stabilisation of PKR dimerisation and leading to increased kinase activity. Phosphorylation of eIF2α then occurs at ser-51, which prevents eIF2α facilitating translation of either host or viral mRNAs.

Given the role PKR plays in the host response to viruses, it is of importance for the successful propagation of the virus to evolve methods of evading PKR. There are numerous ways that viruses have devised to prevent or suppress PKR activation, as shown in Figure 1. Firstly, viruses have evolved ways to sequester dsRNA molecules, which is not surprising considering the importance of dsRNA as a danger signal for viral infection. Vaccinia virus (VACV) is a large DNA virus which contains over 200 open reading frames, many of which encode proteins involved in virus-host interactions which show immune modulatory capabilities. The VACV *E3L* gene encodes the E3 dsRNA-binding protein which is an important virulence factor for the virus. E3 contains a highly conserved dsRNA-binding motif in the C-terminus and that this motif is important to virulence was shown by the fact that VACV expressing dsRNA-binding proteins instead of E3L sustain some

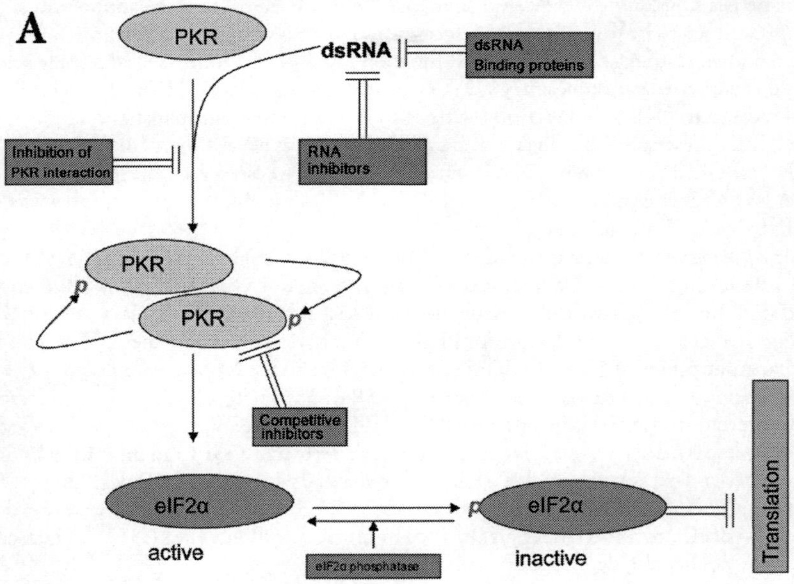

B

Mechanism of inhibition	Virus	Protein
dsRNA binding proteins	VACV IAV EBV HSV	E3 NS1 SM Us11
RNA Inhibitors	HCV EBV AdV	IRES EBER RNA VAI RNA
PKR interaction	HCV IAV VACV EBV	NS5A,E2 P58 E3 SM
PKR degradation	PV	Protease
eIF2α phosphatase	HSV HPV SV-40	γ34.5 E6 Large T antigen

Figure 1. Activation and evasion of PKR signalling. A) schematic of the PKR signalling pathway showing points modulated by viruses. *p*, phosphorylated. B) Viral proteins shown to act on the PKR pathway. See text for details.

of the characteristics of the wild-type virus.[13-16] One property of the multifunctional NS1 protein from influenza A virus (IAV) is its ability,to sequester RNA and thereby block PKR activation.[17,18] Other viral proteins capable of binding dsRNA include the Epstein-Barr virus (EBV) protein SM and the herpes simplex virus (HSV) protein Us11 which both contain the amino acid sequence arg-X-pro, shown to be part of a dsRNA recognition motif.[19,20]

A second viral strategy to disrupt PKR function involves antagonistic viral nucleic acids. For example, the internal ribosome entry site (IRES) of Hepatitis C virus (HCV) has been shown to be able to bind to PKR in competition with dsRNA and prevent autophosphorylation and activation of PKR in vitro.[21] EBV nucleic acid can also antagonise PKR, since EBV-encoded RNA 1 (EBER-1) and EBER-2 can bind PKR in vitro[22-24] and EBER-1 can reverse the inhibitory effects of dsRNA on protein synthesis. AdV-encoded VAI RNA binds to PKR in competition with dsRNA but fails to lead to PKR autophosphorylation or activation.

Thirdly, some viral proteins can bind to and inhibit PKR, such as the HCV protein NS5A which inhibits PKR dimerization and activation.[25,26] The envelope glycoprotein E2 of HCV encodes a second inhibitor of PKR which has sequence similarity to the phosphorylation site of eIF2α,[27] allowing it to act as a pseudosubstrate for PKR.[28] In a variation on this theme, IAV uses a host cellular chaperone protein, p58 as a PKR inhibitor. In uninfected cells p58 is complexed with Hsp40. IAV infection somehow causes disassociation of p58 from Hsp40, allowing p58 to interact with PKR to prevent its dimerization and activation.[26,29] E3 from VACV is not only able to sequester dsRNA as described above, but it is also capable of directly binding to and inhibiting PKR.[30]

A further viral strategy to disable PKR is poliovirus (PV)-induced PKR degradation,[31] while herpes simplex virus (HSV) encodes a protein (the γ34.5 gene product) which promotes dephosphorylation of eIF2α.[32] This strategy is also used by human papilloma virus (HPV), which encodes the protein E6 that also dephosphorylates eIF2α.[33]

Toll-Like Receptors (TLRs)

The second class of PRRs shown to play a role in the anti-viral immune response are the TLRs. The discovery of TLRs began with the identification of the *drosophila* protein Toll, a receptor that is expressed by insects and that was found to be essential for dorsoventral polarity during *drosophila melangoster* embryogenesis.[34] Further studies showed that Toll has an essential role in insect innate immunity with regard to fungal infection.[35] Database searches eventually led to the identification of 13 mammalian TLRs with TLR1-10 found in humans and TLR1-13 found in mice (although TLR10 is defective in mice due to most of the gene being deleted). More recently the Toll pathway has been shown to be required for anti-viral immunity in *drosophila*.[36] TLRs serve to initiate the innate immune response and to orchestrate the ensuing adaptive response.

Recognition of Viral Nucleic Acid by TLR 3, 7, 8 and 9

Following the discovery of PKR, TLR3 was identified as the second PRR receptor for viral dsRNA, based primarily on two lines of evidence. Firstly, PKR-deficient mice could still respond to the synthetic dsRNA poly(I:C).[37,38] Secondly TLR3–/– mice showed impaired production of the cytokines TNFα, IL6 and IL12 and reduced expression of IFN-α and IFN-β when stimulated with poly(I:C) compared to wild type cells.[39] TLR3 has been shown to be present on the cell surface of fibroblasts, however TLR3 was also shown to localize to an intracellular vesicular compartment in myeloid dendritic cells (mDCs) and cannot be detected on their cell surface.[40,41] This compartment has been assumed to be endosomal because inhibition of endosomal acidification abrogates poly(I:C) signalling. TLR3 has been shown to recognise dsRNA viruses such as reovirus, but interestingly also recognises West Nile virus (WNV), a positive ssRNA virus and murine cytomegalovirus (CMV), a DNA virus. This makes sense as both positive ssRNA and DNA viruses have been shown to have dsRNA replication intermediates.[42] However many other positive ssRNA and DNA viruses do not elicit a TLR3 response, the reason for which is not yet clear. It has also been reported that TLR3 is important for cross-presentation during viral infection where uptake of viral antigens from dead or dying cells by dendritic cells are presented to cytotoxic CD8+ T-cells. Thus TLR3 may have a greater contribution to the immune response where cross-presentation is a predominant feature.[43]

TLR9 was first shown to sense unmethylated CpG DNA present in bacterial genomes.[44] This unmethylated DNA is a potent immune stimulator and there is a marked reduction in the frequency of CpG motifs in vertebrate DNA (where cytosine residues are heavily methylated) which prevents host DNA being immune-stimulatory. Viral genomes also contain unmethylated CpG DNA motifs and HSV-1, a DNA virus, was shown to induced Type I IFNs in plasmacytoid dendritic cells (pDCs) through TLR9.[45] Further, TLR9 –/– mice injected with HSV-2 did not produce IFN-α.[46] Like TLR3, TLR9 is located in intracellular endosomal compartments.[47]

TLR7 and TLR8 are more closely related to TLR9 than any other TLRs with respect to their sequence similarity and signalling and have also been shown to reside in endosomal compartments where they detect viral ssRNA. Homodimers consisting of murine TLR7 or human TLR8 have been shown to detect guanosine- and uridine-rich oligonucleotides derived from HIV-1 and induce IFN-α and pro-inflammatory cytokines in DCs and macrophages.[48] Two ssRNA viruses, vesicular stomatitis virus (VSV) and IAV have been shown to elicit an IFN response through a TLR7-mediated pathway.[49] Diebold et al[50] determined that the recognition of RNA by TLR7 requires the presence of both ribose and uridine. Interestingly, both viral and 'self' RNA are capable of activating TLR7 equally efficiently if delivered to endosomes suggesting that TLR7 discriminates between viral and 'self' RNA based on their endosomal accessibility rather than sequence specificity. TLR7 would therefore also be able to detect 'self' RNA endocytosed from damaged tissue. Thus TLR3, TLR7/8 and TLR9 are capable of detecting viral dsRNA, ssRNA and dsDNA respectively (see Fig. 2).

TLR Responses to Viral Proteins

Apart from viral nucleic acids, some viral proteins also seem capable of activating PRR signalling pathways and to date the two TLRs shown to respond to viral proteins are TLR2 and TLR4 (Fig. 2). The first viral protein shown to induce cellular responses through TLRs was the fusion (F) protein of respiratory syncytial virus (RSV), which stimulates cytokine induction via TLR4 and CD14 in mice.[51] RSV causes severe lower respiratory tract infections in children, but the importance the RSV-TLR4 interaction remains unclear as another group failed to see a role for TLR4 mediated cytokine induction in controlling RSV in infected mice.[52] HSV is not only detected by TLR9 via its genomic DNA but also through TLR2, which has been shown to mediate cellular responses to viral glycoproteins. Although TLR2 –/– mice display higher viral titres than wild type mice after HSV infection, they are much less susceptible to challenge with HSV and after infection show reduced cytokine production in the brain, leading to the suggestion that the HSV-TLR2 interaction may lead to encephalitis.[53] The human CMV envelope glycoproteins gB and gH have been shown to mediate NFκB activation and cytokine induction through TLR2 from UV inactivated virions and thus in a replication-independent event. VACV has also been shown to stimulate cytokine production via TLR2, although the VACV ligand for this activation is still unknown. The activation of the innate immune system in response to VACV also requires TLR-independent production of IFN-β.[54]

TLR Signalling Pathways Activated by Viruses

TLRs contain a highly conserved intracellular region of around 200 amino acids termed the Toll/interleukin-1 receptor (TIR) domain. It is the TIR domain that has been shown to be responsible for downstream signalling via homotypic interactions. Ligand binding by the TLRs causes receptor dimerisation resulting in the recruitment of TIR domain-containing adaptors, of which there are five: myeloid differentiation factor 88 (MyD88), MyD88 adaptor-like (MAL), TIR domain-containing adaptor inducing IFNβ (TRIF), TRIF-related adaptor molecule (TRAM) and sterile alpha and HEAT/Armadillo motif (SARM).[55] The adaptors mediate downstream signalling in the cytoplasm (see Fig. 2). All TLRs signal through MyD88, with the exception of TLR3, which utilises only TRIF. Interestingly, TLR4 signals through four adaptors activating a MyD88-dependent pathway via MAL and a TRIF-dependent pathway via TRAM. As such TRIF is involved in TLR3 and TLR4 signalling. SARM has been shown to play a negative role in TLR3 and TLR4 TRIF-dependent signalling by directly binding to TRIF and inhibiting downstream

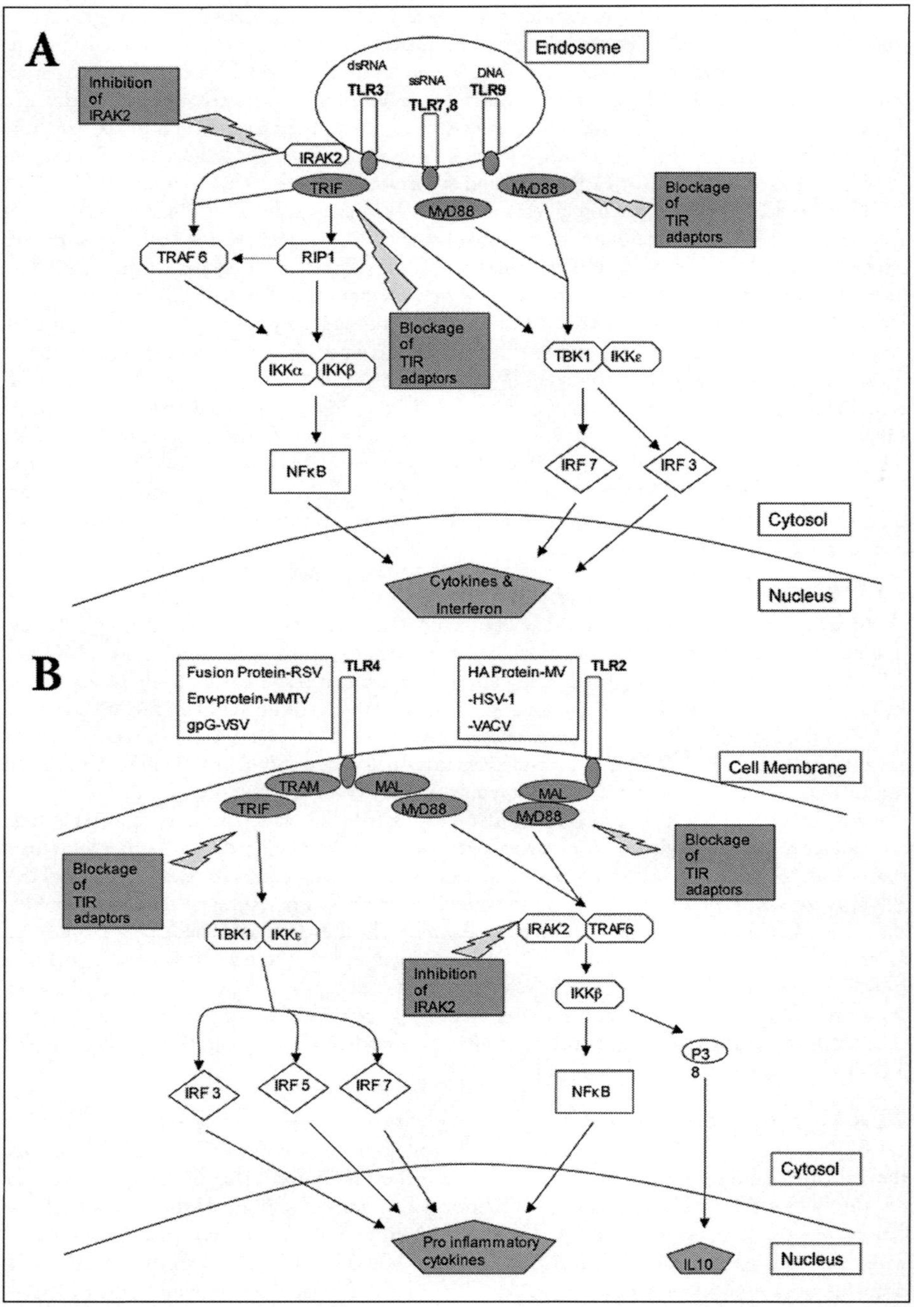

Figure 2. Activation and evasion of TLR signalling. Shown is a schematic of endosomal TLR (A) and membrane TLR (B) signalling pathways to NFκB and IRF activation showing points modulated by viruses (Previous page). Figure continued on next page.

C

Mechanism of inhibition	Virus	Protein
Antagonising TLR adaptor proteins	VACV HCV	A46 NS3/4A
Inhibition of IRAK2	VACV	A52

Figure 2, continued. C) Viral proteins shown to act on the TLR pathway. See text for details.

signals.[56] Differential utilisation of the various adaptor proteins by the TLRs results in the activation of multiple transcription factors including NFκB, IRF3, IRF7 and cJun/ATF2. Of note, many cytokine promoters such as TNF are NFκB-dependent, while Type I IFN induction requires IRF7 in the case of IFNα and IRF3, NFκB and cJun/ATF2 in the case of IFNβ. Thus TLR3 recruitment of TRIF activates NFκB, IRF3, IRF7 and cJun/ATF2 resulting in the upregulation of IFNα/β and of the cytokines TNFα, IL6 and IL12. TLR2 activates NFκB via MyD88 and Mal, while TLR4 can activate NFκB via the MyD88 pathway and both NFκB and IRF3 via TRIF. Finally, TLR7, 8 and 9 activate both NFκB and IRF7 via MyD88.[55]

Viral Evasion of TLR Signalling Pathways

For all the TLRs except TLR3, NFκB activation can proceed via MyD88.[55] MyD88 dependent signalling results in the recruitment of the IL-1R associated kinases (IRAK1, 2, 4 and IRAK-M). IRAK1 and IRAK4 are active kinases whereas IRAK2 and IRAK-M do not possess kinase activity. IRAK4 seems to be the first IRAK to interact with MyD88, leading to the recruitment and activation of IRAK1. IRAK-M appears to have an inhibitory role in that it prevents the dissociation of IRAK1 and IRAK4 from MyD88 and therefore their ability to interact with downstream molecules. TNF receptor-associated factor 6 (TRAF6) is located downstream of IRAKs. TRAF6 is crucial for NFκB activation and is upstream of the inhibitor of NFκB (IκB) kinase (IKK) complex, which is the central NFκB activating kinase complex. Upstream TLR engagement triggers the Lys-63 linked ubiquitination of TRAF6, which results in the recruitment and activation of TGF-β activated protein kinase 1 (TAK1). TAK1 is the kinase responsible for phosphorylating and activating IKKβ which together with IKKα and NFκB essential modulator (NEMO) comprises the IKK complex. IKK then phosphorylates IκB. IκB sequesters NF-κB dimers in the cytoplasm in unstimulated cells and phosphorylation of IκB by IKK targets it for lys-48-linked ubiquitination and subsequent degradation by the proteasome. The IKK complex also phosphorylates the NFκB subunit p65, which is necessary for transactivation of genes. Thereafter, NFκB translocates to the nucleus and stimulates gene transcription. TAK1 also simultaneously activates the JNK and p38 MAP kinase pathways leading to the phosphorylation and activation of ATF2 and cJun.

IRAK1 was initially thought to be the critical kinase for NFκB activation in this MyD88 pathway to TAK1 and IKK via TRAF6. However recently IRAK2 was shown to have a more central role in TRAF6 activation than IRAK1.[57] In fact it was studies of how viruses modulate TLR signalling which led to this conclusion. The VACV protein A52 was found to be an inhibitor of all TLR pathways to NFκB activation.[57-59] This suggested that A52 would target a signalling molecule common to all TLR pathways to NFκB and since A52 interacted with IRAK2 and not IRAK1, this suggested an important role for IRAK2 in NFκB activation (A52 also interacts with TRAF6 directly but surprisingly this interaction does not contribute to NFκB inhibition, but rather leads to more subtle manipulation of the TLR pathway,[57,60] as discussed below). Further work then showed that suppression of IRAK2 expression by siRNA inhibited the TLR-NFκB axis and that expression of IRAK2, but not IRAK1, triggered TRAF6 ubiquitination.[57] IRAK2 can interact with both MyD88 and MAL and has been shown to act downstream of these adaptors.[57] TLR3 however does not utilize MyD88 or MAL but interestingly, A52 is a very potent inhibitor of the TLR3-NFκB axis when compared to other TLRs. It was this observation that led to the discovery that IRAK2 can directly interact with TLR3 to facilitate NFκB activation.[57] Consistent with this, IRAK1 was previously shown to have no role in TLR3 signalling.[11] The study of A52 and IRAK2 provides an excellent example of how understanding viral modulation of innate immunity can reveal important information about host mechanisms.

It is still unclear however, how IRAK2 and TRIF work together on the TLR3 pathway. The N-terminus of TRIF contains a TRAF6 binding motif which has been proposed to integrate TRAF6 into TLR3 mediated NFκB activation. Further, the C-terminus of TRIF mediates a second method of NFκB activation by recruiting receptor-interacting protein-1 (RIP-1), a member of the RIP family involved in TNF receptor mediated NFκB activation. This leads to NFκB but not IRF activation.[61] RIP1, like TRAF6, undergoes lys-63 linked ubiquitination and forms a complex with TAK1.[62] The recruitment of both TRAF6 and RIP1 facilitates TAK1 activation, resulting in maximal NFκB and MAP kinase activation.

In contrast to its inhibitory effect on the TLR pathways to NFκB, A52 has no effect on IRF3 or IRF7 activation, suggesting no role for IRAK2 in these axes.[57] There are a number of now well-defined TLR pathways to IRF activation via TRIF for TLR3 and TLR4 and via MyD88 for TLR7, 8 and 9. In the case of the TRIF pathway, IRF3 activation requires the N-terminal region of TRIF and involves recruitment of TRAF3 and NAK-associated protein 1 (NAP1).[55] This leads to the recruitment and activation of the active IRF3/7 kinase complex containing TANK-binding kinase 1 (TBK1) and IKKε. Once phosphorylated IRF3 and IRF7 form homodimers and translocate into the nucleus to initiate transcription of target genes. For TLR7 and TLR9 signalling, IRAK1 is the required for IRF7 activation.[63,64] In contrast to the TRIF pathway, TBK1 and IKKε are not involved here. Rather it seems that MyD88 and IRF7 directly interact while TRAF6, TRAF3 and IKKα are recruited to the complex and have all been shown to be essential to IRF7 activation.[65] A second VACV protein, A46, can inhibit these TLR-IRF axes.[66] From database searches A46 was shown to contain a TIR domain similar to the adaptor molecules used by the TLRs. Interestingly a number of bacterial pathogens have been subsequently shown to express proteins containing TIR domains.[67] Stack et al[66] demonstrated that A46 was capable of binding to the TIR adaptors MyD88, MAL TRIF and TRAM resulting in the inhibition of both NFκB and IRF activation in all TLR pathways. A46 failed to interact with SARM, which is in fitting with the role for SARM as a negative regulator of TLR signalling. Although A46 and A52 appear to have some overlapping effects on TLR signalling, both proteins independently contribute to the virulence of the virus.[59,66] Even though A46 is still the only viral TIR domain-containing protein identified to date, other viruses have been shown to utilise the strategy of disabling TIR adaptor proteins: HCV encodes a serine protease, NS3/4A, which has been demonstrated to cleave TRIF, thus inhibiting TLR3 signalling.[68]

Viral Subversion of TLR Signalling: More Than Just Inhibition

It has become apparent that viruses not only inhibit TLR signalling to avoid and suppress detection, but also actively subvert the signalling processes to manipulate and direct gene induction to favour the viral life cycle. For example, WNV appears to actually require TLR3 stimulation in order to cause disruption of the blood brain barrier, resulting in enhanced brain infection and encephalitis in mice.[69] This was shown since TLR3 knockout mice do not exhibit this encephalitis and contain a reduced virus titre in the brain, but a greater virus titre in the periphery, compared to wild type mice.[69] Bizarrely, even though VACV encodes inhibitors of TLR3 signalling (see previous section), TLR3 −/− mice infected with VACV showed substantially lower viral replication and have reduced disease morbidity relative to wild-type mice.[70] Thus it appears some viruses use

TLR3 stimulation to enhance viral replication and to progress infection. Another manipulation of the host TLR response involves the selective stimulation of certain TLR pathways for IL-10 production. IL10 inhibits a number of macrophage and monocyte functions including IL12 synthesis, production of nitric oxide and expression of MHC class II costimulatory molecules.[71] Further it has been shown that for some viruses at least, IL-10 production is actually required for viral persistence.[72,73] Not only does VACV A52 inhibit NFκB activation via interacting with IRAK-2 (see previous section), but this protein also interacts with TRAF6 leading to p38 MAP kinase activation and IL-10 promoter induction.[60] Thus during TLR stimulation A52 can switch off NFκB-dependent genes such as IL-8,and enhance IL-10 production instead.[60] A52-mediated enhancement of TLR-induced IL-10 may be important in explaining the contribution of A52 to virulence.[59] Further, mouse mammary tumour virus (MMTV) was shown to induce IL10 through stimulation of TLR4 in B-cells and this was necessary for viral persistence.[74]

Thus there is a complex relationship between TLRs and viruses and viral immunomodulation of TLRs encompasses not only inhibition of signalling pathways for evasion, but also their hijacking and subversion.

RIG-I-Like Helicases (RLHs)

The third and most recently discovered viral PRR family are the retinoic acid-inducible protein –I (RIG-I)-like helicases (RLHs). A role for the RNA helicase RIG-I as a viral PRR was discovered by Yoneyama et al who proposed a novel cytosolic sensor for dsRNA.[75] By screening an expression cDNA library from poly(I:C) stimulated cells for the ability to enhance poly(I:C)-induced IFNβ induction, RIG-I was revealed.[75] The related RLH melanoma differentiation-associated gene 5 (MDA5) was found as a target for the V proteins from paramyxoviruses which inhibit IFN-β production by dsRNA.[76] LGP2 is also a RLH, but may play a negative role in signalling.

Recognition of Viral RNA by RLHs

Both RIG-I and MDA5 have an N-terminal caspase recruitment and activation domain (CARD) and a C-terminal DExD/H box RNA helicase domain.[75,77] LGP2 has the helical domain, but lacks the CARD domain.[78,79] From studies using mice deficient in either RIG-I or MDA5 it was found that the helicases are not redundant in terms of the viruses they respond to. RIG-I was found to be vital for the antiviral response to a number of negative stranded RNA viruses such as Sendai virus, VSV, Japanese encephalitis virus (JEV) and IAV, while MDA5 was found to be essential for the recognition of members of the picornavirus family such as encephalomyocarditis virus (EMCV), Mengo virus and Theiler's virus.[80]

Synthetic RNAs are also differentially recognised by RIG-I and MDA5. Using cDCs from MDA5$^{-/-}$mice it was shown that there was no production of Type I IFNs in response to poly(I:C). In addition the administration of polyI:C to these mice showed impaired production of Type I IFNs.[80,81] This illustrated that MDA5 is the PRR for polyI:C. In contrast it was found that both ssRNA and dsRNA with a 5'triphosphate can activate Type I IFN production via RIG-I.[82,83] This may provide a rationale as to how RIG-I discriminates between self and viral RNA since ribosomal RNA (rRNA) and transfer RNA (tRNA) possess a 5'monophosphate and thus would not be detected by RIG-I. In addition the 5' end of mature messenger RNAs (mRNAs) and small nuclear RNAs (snRNAs) are modified with a methylguanosine cap.

Signalling Pathways Activated by RLHs

Both RIG-I and MDA5 use a common adaptor protein to signal downstream (see Fig. 3). This adaptor was identified by four groups around the same time. Kawai et al referred to this adaptor as IFN-β promoter stimulator-1 (IPS-1), Seth et al named it as mitochondrial antiviral signalling protein (MAVS), Xu et al called the protein virus-induced signalling adaptor (VISA) and Meylan et al termed the protein CARD adaptor inducing INF-β (Cardif).[84-87] Binding of RNA to RLHs causes a conformational change which exposes the CARD domain. IPS-1 contains a CARD domain which interacts with the CARD domains of RIG-I and MDA5. IPS-1 mediates the activation of similar downstream pathways to TRIF and overexpression of IPS-1 induces the

A

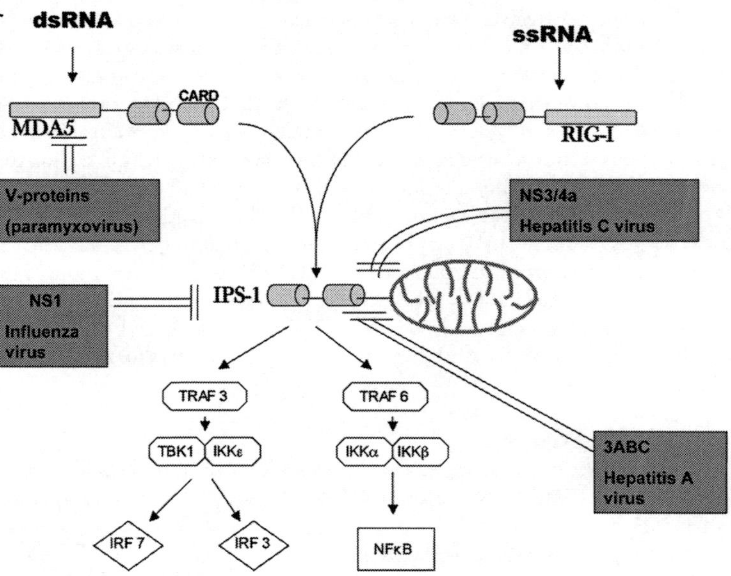

B

Mechanism of inhibition	Virus	Protein
Interaction and inhibition of IPS-1 and RIG-I	Influenza virus	NS1
Interaction and inhibition of MDA5	Various paramyxoviruses	V-proteins
Cleavage of IPS-1 for the mitochondra	Hepatitis C virus	NS3/4A
Degradation of IPS-1	Hepatitis A virus	3ABC protein

Figure 3. Activation and evasion of RLH signalling. A) schematic of the RIG-I and MDA5 signalling pathways showing points modulated by viruses. B) Viral proteins shown to act on the RLH pathway. See text for details.

activation of the IFN-β promoter and NFκB.[88,89] IPS-1 is localised to the outer membrane of the mitochondria. This localization seems to be essential for the function of IPS-1 and in fact HCV employs the NS3/4A protease to block IPS-1 signalling by cleaving it and thereby causing its dissociation from the mitochondria (see Fig. 3).[90] IPS-$1^{-/-}$ mice were more susceptible to both VSV and EMCV infection than wild type mice, which demonstrates that IPS-1 is essential for both RIG-I and MDA5 signalling. Similar to TRIF, IPS-1 signals to IRF3 and IRF7 activation via TRAF3, TBK1 and IKKε.[91,92] IPS-1 can also interact with TRAF2, TRAF 6, FADD and RIP-1 all of which have been proposed to contribute to NFκB activation.[92]

Apart from HCV NS3/4A, which cleaves IPS-1 and paramyxovirus V proteins which bind to and inhibit MDA5, other viral strategies to evade detection by RLHs are becoming apparent. For example, NS1 from IAV inhibits signalling via direct interactions with both RIG-I and IPS-1,[82] while the Hepatitis A virus targets its 3ABC precursor of its $3C^{pro}$ cysteine protease to mitochondria where it colocalizes with and cleaves IPS-1, thereby disrupting activation of IRF3 through the MDA5 pathway.[93]

Other novel cytosolic PRR detectors of viral nucleic acid are have also been discovered recently. Takaoka et al characterised the Z-DNA binding protein (ZBP-1) as a cytoplasmic DNA receptor capable of IRF3 activation and IFNβ induction and as such renamed it as DNA-dependent activator of IRFs (DAI).[94] DAI associates with IRF3 and TBK1 upon stimulation with DNA,[94] but further study is needed to determine all the other signalling components of this pathway and how viruses might evade it. Of interest is the fact that the poxvirus virulence factor E3 which binds dsRNA (see Section 2) also has a DNA-binding motif shown to contribute to virulence.[95] Thus E3 might bind to poxvirus DNA and prevent its detection by DAI or a related cytosolic DNA PRR.

Viral Inhibition Proximal to Transcription Factors

Given that most viruses likely stimulate more than one PRR pathway, an effective viral evasion strategy is to target innate signalling proximal to transcription factor activation, in order to mediate and effect at a common shared downstream point in signalling. There are numerous examples of this.[96] For example rabies virus and borna disease virus express phosphoproteins which inhibit TBK1 activity and thus prevent IRF3/7 activation. The borna disease virus phosphoprotein is phosphorylated by TBK-1 and thus acts as a viral decoy substrate preventing TBK-1 phosphorylating IRF3/7.[97] Direct binding to transcription factors or their coactivator targets is another key method of viral interference. Human herpes virus-8 blocks IRF3 dimerisation via a virally encoded dominant negative IRF termed v-IRF1. This protein inhibits IRF3 (but not IRF7) activation by competing for binding to the IRF3 binding partner CBP/p300.[98] The ICP0 protein from bovine herpes virus binds to IRF3 and leads to its proteosomal degradation. Interestingly, the exact mechanism of ICP0 is not conserved among other herpes viruses, as HSV ICP0 also interacts with IRF3 but instead of causing its degradation, it acts in a similar manner to v-IRF-1 by binding to CBP/p300 and sequestering them in specialised nuclear bodies, thus preventing IFN gene transcription.[96] IRF7 is not as well studied as IRF3 but has more recently been shown to be the target of a number of viruses. For example, EBV encodes an early protein, BZLF-1, which binds to IRF7 and prevents IFN production in virally infected cells, while the rotavirus protein NSP1 promotes the proteosomal degradation of IRF3, 5 and 7.[96]

Similar to the IRFs, NFκB is also targeted for viral evasion and this has been dealt with extensively elsewhere.[99] One striking manipulation of this transcription factor is seen by the number of viruses which contain NFκB DNA-binding motifs in their genome, such as HIV-1, CMV, HSV and EBV. NFκB is not only a fundamental transcription factor in mounting the innate immune response but is also an anti-apoptotic protein, promoting cell survival. Hence NFκB DNA binding motifs are found in the promoter regions of genes encoding many anti-apoptotic proteins such as Bcl-2, Bfl1 and A20 along with pro-survival genes including c-FLIP and IAPs. Thus many viruses not only inhibit NFκB in early infection, but activate NFκB in later stages of viral infection in order to promote the survival of infected cells. For example both EBV and Kaposi's Sarcoma virus activate NFκB.[99] There is mounting evidence showing the role of NFκB in tumorigenesis and EBV is associated with the development of several human malignancies including Burkitt's lymphoma and Hodgkin's lymphoma.[100] EBV encodes an integral transmembrane protein, LMP-1, that functions as an activated CD40 cell surface receptor in B-cells promoting growth and lymphocyte survival. This LMP-1 protein actives the NFκB subunit c-Rel and is essential for the survival of EBV-transformed cells.[99] Thus for a given virus, manipulation of NFκB can be crucial not only for immune evasion but also for the survival of virally infected cells allowing viral replication to continue.

Conclusions

This chapter has shown that the host has multiple mechanisms for detecting viruses, while viruses have evolved numerous immunomodulatory strategies in order to achieve a more successful life cycle. Most components of the innate immune system are targeted by multiple viruses for evasion or subversion and discoveries in this field continue at a pace which sometimes outruns and even informs the discovery of innate immune mechanisms. With the on going exploration of immunomodulation by viruses much more will be unearthed about the unknown components of antiviral signalling. In addition these studies could potentially herald more antiviral treatments.

References

1. Garcia MA, Meurs EF, Esteban M. The dsRNA protein kinase PKR: Virus and cell control. Biochimie 2007; 89(6-7):799-811.
2. Metz DH, Esteban M. Interferon inhibits viral protein synthesis in L-cells infected with vaccinia virus. Nature 1972; 238(5364):385-388.
3. Katze MG, DeCorato D, Safer B et al. Adenovirus VAI RNA complexes with the 68,000 Mr protein kinase to regulate its autophosphorylation and activity. EMBO J 1987; 6(3):689-697.
4. Maitra RK, McMillan NA, Desai S et al. HIV-1 TAR RNA has an intrinsic ability to activate interferon-inducible enzymes. Virology 1994; 204(2):823-827.
5. Ito T, Yang M, May WS. RAX, a cellular activator for double-stranded RNA-dependent protein kinase during stress signaling. J Biol Chem 1999; 274(22):15427-15432.
6. Patel RC, Sen GC. PACT, a protein activator of the interferon-induced protein kinase, PKR. EMBO J 1998; 17(15):4379-4390.
7. Nallagatla SR, Hwang J, Toroney R et al. 5′-triphosphate-dependent activation of PKR by RNAs with short stem-loops. Science 2007; 318(5855):1455-1458.
8. Garcia MA, Gil J, Ventoso I et al. Impact of protein kinase PKR in cell biology: From antiviral to antiproliferative action. Microbiol Mol Biol Rev 2006; 70(4):1032-1060.
9. Williams BR. Signal integration via PKR. Sci STKE 2001; 2001(89):RE2.
10. Verma IM, Stevenson JK, Schwarz EM et al. Rel/NF-kappa B/I kappa B family: Intimate tales of association and dissociation. Genes Dev 1995; 9(22):2723-2735.
11. Jiang Z, Zamanian-Daryoush M, Nie H et al. Poly(I-C)-induced Toll-like receptor 3 (TLR3)-mediated activation of NFkappa B and MAP kinase is through an interleukin-1 receptor-associated kinase (IRAK)-independent pathway employing the signaling components TLR3-TRAF6-TAK1-TAB2-PKR. J Biol Chem 2003; 278(19):16713-16719.
12. Clemens MJ. PKR—A protein kinase regulated by double-stranded RNA. Int J Biochem Cell Biol 1997; 29(7):945-949.
13. Beattie E, Paoletti E, Tartaglia J. Distinct patterns of IFN sensitivity observed in cells infected with vaccinia K3L- and E3L- mutant viruses. Virology 1995; 210(2):254-263.
14. Langland JO, Pettiford S, Jiang B et al. Products of the porcine group C rotavirus NSP3 gene bind specifically to double-stranded RNA and inhibit activation of the interferon-induced protein kinase PKR. J Virol 1994; 68(6):3821-3829.
15. Shors T, Kibler KV, Perkins KB et al. Complementation of vaccinia virus deleted of the E3L gene by mutants of E3L. Virology 1997; 239(2):269-276.
16. Langland JO, Cameron JM, Heck MC et al. Inhibition of PKR by RNA and DNA viruses. Virus Res 2006; 119(1):100-110.
17. Lu Y, Wambach M, Katze MG et al. Binding of the influenza virus NS1 protein to double-stranded RNA inhibits the activation of the protein kinase that phosphorylates the elF-2 translation initiation factor. Virology 1995; 214(1):222-228.
18. Wang W, Riedel K, Lynch P et al. RNA binding by the novel helical domain of the influenza virus NS1 protein requires its dimer structure and a small number of specific basic amino acids. RNA 1999; 5(2):195-205.
19. Khoo D, Perez C, Mohr I. Characterization of RNA determinants recognized by the arginine- and proline-rich region of Us11, a herpes simplex virus Type 1-encoded double-stranded RNA binding protein that prevents PKR activation. J Virol 2002; 76(23):11971-11981.
20. Poppers J, Mulvey M, Perez C et al. Identification of a lytic-cycle Epstein-Barr virus gene product that can regulate PKR activation. J Virol 2003; 77(1):228-236.
21. Vyas J, Elia A, Clemens MJ. Inhibition of the protein kinase PKR by the internal ribosome entry site of hepatitis C virus genomic RNA. RNA 2003; 9(7):858-870.
22. Clarke PA, Schwemmle M, Schickinger J et al. Binding of Epstein-Barr virus small RNA EBER-1 to the double-stranded RNA-activated protein kinase DAI. Nucleic Acids Res 1991; 19(2):243-248.

23. Sharp TV, Schwemmle M, Jeffrey I et al. Comparative analysis of the regulation of the interferon-inducible protein kinase PKR by Epstein-Barr virus RNAs EBER-1 and EBER-2 and adenovirus VAI RNA. Nucleic Acids Res 1993; 21(19):4483-4490.
24. Clarke PA, Sharp NA, Clemens MJ. Translational control by the Epstein-Barr virus small RNA EBER-1. Reversal of the double-stranded RNA-induced inhibition of protein synthesis in reticulocyte lysates. Eur J Biochem 1990; 193(3):635-641.
25. Gale MJ Jr, Korth MJ, Tang NM et al. Evidence that hepatitis C virus resistance to interferon is mediated through repression of the PKR protein kinase by the nonstructural 5A protein. Virology 1997; 230(2):217-227.
26. Tan SL, Gale MJ Jr, Katze MG. Double-stranded RNA-independent dimerization of interferon-induced protein kinase PKR and inhibition of dimerization by the cellular P58IPK inhibitor. Mol Cell Biol 1998; 18(5):2431-2443.
27. Taylor DR, Shi ST, Romano PR et al. Inhibition of the interferon-inducible protein kinase PKR by HCV E2 protein. Science 1999; 285(5424):107-110.
28. Pavio N, Taylor DR, Lai MM. Detection of a novel unglycosylated form of hepatitis C virus E2 envelope protein that is located in the cytosol and interacts with PKR. J Virol 2002; 76(3):1265-1272.
29. Lee TG, Tomita J, Hovanessian AG et al. Characterization and regulation of the 58,000-dalton cellular inhibitor of the interferon-induced, dsRNA-activated protein kinase. J Biol Chem 1992; 267(20):14238-14243.
30. Romano PR, Zhang F, Tan SL et al. Inhibition of double-stranded RNA-dependent protein kinase PKR by vaccinia virus E3: role of complex formation and the E3 N-terminal domain. Mol Cell Biol 1998; 18(12):7304-7316.
31. Black TL, Barber GN, Katze MG. Degradation of the interferon-induced 68,000-M(r) protein kinase by poliovirus requires RNA. J Virol 1993; 67(2):791-800.
32. Chou J, Chen JJ, Gross M et al. Association of a M(r) 90,000 phosphoprotein with protein kinase PKR in cells exhibiting enhanced phosphorylation of translation initiation factor eIF-2 alpha and premature shutoff of protein synthesis after infection with gamma 134.5- mutants of herpes simplex virus 1. Proc Natl Acad Sci USA 1995; 92(23):10516-10520.
33. Kazemi S, Papadopoulou S, Li S et al. Control of alpha subunit of eukaryotic translation initiation factor 2 (eIF2 α) phosphorylation by the human papillomavirus Type 18 E6 oncoprotein: implications for eIF2 alpha-dependent gene expression and cell death. Mol Cell Biol 2004; 24(8):3415-3429.
34. Hashimoto C, Hudson KL, Anderson KV. The Toll gene of drosophila, required for dorsal-ventral embryonic polarity, appears to encode a transmembrane protein. Cell 1988; 52(2):269-279.
35. Lemaitre B, Nicolas E, Michaut L et al. The dorsoventral regulatory gene cassette spatzle/Toll/cactus controls the potent antifungal response in drosophila adults. Cell 1996; 86(6):973-983.
36. Zambon RA, Nandakumar M, Vakharia VN et al. The Toll pathway is important for an antiviral response in drosophila. Proc Natl Acad Sci USA 2005; 102(20):7257-7262.
37. Yang LFR YL, Pavlovic J, Aguzzi A et al. Deficient signaling in mice devoid of double-stranded RNA-dependent protein kinase. EMBO J 1995; 14(24):6095–6106.
38. Zhou A, Paranjape JM, Der SD et al. Interferon action in triply deficient mice reveals the existence of alternative antiviral pathways. Virology 1999; 258(2):435-440.
39. Lena Alexopoulou ACH, Ruslan Medzhitov, Richard Flavell A. Recognition of double-stranded RNA and activation of NF-B by Toll-like receptor 3. Nature 2001; 413:732-738.
40. Schroder M, Bowie AG. TLR3 in antiviral immunity: Key player or bystander? Trends in Immunology 2005; 26(9):462-468.
41. Matsumoto M, Funami K, Tanabe M et al. Subcellular localization of Toll-like receptor 3 in human dendritic cells. J Immunol 2003; 171(6):3154-3162.
42. Weber F, Wagner V, Rasmussen SB et al. Double-stranded RNA is produced by positive-strand RNA viruses and DNA viruses but not in detectable amounts by negative-strand RNA viruses. J Virol 2006; 80(10):5059-5064.
43. Schulz O, Diebold SS, Chen M et al. Toll-like receptor 3 promotes cross-priming to virus-infected cells. Nature 2005; 433(7028):887-892.
44. Hemmi H, Takeuchi O, Kawai T et al. A Toll-like receptor recognizes bacterial DNA. Nature 2000; 408(6813):740-745.
45. Krug A, Luker GD, Barchet W et al. Herpes simplex virus Type 1 activates murine natural interferon-producing cells through toll-like receptor 9. Blood 2004; 103(4):1433-1437.
46. Lund J, Sato A, Akira S et al. Toll-like receptor 9-mediated recognition of Herpes simplex virus-2 by plasmacytoid dendritic cells. J Exp Med 2003; 198(3):513-520.
47. Latz E, Schoenemeyer A, Visintin A et al. TLR9 signals after translocating from the ER to CpG DNA in the lysosome. Nat Immunol 2004; 5(2):190-198.

48. Heil F, Hemmi H, Hochrein H et al. Species-specific recognition of single-stranded RNA via toll-like receptor 7 and 8. Science 2004; 303(5663):1526-1529.
49. Lund JM, Alexopoulou L, Sato A et al. Recognition of single-stranded RNA viruses by Toll-like receptor 7. Proc Natl Acad Sci USA 2004; 101(15):5598-5603.
50. Diebold SS, Massacrier C, Akira S et al. Nucleic acid agonists for Toll-like receptor 7 are defined by the presence of uridine ribonucleotides. Eur J Immunol 2006; 36(12):3256-3267.
51. Kurt-Jones EA, Popova L, Kwinn L et al. Pattern recognition receptors TLR4 and CD14 mediate response to respiratory syncytial virus. Nat Immunol 2000; 1(5):398-401.
52. Ehl S, Bischoff R, Ostler T et al. The role of Toll-like receptor 4 versus interleukin-12 in immunity to respiratory syncytial virus. Eur J Immunol 2004; 34(4):1146-1153.
53. Kurt-Jones EA, Chan M, Zhou S et al. Herpes simplex virus 1 interaction with Toll-like receptor 2 contributes to lethal encephalitis. Proc Natl Acad Sci USA 2004; 101(5):1315-1320.
54. Zhu J, Martinez J, Huang X et al. Innate immunity against vaccinia virus is mediated by TLR2 and requires TLR-independent production of IFN-beta. Blood 2007; 109(2):619-625.
55. O'Neill LA, Bowie AG. The family of five: TIR-domain-containing adaptors in Toll-like receptor signalling. Nat Rev Immunol 2007; 7(5):353-364.
56. Carty M, Goodbody R, Schroder M et al. The human adaptor SARM negatively regulates adaptor protein TRIF-dependent Toll-like receptor signaling. Nat Immunol 2006; 7(10):1074-1081.
57. Keating SE, Maloney GM, Moran EM et al. IRAK-2 participates in multiple toll-like receptor signaling pathways to NFkappaB via activation of TRAF6 ubiquitination. J Biol Chem 2007; 282(46):33435-33443.
58. Bowie A, Kiss-Toth E, Symons JA et al. A46R and A52R from vaccinia virus are antagonists of host IL-1 and toll-like receptor signaling. Proc Natl Acad Sci USA 2000; 97(18):10162-10167.
59. Harte MT, Haga IR, Maloney G et al. The poxvirus protein A52R targets Toll-like receptor signaling complexes to suppress host defense. J Exp Med 2003; 197(3):343-351.
60. Maloney G, Schroder M, Bowie AG. Vaccinia virus protein A52R activates p38 mitogen-activated protein kinase and potentiates lipopolysaccharide-induced interleukin-10. J Biol Chem 2005; 280(35):30838-30844.
61. Meylan E, Burns K, Hofmann K et al. RIP1 is an essential mediator of Toll-like receptor 3-induced NF-kappa B activation. Nat Immunol 2004; 5(5):503-507.
62. Cusson-Hermance N, Khurana S, Lee TH et al. Rip1 mediates the Trif-dependent toll-like receptor 3- and 4-induced NF-{kappa}B activation but does not contribute to interferon regulatory factor 3 activation. J Biol Chem 2005; 280(44):36560-36566.
63. Kim TW, Staschke K, Bulek K et al. A critical role for IRAK4 kinase activity in Toll-like receptor-mediated innate immunity. J Exp Med 2007; 204(5):1025-1036.
64. Uematsu S, Sato S, Yamamoto M et al. Interleukin-1 receptor-associated kinase-1 plays an essential role for Toll-like receptor (TLR)7- and TLR9-mediated interferon-{alpha} induction. J Exp Med 2005; 201(6):915-923.
65. Hoshino K, Sugiyama T, Matsumoto M et al. IkappaB kinase-alpha is critical for interferon-alpha production induced by Toll-like receptors 7 and 9. Nature 2006; 440(7086):949-953.
66. Stack J, Haga IR, Schroder M et al. Vaccinia virus protein A46R targets multiple Toll-like-interleukin-1 receptor adaptors and contributes to virulence. J Exp Med 2005; 201(6):1007-1018.
67. Cirl C, Wieser A, Yadav M et al. Subversion of Toll-like receptor signaling by a unique family of bacterial Toll/interleukin-1 receptor domain-containing proteins. Nat Med 2008; 14(4):399-406.
68. Li K, Foy E, Ferreon JC et al. Immune evasion by hepatitis C virus NS3/4A protease-mediated cleavage of the Toll-like receptor 3 adaptor protein TRIF. Proc Natl Acad Sci USA 2005; 102(8):2992-2997.
69. Wang T, Town T, Alexopoulou L et al. Toll-like receptor 3 mediates West Nile virus entry into the brain causing lethal encephalitis. Nat Med 2004; 10(12):1366-1373.
70. Hutchens M, Luker KE, Sottile P et al. TLR3 increases disease morbidity and mortality from vaccinia infection. J Immunol 2008; 180(1):483-491.
71. Moore KW, de Waal Malefyt R, Coffman RL et al. Interleukin-10 and the interleukin-10 receptor. Annu Rev Immunol 2001; 19:683-765.
72. Brooks DG, Trifilo MJ, Edelmann KH et al. Interleukin-10 determines viral clearance or persistence in vivo. Nat Med 2006; 12(11):1301-1309.
73. Ejrnaes M, Filippi CM, Martinic MM et al. Resolution of a chronic viral infection after interleukin-10 receptor blockade. J Exp Med 2006; 203(11):2461-2472.
74. Jude BA, Pobezinskaya Y, Bishop J et al. Subversion of the innate immune system by a retrovirus. Nat Immunol 2003; 4(6):573-578.
75. Yoneyama M, Kikuchi M, Natsukawa T et al. The RNA helicase RIG-I has an essential function in double-stranded RNA-induced innate antiviral responses. Nat Immunol 2004; 5(7):730-737.

76. Andrejeva J, Childs KS, Young DF et al. The V proteins of paramyxoviruses bind the IFN-inducible RNA helicase, mda-5 and inhibit its activation of the IFN-beta promoter. Proc Natl Acad Sci USA 2004; 101(49):17264-17269.
77. Saito T, Hirai R, Loo YM et al. Regulation of innate antiviral defenses through a shared repressor domain in RIG-I and LGP2. Proc Natl Acad Sci USA 2007; 104(2):582-587.
78. Rothenfusser S, Goutagny N, DiPerna G et al. The RNA helicase Lgp2 inhibits TLR-independent sensing of viral replication by retinoic acid-inducible gene-I. J Immunol 2005; 175(8):5260-5268.
79. Komuro A, Horvath CM. RNA- and virus-independent inhibition of antiviral signaling by RNA helicase LGP2. J Virol 2006; 80(24):12332-12342.
80. Kato H, Takeuchi O, Sato S et al. Differential roles of MDA5 and RIG-I helicases in the recognition of RNA viruses. Nature 2006; 441(7089):101-105.
81. Takeuchi O, Akira S. Recognition of viruses by innate immunity. Immunol Rev 2007; 220:214-224.
82. Pichlmair A, Schulz O, Tan CP et al. RIG-I-mediated antiviral responses to single-stranded RNA bearing 5′-phosphates. Science 2006; 314(5801):997-1001.
83. Hornung V, Ellegast J, Kim S et al 5′-Triphosphate RNA is the ligand for RIG-I. Science 2006; 314(5801):994-997.
84. Kawai T, Takahashi K, Sato S et al. IPS-1, an adaptor triggering RIG-I- and Mda5-mediated Type I interferon induction. Nat Immunol 2005; 6(10):981-988.
85. Seth RB, Sun L, Ea CK et al. Identification and characterization of MAVS, a mitochondrial antiviral signaling protein that activates NF-kappaB and IRF 3. Cell 2005; 122(5):669-682.
86. Xu LG, Wang YY, Han KJ et al. VISA is an adapter protein required for virus-triggered IFN-beta signaling. Mol Cell 2005; 19(6):727-740.
87. Meylan E, Curran J, Hofmann K et al. Cardif is an adaptor protein in the RIG-I antiviral pathway and is targeted by hepatitis C virus. Nature 2005; 437(7062):1167-1172.
88. Kumar H, Kawai T, Kato H et al. Essential role of IPS-1 in innate immune responses against RNA viruses. J Exp Med 2006; 203(7):1795-1803.
89. Sun Q, Sun L, Liu HH et al. The specific and essential role of MAVS in antiviral innate immune responses. Immunity 2006; 24(5):633-642.
90. Loo YM, Owen DM, Li K et al. Viral and therapeutic control of IFN-beta promoter stimulator 1 during hepatitis C virus infection. Proc Natl Acad Sci USA 2006; 103(15):6001-6006.
91. Saha SK, Pietras EM, He JQ et al. Regulation of antiviral responses by a direct and specific interaction between TRAF3 and Cardif. EMBO J 2006; 25(14):3257-3263.
92. Pichlmair A, Reis e Sousa C. Innate recognition of viruses. Immunity 2007; 27(3):370-383.
93. Yang Y, Liang Y, Qu L et al. Disruption of innate immunity due to mitochondrial targeting of a picornaviral protease precursor. Proc Natl Acad Sci USA 2007; 104(17):7253-7258.
94. Takaoka A, Wang Z, Choi MK et al. DAI (DLM-1/ZBP1) is a cytosolic DNA sensor and an activator of innate immune response. Nature 2007; 448(7152):501-505.
95. Kim YG, Muralinath M, Brandt T et al. A role for Z-DNA binding in vaccinia virus pathogenesis. Proc Natl Acad Sci USA 2003; 100(12):6974-6979.
96. Unterholzner L, Bowie AG. The interplay between viruses and innate immune signaling: Recent insights and therapeutic opportunities. Biochem Pharmacol 2008; 75(3):589-602.
97. Unterstab G, Ludwig S, Anton A et al. Viral targeting of the interferon-{beta}-inducing Traf family member-associated NF-{kappa}B activator (TANK)-binding kinase-1. Proc Natl Acad Sci USA 2005; 102(38):13640-13645.
98. Lin R, Genin P, Mamane Y et al. HHV-8 encoded vIRF-1 represses the interferon antiviral response by blocking IRF-3 recruitment of the CBP/p300 coactivators. Oncogene 2001; 20(7):800-811.
99. Santoro MG, Rossi A, Amici C. NF-kappaB and virus infection: Who controls whom. EMBO J 2003; 22(11):2552-2560.
100. Hiscott J, Nguyen TL, Arguello M et al. Manipulation of the nuclear factor-kappaB pathway and the innate immune response by viruses. Oncogene 2006; 25(51):6844-6867.

CHAPTER 5

Viral TNF Inhibitors as Potential Therapeutics

Masmudur M. Rahman, Alexandra R. Lucas and Grant McFadden*

Abstract

The immune system functions by maintaining a delicate balance between the activities of pro-inflammatory and anti-inflammatory pathways. Unbalanced activation of these pathways often leads to the development of serious inflammatory diseases. TNF (Tumor Necrosis Factor) is a key pro-inflammatory cytokine, which can cause several inflammatory diseases when inappropriately up-regulated. Inhibition of TNF activities by using modulatory recombinant proteins has become a successful therapeutic approach to control TNF activity levels but these anti-TNF reagents also have risks and certain limitations. Biological molecules with a different mode of action in regulating TNF biology might provide a clinically useful alternative to the current therapeutics or in some cases might be efficacious in combination with existing anti-TNF therapies. TNF is also a powerful host defense cytokine commonly induced in the host response against various invading pathogens. Many viral pathogens can block TNF function by encoding modulators of TNF, its receptors or downstream signaling pathways. Here, we review the known virus-encoded TNF inhibitors and evaluate their potential as alternative future anti-TNF therapies.

Introduction

Dynamic interactions are set into motion between the host and pathogens whenever they encounter each other. All successful pathogens, including viruses, bacteria and intracellular parasites have adapted diverse mechanisms to counteract the innate and adaptive responses mounted by the host. During this process many have evolved to express specific pathogen-encoded molecules that have regulatory roles in controlling the immune system of the infected host.[1-3] These pathogen-derived molecules have often been well-honed by evolutionary selection pressures and can be attractive platform candidates as novel therapeutics to regulate the host immune system in diseases where exacerbated immune or inflammatory cascades have become pathologic to the host.[4,5]

Pro-inflammatory cytokines like TNFα (here called TNF) play very important roles in orchestrating host defense against invading pathogens, but uncontrolled expression of these cytokines sometimes creates inflammatory diseases in humans if not properly regulated. Various anti-TNF therapeutics, such as neutralizing monoclonal antibodies or Fc fusions of TNF receptor ectodomains, have now entered into the arena of clinical usage to control inappropriate and excessive elaboration of TNF. Virus-encoded TNF inhibitors or modulators of TNF function can be exceedingly potent inhibitors of TNF pro-inflammatory activities.[6] Here we discuss whether any of these virus-derived inhibitors might have potential clinical utility as an alternative strategy to dampen TNF-mediated pathologies.

*Corresponding Author: Grant McFadden—Department of Molecular Genetics and Microbiology, College of Medicine, University of Florida, 1600 SW Archer Rd., ARB Rm R5-295, PO Box 100266, Gainesville, FL 32610, USA. Email: grantmcf@ufl.edu

Pathogen-Derived Immunomodulatory Molecules, edited by Padraic G. Fallon.

TNF and TNF-Mediated Signaling

TNF is first expressed as a membrane-bound ligand that can be cleaved and secreted as a nonglycosylated trimer of a 17-kDa protein. TNF is predominantly expressed from macrophages, monocytes, $CD4^+$ and $CD8^+$ T-cells, smooth muscle cells, activated NK cells, neutrophils and fibroblasts. TNF production is inducible by a number of diverse stimuli, such as interferons, IL-2 and IL-18. The initial precursor protein, 26 kDa pro-TNF, is translated, translocated to the endoplasmic reticulum, transported to the cell surface via the Golgi apparatus and is then presented on the cell membrane as a homo-trimeric complex. This cell-surface form of TNF can interact with TNF receptors of neighboring cells or it can be cleaved and released from the cell surface as a soluble trimeric ligand by the TNF converting enzyme (TACE). Cleavage and release from the cell surface appears to have some role for the biological properties of the TNF molecule in vivo, but both forms of the ligand can induce potent signaling activities following interaction with the two known TNF receptors in cell culture. For both the cell associated and secreted forms of TNF, ligand trimerization is required for biological activity. Either the cell bound or soluble TNF ligand binds to two structurally distinct receptors: Type I (TNFR1/p55) and Type II (TNFR2/p75), which are present on the membrane of all cell types except erythrocytes. The two receptors differ significantly in their binding affinities with TNF and other TNF-superfamily members, as well as differing in their intracellular signaling pathways. Both receptors have multiple cytoplasmic domains that control their signaling properties but TNFR1 also has an additional intracellular death domain (DD) for its diverse signaling events.[7,8]

The trimeric TNF ligand binds to the extracellular domain of the receptors, via domains referred to as Cysteine-Rich Domains (CRDs), which induces conformational changes in the receptor and activates the intracellular signaling pathway, which itself can vary according to the cell type. Binding of TNF with TNFR1 leads to the release of the inhibitory protein silencer of death domains (SODD) from TNFR1 intracellular DD.[9] Release of SODD allows binding of TRADD (TNFR1-associated death domain protein) to the DD, which can further activate either the apoptotic pathway, via the Fas-associated death domain (FADD) protein, or the pro-inflammatory pathway, via TNF receptor-associated factor 2 (TRAF2) and receptor-interacting protein (RIP), resulting in the activation of nuclear factor-κB (NF-κB) (Fig. 1). In contrast to TNF-R1, TNF-R2 is unable to activate the TRADD/FADD pathway and signals only through the TRAF2-associated pathway. Some studies have indicated the presence of cross-talk between the two receptors, which is likely to be responsible for the net response of a cell upon TNF stimulation.[7] It is also possible that other cellular receptors can form complexes with TNF-receptors and thus add yet more levels of complexity in TNF-induced signaling.

PLAD Domain of TNFRs

The TNFR superfamily members are all Type I transmembrane proteins characterized by the presence of one to six hallmark CRDs. Many members of the TNFR superfamily (e.g., FAS, TNFR1 and TNFR2) exist as pre-assembled oligomers on the cell surface. This preligand assembly of TNFR oligomers is mediated by the preligand assembly domain (PLAD), which resides within the N-terminal cysteine-rich domain of the receptors and is not directly involved in ligand binding.[10] PLAD-mediated preligand assembly has also been reported for TRAIL receptors and viral TNFR homologues.[11,12] The PLAD domain of TNFR1 is critical in TNF responses, because mutation in the PLAD region reduces NF-κB activation and results in the TNFR-associated periodic syndrome, an autoinflammatory syndrome in man.[13] Also, mutation in the PLAD region of FAS has been found to participate in pathogenesis of autoimmune lymphoproliferative syndrome (ALPS), a human genetic disease involving defective apoptosis, lymphocyte accumulation and autoimmunity.[14] The mutant form of PLAD appears to inhibit the pre-assembly of FAS chain, thereby blocking the FAS intracellular signaling pathway.[14] Recent evidence indicates that PLAD-mediated receptor association regulates cellular responses to TNF-like cytokines, especially in cells of the immune system such as $CD4^+$ and $CD8^+$ T-cells.[15] Thus, targeting preligand assembly itself may offer new possibilities for therapeutic intervention in different pathological conditions involving hyperactive

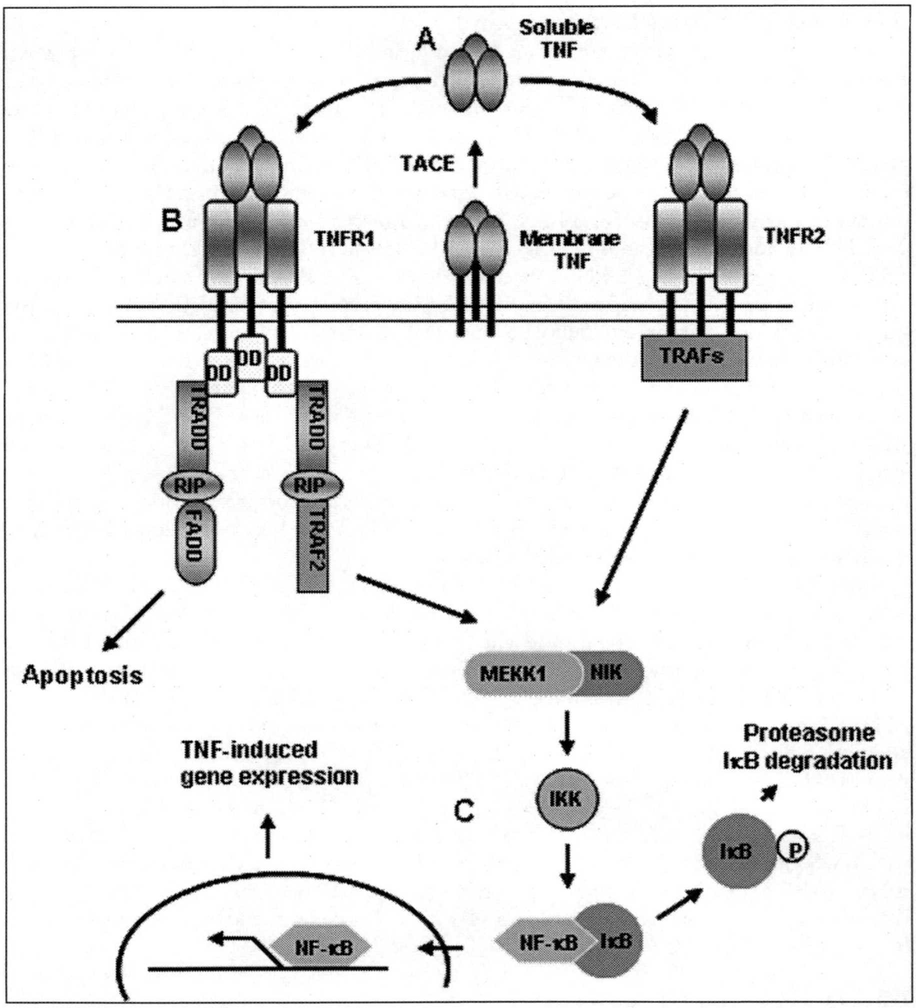

Figure 1. TNF signaling pathways and potential targets for viral inhibition. Binding of TNF with TNFR1 can lead to the activation of either pro-apoptotic signaling or NF-κB-mediated TNF-induced gene expression. TNFR1 or TNFR2 can activate NF-κB through several downstream proteins like TNFR1-associated death domain-containing protein (TRADD), receptor-interacting protein (RIP), TNFR-associated factor 2 (TRAF2), mitogen-activated protein kinase kinase 1 (MEKK1), nuclear factor-κB-inducing kinase (NIK), inhibitor of nuclear factor-κB (IκB) kinase kinase (IKK). Virus-encoded proteins inhibit TNF-mediated responses by directly binding to TNF (A), down-regulating or binding to TNF receptors and associated factors (B) and inhibiting the proteins involved in activation of NF-κB (C). DD, death domain; TACE, TNF-α-converting enzyme.

TNF signaling. PLAD domain proteins can also effectively prevent TNFR signaling and potently inhibit inflammatory arthritis.[16]

TNF-Mediated Diseases

The immune system constantly maintains a delicate balance between the pro-inflammatory and anti-inflammatory mediators or cytokines. During many disease states, this balance is lost and the pro-inflammatory cytokines like TNF can become inappropriately upregulated. This, in turn, ultimately induces excessive levels of adhesion molecules on the endothelium, stimulates fibroblast proliferation and recruits leukocytes from the circulation into tissues or sites where they can be pathologic, such as the synovial fluid.[17] Enhanced levels of TNF are associated with the development of a variety of inflammatory conditions, like Rheumatoid Arthritis (RA), juvenile RA, Crohn's disease (CD), Ankylosing Spondylitis (AS), Psoriatic Arthritis (PsA), Inflammatory Bowel disease (IBD) and asthma. TNF also can play an indirect role in other inflammatory conditions, as reported in the case of ocular inflammatory development.[18] The recent findings that hyperactive systemic inflammation contributes to the development of atherosclerosis and Type 2 diabetes mellitus constitutes a major breakthrough in understanding the mechanisms underlying these conditions. Thus, TNF has been shown to play a key role in many human inflammatory disorders and is considered to be a prime therapeutic target for drug development.

Apart from its direct pro-inflammatory role, TNF also stimulates production of other pro-inflammatory cytokines (such as IL-1 and IL-6), chemokines, reactive oxygen intermediates, nitric oxide and prostaglandins and increases the rate of tissue remodeling by matrix-degrading proteases.[19] TNF promotes angiogenesis and osteoclast differentiation and activates osteoclasts that stimulate bone lysis, leading to joint erosions particularly at the marginal surfaces.[20] TNF also increases the rate of tissue remodeling by matrix-degrading proteases and directly mediates pain, fever and cachexia. Additionally, TNF has a significant role in lipid metabolism. In animal models, administration of exogenous TNF leads to severe impairment of glucose tolerance and insulin sensitivity.[21] Thus, TNF might serve as a therapeutic target in these disorders as well.

Current Anti-TNF Therapies in Humans

Several recombinant protein-based inhibitors of TNF activity have been developed for clinical therapy. Three of these, Etanercept (Enbrel), infliximab (Remicade) and adalimumab (Humira) have already been approved for clinical use in various inflammatory diseases. Etanercept is a dimeric fusion protein consisting of two extracellular domains of the human TNFR2, linked to the Fc portion of a Type 1 human immunoglobulin (IgG1). The Fc portion helps to maintain the molecule in the circulation[22] and it has a relatively short half life of 4-5 days. Infliximab is a chimeric mAb, composed of human constant regions of IgG1, with murine variable regions. It binds to both soluble and membrane TNF with high affinity and exhibits lower nonspecific effects on other pathways.[23] Infliximab has a circulating half life of 8-10 days. Adalimumab is a complete human IgG1 anti-TNF mAb that binds to both soluble and membrane bound TNF with high affinity. It has a relatively longer half life of 12-14 days. Several other protein based anti-TNF therapeutics are also currently in clinical trials. Numerous studies have been conducted to understand the mechanism of action of the various anti-TNF therapeutics in immune-mediated diseases. In addition to neutralization of soluble TNF, these drugs clearly affect diverse intracellular signaling which regulate cell cycle arrest, apoptosis, or suppression of cytokine production.[24]

Although the current protein-based TNF inhibitors have demonstrated ligand-inhibitory efficacy, they can also exhibit potentially serious adverse effects such as a greater predisposition towards secondary infections, congestive heart failure, neurologic changes (demyelination), lymphomas, re-exacerbation of latent tuberculosis and problems related to autoimmunity such as lupus-like syndrome.[25] With infliximab, acute allergic reactions are seen in approximately 5% of intravenous infusions as well.

Safety Issues with Current Anti-TNF Therapies

Some of the major safety considerations regarding the long term use of TNF antagonists include infections, autoimmune disease, demyelinating disease, malignancies and congestive heart failure. Because TNF has significant role in host defense to both bacterial and viral invasion, one of the main concerns with anti-TNF therapy is risk of infection. Nevertheless, patients with RA who received anti-TNF agents do not show a significant increase in the overall rate of infections during treatment but there are case reports of pulmonary listeriosis,[26] pulmonary aspergillosis,[27] *Pneumocystis carinii* pneumonia[28] and reactivated histoplasmosis[29] in some patients. Tuberculosis (TB) recrudescence has been the most common serious infection observed in patients receiving TNF antagonists.[30] However, the incidence of TB is also influenced by age, concomitant immunosuppressive regimens, socioeconomic status and geography.[31] Screening of patients for the possibility of latent TB and other related pathogens before treatment would further reduce the rate of infections exacerbated by treatment with TNF inhibitors.

The regulatory role of TNF with antigen-presenting cell function may be associated with autoimmune diseases arising in some patients treated with TNF antagonists. Anti-TNF treatment very often results in the development of antinuclear antibodies and anti- double-stranded DNA antibodies.[32] Some of the reported autoimmune diseases associated with anti-TNF therapy are systemic lupus erythematosus (SLE) and different forms of vasculitis.[33] Among the neurological diseases reported, demyelinating disease is most common in patients with inflammatory arthritis. Congestive heart failure has been reported, mostly in RA patients. Although not common, anti-TNF drugs may also increase incidence of solid tumor development in patients with RA.[33] In many cases the complications from anti-TNF treatment depends on the specific biological agent used.

Recently and for the first time, it has been reported that perforating folliculitis (a type of perforating dermatosis characterized by trans-epithelial elimination of dermal structures) is associated with the administration of the TNF inhibitors, infliximab and etanercept, in a patient suffering from RA and pulmonary fibrosis.[34] The possible reason is that TNF directly inhibits fibronectin production and promotes its degradation through stimulation of several metalloproteinases.[35] Therefore, the blockade of TNF may induce fibronectin accumulation, favoring in some way the perforating phenomenonobserved in these patients.

Viral TNF Inhibitors as Alternative Therapeutics

Viruses have adapted diverse strategies to neutralize TNF and TNF-mediated responses by targeting almost every step of the TNF response pathway.[6,36,37] The virus-encoded modulators can either directly bind the ligand and/or receptor, or components of the TNFR signaling pathway, to inhibit the TNF response (Fig. 1). Some viral regulators have adapted a strategy of down-regulating the expression of cell surface TNFRs. Components of the TNF signal transduction pathway are also a recurrent target for viral immune evasion because of their central roles in mounting innate and adaptive immune responses.[36,37] Some of the known viral immune modulating molecules directly bind and inhibit the function of TNF ligand (Table 1) whereas others modulate downstream TNFR functions (Table 2).

So far, poxviruses are the only viruses that express viral modulators that can bind and sequester extracellular TNF prior to its TNFR engagement.[6] Two distinct classes of extracellular poxvirus TNF-binding proteins have been identified: those that resemble the mammalian TNFRs, termed vTNFRs and a recently characterized family of proteins that resemble the mammalian MHC class I heavy chain, termed vTNF-BPs, that bind and inhibit TNF with unusually tight affinity.[38,39]

Viral TNFR Homologues

Among these vTNFRs are the T2-like family members encoded by Leporipoxviruses and the cytokine response modifier (Crm) family members encoded by Orthopoxviruses.[40,41] Like their mammalian counterparts, poxvirus vTNFR superfamily proteins comprise a tandem array of CRDs. These CRDs are each composed of two distinct structural modules that can adopt one of a number of conserved conformations (primarily A or B), can contain zero to two disulphide bonds and are named according to topology and disulphide bond count (A1, A2, etc.). Individual poxviruses can express different numbers of TNFR-like molecules: cowpox virus (CPXV) expresses up to five TNFR superfamily molecules, cytokine response modifier B (CrmB), CrmC, CrmD, CrmE and vCD30,[42-46] while variola virus (VARV) and monkeypox virus (MPXV), which causes

Table 1. Poxvirus-encoded TNF inhibitors

Gene/ORF	Virus	Characteristics	Binding Proteins	Bioactivity/ Function	Reference
CrmB	Cowpox	Secreted (early gene)	Human and mouse TNF and LTα	Blocks human and mouse TNF and LTα function	42
CrmB	Variola	Secreted	Human, mouse, rat TNF; human LTα and chemokines	Blocks human, mouse and rat TNF function	47
CrmB	Monkeypox	secreted	Human, mouse and rabbit TNF	Blocks human, mouse and rabbit TNF function	48
CrmC	Cowpox	Secreted (late gene)	Human and mouse TNF	Blocks human and mouse TNF function	43
CrmD	Cowpox and Ectromelia	Secreted (late gene)	Human, mouse and rat TNF; human LTα	Blocks human TNF and LTα function	44
CrmE	Cowpox	Secreted	Human, mouse and Rat TNF	Blocks human TNF function	45
CrmE	Vaccinia	Secreted and cell surface	Human, mouse and rat TNF	Blocks human TNF function	51
M-T2	Myxoma	Secreted (early gene)	Rabbit TNF; human TNFR	Blocks rabbit TNF, human TNF function	12,55
TPV-2L	Tanapox	Secreted (early)	Human, monkey, canine and rabbit TNF	Blocks human, monkey and canine TNF function	38,39
YMTV-2L	Yabamonkey tumor virus	secreted	Human, monkey and rabbit TNF	Blocks human and monkey TNF function	39

a smallpox-like disease in humans, express only one, CrmB,[47,48] as does Ectromelia virus, a mouse pathogen that cause mousepox, which expresses CrmD only.[49] These CrmB, C, D, E proteins bear no structural similarity to CrmA which is a cross-class serpin. Poxviral TNFR superfamily members differ from their mammalian counterparts in that they lack a transmembrane region and cytoplasmic domain for signaling and are transported to the extracellular millieu. Vaccinia virus (VACV) strains USSR, Lister and Evans encode two functional vTNFR superfamily members: CrmC and CrmE.[50] These are the smallest of the known poxvirus vTNFRs, containing an N-terminal signal peptide for secretion into the extracellular matrix, three CRDs and a short C-terminal extension.

In some cases, poxvirus-encoded vTNFR homologues have additional properties in addition to the binding and inhibiting of TNF functions. This has been demonstrated in case of myxoma virus encoded M-T2 protein, which has a second anti-apoptotic role.[12] In case of CrmE, it has recently been shown that the protein also possesses chemokine binding properties.[47] Thus, unlike the engineered commercial TNF inhibitors currently used in humans, the virus-encoded modulators sometime possess additional anti-inflammatory properties above and beyond just TNF inhibition.

Table 2. Virus-encoded TNF modulators

Viral Proteins	Virus	Mode of Action	Effects on TNF	Reference
RID complex	Adenovirus	Downregulates surface TNFR1 level	Inhibits TNF signaling	56,57
E6 and E7	HPV16	Binds to TNFR1 and inhibits DISC formation	Inhibits TNF signaling	66
3A	Poliovirus	Reduces the surface level of TNFRs	Inhibits TNF signaling	71
NS5A	HCV	Interacts with TRAF2 and TRADD	Inhibits TNF signaling	73
NS5B	HCV	Interacts with IKKα	Inhibits TNF signaling	86
Early proteins	HCMV	Reduces the surface level of TNFR1	Inhibits TNF signaling	68
Early proteins	MCMV	Reduces the surface level of TNFR1 and TNFR2	Inhibits TNF signaling	69
N1L	Vaccinia	Inhibits TRAF2, TRAF6, IKK-α and IKK-β	Inhibits activation of NF-κB by TNF	78
K1L	Vaccinia	Inhibit IκBα degradation	Inhibits TNF-induced gene expression	82
A238L (ASFV-IκB)	ASFV	Binds to NF-κB	Inhibits TNF-induced gene expression	84,85

CrmE

All members of the poxvirus-encoded vTNFR superfamily of proteins lack a C-terminal transmembrane domain like the ones in TNFR1 or TNFR2. CrmE from CPXV is expressed as a secreted 18kDa protein. CrmE binds to human, mouse and rat TNF but not to any other member of TNF superfamily.[45] Biologically, it protects cells from the cytolytic activity of human TNF by binding to it with high affinity. The CrmE ortholog from vaccinia virus has also the same properties as CPXV ortholog.[51] Cells infected with VACV strains that encode CrmE (e.g., strain Lister) display both soluble and cell surface associated CrmE activity.[51] VACV CrmE inhibited the cytotoxic and apoptotic activity of only human TNF in vitro. The expression of CrmE by a VACV strain (VACV WR) that does not normally encode a viral TNFR enhances virus virulence.[51] It was predicted that the CrmE members structurally resemble TNFR2. The structure of CrmE protein from VACV strain Lister has been solved recently[52] which shares significant sequence similarity with mammalian Type 2 TNF receptors (TNFSFR1B, p75; TNFR2). The bacterially expressed and purified CrmE is a monomer in solution and forms a stable complex with recombinant human TNF to form a heterohexamer. The structure confirms that CrmE adopts the canonical TNFR fold but only one of the two "ligand binding" loops of TNFR1 is conserved in CrmE, suggesting a mechanism for the higher affinity of poxvirus vTNFRs for TNF over LT-α.[52]

CrmB and CrmD

The CrmB from VARV, MPXV and CPXV differ in their efficiencies of inhibition of cytotoxic effects of human, mouse or rabbit TNF.[48] The CrmB and CrmD, in addition to their CRD, have an extended C terminus. The TNF binding activity is exclusively located in the N-terminal CRDs of these proteins. Surface Plasmon Resonance (SPR) screening of CrmB with different cytokines and chemokines identified that it also binds to human chemokines. The chemokine binding region is exclusively located to the C-terminal CTD. This suggests that CrmB and CrmD have distinct and independent binding sites for TNF and chemokines and might thus have a broader role as an anti-inflammatory modulator. The CTD of the variola version has been named the SECRET (smallpox virus-encoded chemokine receptor) domain.[47] The CrmB proteins from VARV, MPXV and CPXV were tested for their ability to protect BALB/c mice against LPS induced endotoxic shock. VARV-CrmB protein exhibited a more efficient protective effect with an increase in the animal survival rate. This protection could be the combined effect of TNF-binding and chemokine binding activity of the N-terminal CRDs and C-terminal SECRET domain respectively.[48] However, further in vivo studies are required to understand the pharmacologic properties of these Crm proteins.

Viral PLAD Like Domain

The role of PLAD domain in regulating TNF-like cytokine signaling has been recently studied in the viral TNFR homologue from myxoma virus. The myxoma virus encoded protein T2 (M-T2) binds and inhibits only rabbit TNF.[53] Rabbits infected with M-T2 knockout myxoma virus exhibited a markedly attenuated disease progression compared to rabbits infected with the wild type myxoma virus, suggesting that M-T2 is a viral virulence factor. Another function of M-T2 is to prevent apoptosis of myxoma virus infected rabbit $CD4^+$ RL5 T-cells.[54] RL5 cells infected with T2 KO virus undergo apoptosis and prevent optimal virus replication as compared to WT myxoma virus infected cells. Deletion analysis of M-T2 demonstrated that the intracellular part of the protein possesses anti-apoptotic activity because active purified M-T2 protein added to the culture supernatants of vMyxT2ko-infected RL5 cells failed to rescue these cells from virus-induced apoptosis.[55] Thus, M-T2 has dual roles in two different forms; secreted M-T2 binds and inhibits rabbit TNF, whereas a different N-terminal domain of the intracellular M-T2 protein blocks virus-infected lymphocyte apoptosis.[41]

The anti-apoptotic role of M-T2 as inhibitor of TNFR-mediated cell death has been further investigated recently. The first evidence came from the observation that human Jurkat T-cells expressing M-T2 were resistant to TNF and TNFR-induced cell death.[12] Note that intact M-T2 protein neither binds nor inhibits human TNF. This M-T2-mediated inhibition of TNFR-induced cell death requires the PLAD domain located in the N-terminus of the M-T2 protein. Further, biochemical and colocalization studies using fluorescently tagged receptors demonstrated that M-T2 interacts with human TNFR1 and TNFR2 via the PLAD.[12] Thus, the M-T2 CRD domain interacts with TNF in a species-dependent fashion whereas the M-T2 PLAD domain binds both TNFRs in a species-independent manner. A version of the PLAD domain is present and conserved in all poxvirus T2-like proteins. It is entirely possible that the viral PLAD domain might also interact with other host proteins, in order to mediate its anti-apoptotic role. Viral PLAD domains in theory could be utilized as components of future anti-TNF therapeutics if the issue of protein delivery to the appropriate intracellular location could be addressed.

Viral TNF-Binding Proteins Unrelated to Host TNFRs

Another class of TNF inhibitors has been identified from certain members of poxviruses which do not encode conventional vTNFR orthologs. The viral TNF binding protein (vTNF-BP) was first identified in Tanapox virus (TPV), a member of the Yatapoxvirus.[38] Functional orthologs are also present in other member of Yatapoxviruses, Yaba-like disease virus (YLDV) and Yaba monkey tumor virus (YMTV) and a version is also present in swinepox virus.[39] Unlike poxvirus vTNFRs, only one copy of these vTNF-BPs inhibitors are present per genome. This class of vTNF-BPs exhibit closest amino acid sequence similarity to MHC class I molecules but, unlike the cellular counterpart, the viral protein lacks a transmembrane domain. The vTNF-BPs tested to date bind to TNFs from different species with diverse affinities. The TPV-encoded vTNF-BP, called TPV-2L, binds to human TNF with very high affinity (Kd, 43pM) which is the highest affinity reported to date among the known protein TNF inhibitors.[38] TPV-2L can also bind and inhibit monkey and canine TNF with high affinity. The vTNF-BP encoded by YMTV, called YMTV-2L, can also bind and inhibit human and monkey TNF.[39] Binding of TPV-2L and YMTV-2L with TNF also inhibit the TNF-mediated signaling pathway as measured by the degradation of IκBα.[39] Interestingly, the orthologous vTNF-BP from Swinepox virus exclusively binds and inhibits only porcine TNF.[39] TPV-2L and YMTV-2L have very good potential as anti-TNF agents, in part because their very high affinity for human TNF suggests they could achieve TNF inhibition at pharmacologically lower doses than the currently used TNF inhibitors. Although vTNF-BPs share no sequence relationship with TNFRs, the folding clearly results in very high affinity binding with human TNF. Using human TNF mutants which possess differential binding for TNFR1 and TNFR2, it was demonstrated that the interacting residues for human TNF are different for TNFRs compared to these vTNF-BPs.[39] However, solving the crystal structure of these class of TNF inhibitors complexed with TNF would shed more light on their function and future potential as therapeutics.

Viral Proteins That Modulate TNF Receptors and Regulate Downstream Signaling

Viral proteins which modulate the TNF receptor or downstream components of the signaling pathway can also in theory be developed as potential anti-TNF therapeutics. Down-regulation of TNFR on the cell surface is another anti-TNF mechanism adapted by many viral proteins. The adenovirus encoded receptor internalization and degradation (RID) complex associated proteins down-regulate the surface levels of TNFR1 and thereby inhibit NF-κB activation. The adenovirus early transcription region 3 (Ad E3) encodes at least seven proteins, five of which block the acquired or innate immune response. Three of these, Ad E3-14.7K, Ad E3-10.4K and Ad E3-14.5K, impose inhibitory effects on the TNF pathway.[56] Two of these proteins, 10.4K (RIDα) and 14.5K (RIDβ), form the hetero-trimeric complex RID in the plasma membrane which inhibits signaling through TNFR1. RID down-regulates surface TNFR1 levels by reducing the assembly of TNFR1 signaling complex and thus inhibiting TNF-induced activation of NF-κB. In terms of the NF-κB pathway, RID blocks the association of members of the IKK complex, as well as the protein kinase RIP, with the TNFR1.[56] From the RID complex, RIDβ directly interacts with TNFR1 and its tyrosine sorting motif plays a major role in the down-regulation of TNFR1 by a clathrin-dependent process where TNFR1 is degraded by an endosomal/lysosomal pathway.[57] In addition to TNFR1, RID can also down-regulate the expression of FAS,[58] TNF-related apoptosis-inducing ligand receptor 1 (TRAIL-R1 and R2)[59,60] as well as epidermal growth factor receptor (EGFR)[61] in several cell lines.

The down-regulation of TNFR1 by RID also inhibits TNF-induced secretion of chemokines. The anti-inflammatory role of RID has been demonstrated using LPS treatment as the pro-inflammatory stimulus. RID can inhibit LPS-induced signaling pathway without affecting the expression of the LPS receptor, Toll-like receptor 4 (TLR4).[62] This suggests that RID also has intracellular targets that inhibit signal transduction and chemokine expression without receptor down-regulation. The potential use of RID as therapeutic immunomodulator has been tested in vivo where RID facilitates transplantation of allogenic pancreatic cells[63] and decreases autoimmune Type I diabetes incidence in NOD mice.[64,65]

Human Papillomaviruses (HPVs), the major cause of cervical cancer, can infect various human epithelial tissues where they inhibit TNF induced apoptosis. HPV16 encodes two oncogene products, E6 and E7, that associate with other cellular proteins involved in cell proliferation and apoptosis and thereby modulate their function. HPV16 E6 protein selectively binds to TNFR1 and affects the transmission of pro-apoptotic signals triggered by TNF.[66] E6 binds to the C-terminal 41 amino acids of TNFR1 and inhibits binding of TRADD to TNFR1 and thereby blocks formation of the death-inducing signaling complex (DISC). This inhibition subsequently blocks transmission of apoptotic signals by inhibiting the activation of initiator caspases such as caspase 8. E6-mediated protection against TNF-induced apoptosis occurs in cells of different species (mouse and human) and tissues (fibroblast, osteosarcoma and histiocyte/monocyte). Both E6 and E7 of HPV16 increased the transcription of cIAP1 and cIAP2 by upregulation of NF-κB–expression and confer resistance to TNF in human keratinocytes.[67]

Cytomegalovirus (CMV) encodes proteins with diverse immunomodulatory functions. Human and murine CMV have developed mechanisms to evade the TNF-induced antiviral state by dysregulating TNFRs. HCMV infection of THP1 cells reduced the level of TNFR1 on the cell surface by accumulating the receptor pool in the trans-Golgi network.[68] Time course analysis and drug inhibition studies suggest that viral early gene products may target trafficking of TNFR1.[68] MCMV infection blocked TNF-induced nuclear translocation of NF-κB, which decreased the level of both TNFR1 and TNFR2 in bone marrow derived macrophages.[69] This was mediated by expression of still-unknown viral immediate early and/or early genes. Identification of the viral molecule(s) responsible could yield a potential therapeutic drug target.

Poliovirus noncapsid protein 3A is a multifunctional viral protein involved in poliovirus RNA replication.[70] One of its functions is suppression of protein trafficking between the ER and Golgi apparatus. In infected cells, it affects the intracellular trafficking of TNFR and induces TNF resistance by eliminating TNFRs from the plasma membrane.[71] This 3A-protein-mediated inhibition of ER to Golgi traffic of TNFR was limited to poliovirus and coxsackievirus B3. Further investigation is required to understand whether this inhibition is selective for TNFR or not.

The hepatitis C virus (HCV) is the major cause of non-A, nonB hepatitis, which often leads to liver cirrhosis and hepatocellular carcinoma. Two HCV-encoded proteins, nonstructural protein 5A (NS5A) and NS5B, have regulatory roles in TNF signaling and both have potential as anti-TNF therapeutics. Both NS5A and NS5B are essential components of the HCV replication complex. NS5A is a multifunctional phosphoprotein and inhibits TNF-induced signaling using multiple mechanisms. In HepG2 cells expression of NS5A protected these cells from TNF-induced cell death.[72] NS5A binds to the TNFR1 signaling complex through its interaction with TRAF2 and subsequently inhibits TRAF2-dependent NF-κB activation, thereby sensitizing the cells to TNF-induced cytotoxicity.[73] In another study using NS5A transgenic mice, it has been demonstrated that NS5A interacts with TRADD, inhibiting its association with FADD and TNF-mediated apoptosis, resulting in persistent infection.[74] The inhibition of intrinsic apoptotic signals is mediated by the putative BH (Bcl-2 homology) domain of NS5A, which allows it to bind to the pro-apoptotic protein Bax, rendering cells refractory to certain pro-apoptotic agonists.[75] The HCV NS5B protein is a membrane-associated phosphoprotein that possesses an RNA dependent RNA polymerase activity.[76] NS5B inhibited TNF-induced NF-κB activation in HEK293 and hepatic cells. This inhibition is mediated by its direct interaction with IKKα.

Cell Signaling Inhibitors from Viruses That Inhibit Activation of NF-κB

The ideal candidate for a virus-derived anti-TNF therapeutic would be more specific and less toxic to the human system than current therapies. Several viral inhibitors can specifically block intracellular signaling, often resulting in reduced TNF production or its TNFR-dependant effects. Activation of NF-κB is an early event that occurs within minutes after exposure to TNF and plays important role in inflammation, regulation of cell proliferation, activation and survival.[77] This activation process has turned out to be an attractive target for viruses to escape immune defenses and many viruses have evolved specific gene products to inhibit the TNF-induced NF-κB activation.

Vaccinia virus encoded protein N1L, a viral virulence factor, inhibits signaling through NF-κB from both TNF and LTα. This N1L-mediated inhibition of NF-κB occurs by association with IKK-γ and inhibition of IKK-α and IKK-β.[78] N1L also inhibits IRF3 signaling and thus might play a broad role as viral immunomodulator of innate immunity. However, N1L might have even additional biologic roles. N1L inhibits apoptosis and interacts with pro-apoptotic Bcl2 proteins like Bid, Bax, Bad, Bim and Bak.[79,80] Recent crystal structure data of N1L suggest that it belongs to Bcl2 family of anti-apoptotic proteins, although no amino acid sequence similarity to these cellular proteins was observed by sequence analysis.[80] The structure also revealed that N1L contains a surface groove that resembles the BH3-binding grooves of other Bcl-2 proteins. In another report, it was shown that N1L also inhibits the release of pro-inflammatory cytokines like TNF, IL-1β, IFNα and IFNβ and the anti-inflammatory cytokine IL-10 from human primary monocytes.[81]

Another vaccinia virus-encoded protein, K1L, inhibits NF-κB activation and could be a potential inhibitor of TNF-mediated responses. K1L inhibits NF-κB by preventing IκBα degradation, probably by interfering directly with IKK to prevent phosphorylation or indirectly by hampering kinases that act upstream of IKK.[82] K1L was originally identified as one of the vaccinia host range genes required for viral replication in certain human cells and rabbit kidney cells RK13. The host range function of K1L is associated with the ankyrin repeats present in this protein.[83]

African swine fever virus (ASFV) A238L, which is an ankyrin-repeat–containing homolog of host IκB (ASFV-IκB), binds to NF-κB following degradation of host IκB and inhibits the nuclear transportation of NF-κB.[84] It also inhibits TNF-induced gene expression. The mechanism of regulation of TNF lies in the promoter region that involves CREB binding protein (CBP)/p300 function. Localization of A238L in the nucleus and binds to the cyclic AMP-responsive element and displace the CBP/p300 coactivators.[85]

Conclusions

Given the sophistication with which viral proteins are currently known to exercise their anti-TNF properties, it would be negligent to ignore the lessons they can impart about how to inhibit TNF and functionally disable pro-inflammatory cascades. Current biotechnology has generated efficient protein-based TNF inhibitors with Kd values in the nanomolar range, but viral TNF-BPs can bind and inhibit human TNF with more than an order of magnitude higher affinity than any commercial TNF inhibitor. At this point, it is unknown if any specific vTNF inhibitors will be more or less immunogenic then the clinical TNF blockade reagents currently in use, but experience has shown that immunogenicity issues are not resolvable with animal models alone. Realistically, the potential of vTNF inhibitors for treating inflammatory disorders will only be settled with properly designed clinical trials.

Acknowledgements

GM is supported by an International Scholarship from HHMI and AL holds the Ethel Smith Endowed Chair in Vasculitis. GM and AL are cofounders of Viron Therapeutics, which is developing viral proteins as anti-inflammatory therapeutics.

References

1. Finlay BB, McFadden G. Anti-immunology: evasion of the host immune system by bacterial and viral pathogens. Cell 2006; 124(4):767-782.
2. Roy CR, Mocarski ES. Pathogen subversion of cell-intrinsic innate immunity. Nat Immunol 2007; 8(11):1179-1187.
3. Unterholzner L, Bowie AG. The interplay between viruses and innate immune signaling: Recent insights and therapeutic opportunities. Biochem Pharmacol 2008; 75(3):589-602.
4. Lucas A, McFadden G. Secreted immunomodulatory viral proteins as novel biotherapeutics. J Immunol 2004; 173(8):4765-4774.
5. Fallon PG, Alcami A. Pathogen-derived immunomodulatory molecules: future immunotherapeutics? Trends Immunol 2006; 27(10):470-476.
6. Rahman MM, McFadden G. Modulation of tumor necrosis factor by microbial pathogens. PLoS Pathog 2006; 2(2):e4.
7. Aggarwal BB. Signalling pathways of the TNF superfamily: a double-edged sword. Nat Rev Immunol 2003; 3(9):745-756.
8. Hehlgans T, Pfeffer K. The intriguing biology of the tumour necrosis factor/tumour necrosis factor receptor superfamily: players, rules and the games. Immunology 2005; 115(1):1-20.
9. Takada H, Chen NJ, Mirtsos C et al. Role of SODD in regulation of tumor necrosis factor responses. Mol Cell Biol 2003; 23(11):4026-4033.
10. Chan FK, Chun HJ, Zheng L et al. A domain in TNF receptors that mediates ligand-independent receptor assembly and signaling. Science 2000; 288(5475):2351-2354.
11. Clancy L, Mruk K, Archer K et al. Preligand assembly domain-mediated ligand-independent association between TRAIL receptor 4 (TR4) and TR2 regulates TRAIL-induced apoptosis. Proc Natl Acad Sci USA 2005; 102(50):18099-18104.

12. Sedger LM, Osvath SR, Xu XM et al. Poxvirus tumor necrosis factor receptor (TNFR)-like T2 proteins contain a conserved preligand assembly domain that inhibits cellular TNFR1-induced cell death. J Virol 2006; 80(18):9300-9309.
13. Siebert S, Fielding CA, Williams BD et al. Mutation of the extracellular domain of tumour necrosis factor receptor 1 causes reduced NF-kappaB activation due to decreased surface expression. FEBS Lett 2005; 579(23):5193-5198.
14. Siegel RM, Frederiksen JK, Zacharias DA et al. Fas preassociation required for apoptosis signaling and dominant inhibition by pathogenic mutations. Science 2000; 288(5475):2354-2357.
15. Chan FK. Three is better than one: preligand receptor assembly in the regulation of TNF receptor signaling. Cytokine 2007; 37(2):101-107.
16. Deng GM, Zheng L, Chan FK et al. Amelioration of inflammatory arthritis by targeting the preligand assembly domain of tumor necrosis factor receptors. Nat Med 2005; 11(10):1066-1072.
17. Choy EH, Panayi GS. Cytokine pathways and joint inflammation in rheumatoid arthritis. N Engl J Med 2001; 344(12):907-916.
18. Theodossiadis PG, Markomichelakis NN, Sfikakis PP. Tumor necrosis factor antagonists: preliminary evidence for an emerging approach in the treatment of ocular inflammation. Retina 2007; 27(4):399-413.
19. Tak PP, Taylor PC, Breedveld FC et al. Decrease in cellularity and expression of adhesion molecules by anti-tumor necrosis factor alpha monoclonal antibody treatment in patients with rheumatoid arthritis. Arthritis Rheum 1996; 39(7):1077-1081.
20. Goldring SR, Gravallese EM. Pathogenesis of bone lesions in rheumatoid arthritis. Curr Rheumatol Rep 2002; 4(3):226-231.
21. Hotamisligil GS, Shargill NS, Spiegelman BM. Adipose expression of tumor necrosis factor-alpha: direct role in obesity-linked insulin resistance. Science 1993; 259(5091):87-91.
22. Mohler KM, Torrance DS, Smith CA et al. Soluble tumor necrosis factor (TNF) receptors are effective therapeutic agents in lethal endotoxemia and function simultaneously as both TNF carriers and TNF antagonists. J Immunol 1993; 151(3):1548-1561.
23. Harriman G, Harper LK, Schaible TF. Summary of clinical trials in rheumatoid arthritis using infliximab, an anti-TNFalpha treatment. Ann Rheum Dis 1999; 58(Suppl 1):I61-64.
24. Wong M, Ziring D, Korin Y et al. TNF α blockade in human diseases: Mechanisms and future directions. Clin Immunol 2007; 126(2): 121-136.
25. Palladino MA, Bahjat FR, Theodorakis EA et al. Anti-TNF-α therapies: the next generation. Nat Rev Drug Discov 2003; 2(9):736-746.
26. Slifman NR, Gershon SK, Lee JH et al. Listeria monocytogenes infection as a complication of treatment with tumor necrosis factor alpha-neutralizing agents. Arthritis Rheum 2003; 48(2):319-324.
27. De Rosa FG, Shaz D, Campagna AC et al. Invasive pulmonary aspergillosis soon after therapy with infliximab, a tumor necrosis factor-alpha-neutralizing antibody: a possible healthcare-associated case? Infect Control Hosp Epidemiol 2003; 24(7):477-482.
28. Tai TL, O'Rourke KP, McWeeney M et al. Pneumocystis carinii pneumonia following a second infusion of infliximab. Rheumatology (Oxford) 2002; 41(8):951-952.
29. Lee JH, Slifman NR, Gershon SK et al. Life-threatening histoplasmosis complicating immunotherapy with tumor necrosis factor alpha antagonists infliximab and etanercept. Arthritis Rheum 2002; 46(10):2565-2570.
30. Gomez-Reino JJ, Carmona L, Angel Descalzo M. Risk of tuberculosis in patients treated with tumor necrosis factor antagonists due to incomplete prevention of reactivation of latent infection. Arthritis Rheum 2007; 57(5):756-761.
31. Ellerin T, Rubin RH, Weinblatt ME. Infections and anti-tumor necrosis factor alpha therapy. Arthritis Rheum 2003; 48(11):3013-3022.
32. Charles PJ, Smeenk RJ, De Jong J et al. Assessment of antibodies to double-stranded DNA induced in rheumatoid arthritis patients following treatment with infliximab, a monoclonal antibody to tumor necrosis factor alpha: findings in open-label and randomized placebo-controlled trials. Arthritis Rheum 2000; 43(11):2383-2390.
33. Hyrich KL, Silman AJ, Watson KD et al. Anti-tumour necrosis factor alpha therapy in rheumatoid arthritis: an update on safety. Ann Rheum Dis 2004; 63(12):1538-1543.
34. Gilaberte Y, Coscojuela C, Vazquez C et al. Perforating folliculitis associated with tumour necrosis factor-alpha inhibitors administered for rheumatoid arthritis. Br J Dermatol 2007; 156(2):368-371.
35. Morgan MB, Truitt CA, Taira J et al. Fibronectin and the extracellular matrix in the perforating disorders of the skin. Am J Dermatopathol 1998; 20(2):147-154.
36. Benedict CA. Viruses and the TNF-related cytokines, an evolving battle. Cytokine Growth Factor Rev 2003; 14(3-4):349-357.

37. Benedict CA, Banks TA, Ware CF. Death and survival: viral regulation of TNF signaling pathways. Curr Opin Immunol 2003; 15(1):59-65.
38. Brunetti CR, Paulose-Murphy M, Singh R et al. A secreted high-affinity inhibitor of human TNF from Tanapox virus. Proc Natl Acad Sci USA 2003; 100(8):4831-4836.
39. Rahman MM, Barrett JW, Brouckaert P et al. Variation in ligand binding specificities of a novel class of poxvirus-encoded tumor necrosis factor-binding protein. J Biol Chem 2006; 281(32):22517-22526.
40. Cunnion KM. Tumor necrosis factor receptors encoded by poxviruses. Mol Genet Metab 1999; 67(4):278-282.
41. Xu X, Nash P, McFadden G. Myxoma virus expresses a TNF receptor homolog with two distinct functions. Virus Genes 2000; 21(1-2):97-109.
42. Hu FQ, Smith CA, Pickup DJ. Cowpox virus contains two copies of an early gene encoding a soluble secreted form of the Type II TNF receptor. Virology 1994; 204(1):343-356.
43. Smith CA, Hu FQ, Smith TD et al. Cowpox virus genome encodes a second soluble homologue of cellular TNF receptors, distinct from CrmB, that binds TNF but not LT alpha. Virology 1996; 223(1):132-147.
44. Loparev VN, Parsons JM, Knight JC et al. A third distinct tumor necrosis factor receptor of orthopoxviruses. Proc Natl Acad Sci USA 1998; 95(7):3786-3791.
45. Saraiva M, Alcami A. CrmE, a novel soluble tumor necrosis factor receptor encoded by poxviruses. J Virol 2001; 75(1):226-233.
46. Panus JF, Smith CA, Ray CA et al. Cowpox virus encodes a fifth member of the tumor necrosis factor receptor family: a soluble, secreted CD30 homologue. Proc Natl Acad Sci USA 2002; 99(12):8348-8353.
47. Alejo A, Ruiz-Arguello MB, Ho Y et al. A chemokine-binding domain in the tumor necrosis factor receptor from variola (smallpox) virus. Proc Natl Acad Sci USA 2006; 103(15):5995-6000.
48. Gileva IP, Nepomnyashchikh TS, Antonets DV et al. Properties of the recombinant TNF-binding proteins from variola, monkeypox and cowpox viruses are different. Biochim Biophys Acta 2006; 1764(11):1710-1718.
49. Smith VP, Alcami A. Expression of secreted cytokine and chemokine inhibitors by ectromelia virus. J Virol 2000; 74(18):8460-8471.
50. Alcami A, Khanna A, Paul NL et al. Vaccinia virus strains Lister, USSR and Evans express soluble and cell-surface tumour necrosis factor receptors. J Gen Virol 1999; 80(Pt 4):949-959.
51. Reading PC, Khanna A, Smith GL. Vaccinia virus CrmE encodes a soluble and cell surface tumor necrosis factor receptor that contributes to virus virulence. Virology 2002; 292(2):285-298.
52. Graham SC, Bahar MW, Abrescia NG et al. Structure of CrmE, a virus-encoded tumour necrosis factor receptor. J Mol Biol 2007; 372(3):660-671.
53. Schreiber M, McFadden G. The myxoma virus TNF-receptor homologue (T2) inhibits tumor necrosis factor-alpha in a species-specific fashion. Virology 1994; 204(2):692-705.
54. Macen JL, Graham KA, Lee SF et al. Expression of the myxoma virus tumor necrosis factor receptor homologue and M11L genes is required to prevent virus-induced apoptosis in infected rabbit T-lymphocytes. Virology 1996; 218(1):232-237.
55. Schreiber M, Sedger L, McFadden G. Distinct domains of M-T2, the myxoma virus tumor necrosis factor (TNF) receptor homolog, mediate extracellular TNF binding and intracellular apoptosis inhibition. J Virol 1997; 71(3):2171-2181.
56. Fessler SP, Chin YR, Horwitz MS. Inhibition of tumor necrosis factor (TNF) signal transduction by the adenovirus group C RID complex involves downregulation of surface levels of TNF receptor 1. J Virol 2004; 78(23):13113-13121.
57. Chin YR, Horwitz MS. Mechanism for removal of tumor necrosis factor receptor 1 from the cell surface by the adenovirus RIDalpha/beta complex. J Virol 2005; 79(21):13606-13617.
58. Shisler J, Yang C, Walter B et al. The adenovirus E3-10.4K/14.5K complex mediates loss of cell surface Fas (CD95) and resistance to Fas-induced apoptosis. J Virol 1997; 71(11):8299-8306.
59. Benedict CA, Norris PS, Prigozy TI et al. Three adenovirus E3 proteins cooperate to evade apoptosis by tumor necrosis factor-related apoptosis-inducing ligand receptor-1 and -2. J Biol Chem 2001; 276(5):3270-3278.
60. Tollefson AE, Toth K, Doronin K et al. Inhibition of TRAIL-induced apoptosis and forced internalization of TRAIL receptor 1 by adenovirus proteins. J Virol 2001; 75(19):8875-8887.
61. Tollefson AE, Stewart AR, Yei SP et al. The 10,400- and 14,500-dalton proteins encoded by region E3 of adenovirus form a complex and function together to down-regulate the epidermal growth factor receptor. J Virol 1991; 65(6):3095-3105.
62. Delgado-Lopez F, Horwitz MS. Adenovirus RIDalphabeta complex inhibits lipopolysaccharide signaling without altering TLR4 cell surface expression. J Virol 2006; 80(13):6378-6386.

63. Efrat S, Fejer G, Brownlee M et al. Prolonged survival of pancreatic islet allografts mediated by adenovirus immunoregulatory transgenes. Proc Natl Acad Sci USA 1995; 92(15):6947-6951.
64. Efrat S, Serreze D, Svetlanov A et al. Adenovirus early region 3(E3) immunomodulatory genes decrease the incidence of autoimmune diabetes in NOD mice. Diabetes 2001; 50(5):980-984.
65. Pierce MA, Chapman HD, Post CM et al. Adenovirus early region 3 antiapoptotic 10.4K, 14.5K and 14.7K genes decrease the incidence of autoimmune diabetes in NOD mice. Diabetes 2003; 52(5):1119-1127.
66. Filippova M, Song H, Connolly JL et al. The human papillomavirus 16 E6 protein binds to tumor necrosis factor (TNF) R1 and protects cells from TNF-induced apoptosis. J Biol Chem 2002; 277(24):21730-21739.
67. Yuan H, Fu F, Zhuo J et al. Human papillomavirus Type 16 E6 and E7 oncoproteins upregulate c-IAP2 gene expression and confer resistance to apoptosis. Oncogene 2005; 24(32):5069-5078.
68. Baillie J, Sahlender DA, Sinclair JH. Human cytomegalovirus infection inhibits tumor necrosis factor alpha (TNF-α) signaling by targeting the 55-kilodalton TNF-α receptor. J Virol 2003; 77(12):7007-7016.
69. Popkin DL, Virgin HWt. Murine cytomegalovirus infection inhibits tumor necrosis factor alpha responses in primary macrophages. J Virol 2003; 77(18):10125-10130.
70. Doedens JR, Giddings TH Jr, Kirkegaard K. Inhibition of endoplasmic reticulum-to-Golgi traffic by poliovirus protein 3A: genetic and ultrastructural analysis. J Virol 1997; 71(12):9054-9064.
71. Neznanov N, Kondratova A, Chumakov KM et al. Poliovirus protein 3A inhibits tumor necrosis factor (TNF)-induced apoptosis by eliminating the TNF receptor from the cell surface. J Virol 2001; 75(21):10409-10420.
72. Ghosh AK, Majumder M, Steele R et al. Hepatitis C virus NS5A protein protects against TNF-α mediated apoptotic cell death. Virus Res 2000; 67(2):173-178.
73. Park KJ, Choi SH, Choi DH et al. 1 Hepatitis C virus NS5A protein modulates c-Jun N-terminal kinase through interaction with tumor necrosis factor receptor-associated factor 2. J Biol Chem 2003; 278(33):30711-30718.
74. Majumder M, Ghosh AK, Steele R et al. Hepatitis C virus NS5A protein impairs TNF-mediated hepatic apoptosis, but not by an anti-FAS antibody, in transgenic mice. Virology 2002; 294(1):94-105.
75. Chung YL, Sheu ML, Yen SH. Hepatitis C virus NS5A as a potential viral Bcl-2 homologue interacts with Bax and inhibits apoptosis in hepatocellular carcinoma. Int J Cancer 2003; 107(1):65-73.
76. Behrens SE, Tomei L, De Francesco R. Identification and properties of the RNA-dependent RNA polymerase of hepatitis C virus. EMBO J 1996; 15(1):12-22.
77. Karin M, Lin A. NF-kappaB at the crossroads of life and death. Nat Immunol 2002; 3(3):221-227.
78. DiPerna G, Stack J, Bowie AG et al. Poxvirus protein N1L targets the I-kappaB kinase complex, inhibits signaling to NF-kappaB by the tumor necrosis factor superfamily of receptors and inhibits NF-kappaB and IRF3 signaling by toll-like receptors. J Biol Chem 2004; 279(35):36570-36578.
79. Aoyagi M, Zhai D, Jin C et al. Vaccinia virus N1L protein resembles a B-cell lymphoma-2 (Bcl-2) family protein. Protein Sci 2007; 16(1):118-124.
80. Cooray S, Bahar MW, Abrescia NG et al. Functional and structural studies of the vaccinia virus virulence factor N1 reveal a Bcl-2-like anti-apoptotic protein. J Gen Virol 2007; 88(Pt 6):1656-1666.
81. Zhang Z, Abrahams MR, Hunt LA et al. The vaccinia virus N1L protein influences cytokine secretion in vitro after infection. Ann N Y Acad Sci 2005; 1056:69-86.
82. Shisler JL, Jin XL. The vaccinia virus K1L gene product inhibits host NF-kappaB activation by preventing I kappa B alpha degradation. J Virol 2004; 78(7):3553-3560.
83. Bradley RR, Terajima M. Vaccinia virus K1L protein mediates host-range function in RK-13 cells via ankyrin repeat and may interact with a cellular GTPase-activating protein. Virus Res 2005; 114(1-2):104-112.
84. Tait SW, Reid EB, Greaves DR et al. Mechanism of inactivation of NF-kappa B by a viral homologue of I kappa B alpha. Signal-induced release of i kappa b alpha results in binding of the viral homologue to NF-kappa B. J Biol Chem 2000; 275(44):34656-34664.
85. Granja AG, Nogal ML, Hurtado C et al. The viral protein A238L inhibits TNF-α expression through a CBP/p300 transcriptional coactivators pathway. J Immunol 2006; 176(1):451-462.
86. Choi SH, Park KJ, Ahn BY et al. Hepatitis C virus nonstructural 5B protein regulates tumor necrosis factor alpha signaling through effects on cellular I kappa B kinase. Mol Cell Biol 2006; 26(8):3048-3059.

Chapter 6

Lipoxins as an Immune-Escape Mechanism

Fabiana S. Machado and Julio Aliberti*

Here, we discuss the mechanisms of repression of signaling pathways that are triggered by Lipoxin (LX) and are responsible for control of pro-inflammatory response during chronic phase of *Toxoplasma gondii* infection. We also discuss this mechanism from the perspective of the pathogen, which pirates the host's lipoxygenase machinery to its own advantage as a probable immune-escape mechanism. Pro-inflammatory mediators such as IL-12, IFN-γ and TNF are essential in controlling parasite growth during *T. gondii* infection. However, it is clear that exacerbated production of these cytokines results in host tissue damage. LX, an anti-inflammatory eicosanoid, plays an important role in regulation of immune response to *T. gondii*.

Introduction

Toxoplasmosis is a parasitic disease, usually asymptomatic, caused by *Toxoplasma gondii*. *Toxoplasma* infection can become a serious disease in the setting of congenital infection and in immunodeficient (AIDS patients and transplant recipients[1]). After infection with *T. gondii*, the released parasites from ruptured cysts actively enter host cells[2] including macrophages and dendritic cells (DCs). Once intracellular the tachyzoites quickly replicate and in immunocompetent individuals, the acute phase of infection resolves when the remaining fast-replicating parasites switch to bradyzoites form of *T. gondii*, probably as a response to immune attack and seclude themselves in cysts within tissues allowing the parasite escape of the immune responses and this phase is known as chronic or persistent infection.[3]

The main pathology resulting from toxoplasma infection occurs during chronic disease, when immune suppression caused by drugs or infections, such as HIV, can lead to reversion from the dormant encysted bradyzoite stage back to the fast replicating tachyzoites, which rupture cysts within heart or skeletal muscle and the central nervous system (CNS) causing local tissue necrosis, in the CNS it is often lethal. Intracellular parasites, including *T. gondii*, are capable of modulating numerous immune response functions in order to promote the host and themselves survival, consequently achieving successful transmission to a new host.

It is clear that *T. gondii* has evolved various strategies to take advantage of host cell signaling regulatory mechanism by distorting this balance between positive and negative influences, which allow the parasite to persist through the chronic phase of the disease and prevents host death.[4] Here, we discussed how hosts and *T. gondii*, make use of some immune-regulatory pathways to promote host survival.

*Corresponding Author: Julio Aliberti—Division of Molecular Immunology, Cincinnati Children's Hospital Medical Center and University of Cincinnati College of Medicine, Cincinnati, Ohio, 45229, USA. Email: julio.aliberti@cchmc.org

Pathogen-Derived Immunomodulatory Molecules, edited by Padraic G. Fallon.

Lipoxins

LXA_4 or (5S,6S,15S)-5,6,15-trihydroxy-7,9,13-trans-11-cis-eicosatentraenoic acid, is an eicosanoid derived from the lypoxigenase (LO)-metabolism of arachidonic acid (AA). Initial studies found that LXA_4 was secreted by neutrophils and inhibited the activating effects of LTB_4 on platelets.[5] Since then, a growing list of anti-inflammatory effects have been associated with LXA_4 and its analogs.[6] LO-catalyzed oxygenation of unsaturated fatty acids initiates the formation of a variety of compounds with diverse biological activities. The biosynthetic pathways involved in LXA_4 formation are complex, involve the actions of at least two independent LOs (Fig. 1) and can occur through trans-cellular cascades. However, the activity of 5-LO seems to be a common step in LXA_4 synthesis.[5] 5-LO is produced as a propeptide that is activated by cleavage. Low levels of active 5-LO are found in different cell types, including macrophages, platelets, DCs and neutrophils.[7] The expression of a 5-LO-activating protein (FLAP) is a key signal for induction of 5-LO activity. Of note, Aspirin acetylation of cyclo-oxygenase (COX),[5] leads to a change in functionality that can drive synthesis of the "Aspirin-triggered LXA_4" (ATL): 15-epi-LXA_4. ATL has the same activity, but is more metabolically stable than LXA_4.[5]

LXA_4 is thought to bind to two types of receptors[8]: (a) a surface membrane seven-transmembrane G-protein-coupled receptor, FPRL-1 or ALX[9]; and (b) the nuclear ligand-activated transcription factor, the Aryl hydrocarbon receptor (AhR).[10] The binding of ligands (i.e., LX) to AhR results in the formation of an active transcription factor that binds to DNA domains—dioxin-responsive elements (DRE)—that activate the expression of a specific panel of genes.[11] In the absence of ligand, AhR is located in the cytosol in association with several molecules, Hsp90, the immunophilin-like protein XAP2 and the cochaperone p23. Ligand binding allows for release of the AhR, its translocation to the nucleus and heterodimerization with the AhR nuclear translocator (ARNT; HIF-1β). The AhR/ARNT heterodimer binds to specific DNA sequences (aryl hydrocarbon elements [AHREs], also known as DRE or XRE), where, in concert with the co-activators CBP/p300 (ARNT) and RIPI40 (AhR), they lead to target-gene transcription leading to expression of target genes.[11](Fig. 2).

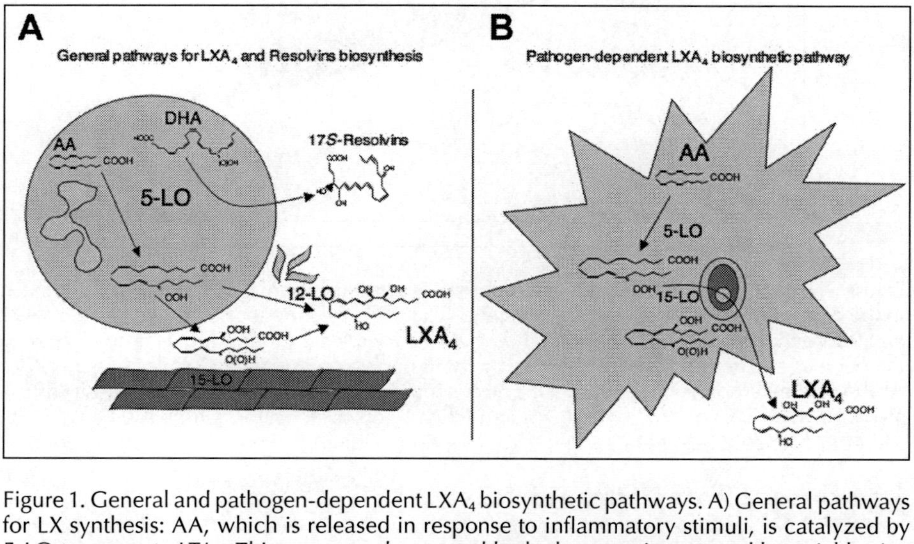

Figure 1. General and pathogen-dependent LXA_4 biosynthetic pathways. A) General pathways for LX synthesis: AA, which is released in response to inflammatory stimuli, is catalyzed by 5-LO to generate LTA_4. This compound, secreted by leukocytes, is captured by neighboring platelets or endothelial cells and, through the actions of 12-LO or 15-LO, respectively, is converted to LXA_4. B) Pathogen-dependent LXA_4 biosynthetic pathway: After the generation of LTA_4 in a 5-LO-dependent manner, it is catalyzed by pathogen-secreted 15-LO into LXA_4, which is then secreted by the infected cell.

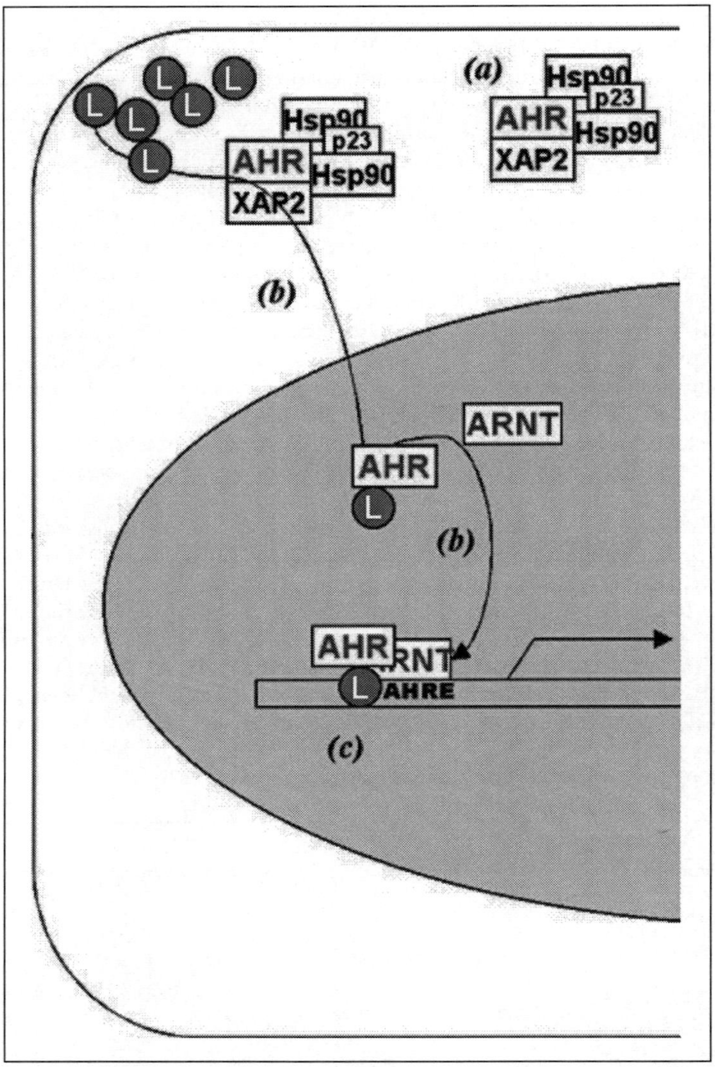

Figure 2. Ligand-induced AhR activation pathway. a) In the absence of ligand, AhR is located in the cytosol in association with 2 molecules of Hsp90, the immunophilin-like protein XAP2 and the cochaperone p23. b) Ligand binding allows for release of the AhR, its translocation to the nucleus and heterodimerization with the AHR nuclear translocator (ARNT; HIF-1b). c) The AhR/ARNT heterodimer binds to specific DNA sequences (aryl hydrocarbon elements [AhREs], also known as DRE or XRE), where, in concert with the co-activators CBP/p300 (ARNT) and RIPI40 (AhR), they lead to target-gene transcription leading to expression of target genes.

Lipoxins and Toxoplasma Infection

5-LO/LXA_4 plays an important counter-regulatory role in mouse models of toxoplasmosis. During infection with *T. gondii*, serum levels of LXA_4 increase steadily over the course of the acute phase and remain at high levels during chronic diseases.[12] Such high levels in the serum of chronically infected animals indicated that this mediator might exert relevant biological functions during this stage of the disease. Despite the fact that DCs are one of the main targets for the immune modulatory actions of LXA_4 during *T. gondii* infection, this cell population does not produce detectable levels of the eicosanoid.[12] Instead, resident splenic macrophages up-regulate 5-LO expression after in vivo stimulation with parasite extract[12] indicating the participation of macrophages in the generation of LXs during experimental toxoplasmosis.

The action of 5-LO on AA results in the formation of LTA_4, which can be rapidly converted to LXA_4 through the actions of a second enzyme, 15-LO (Fig. 1A). Although the 5-LO activity after *T. gondii* infection was known to be associated with splenic macrophages,[12] the 15-LO-expressing cell population was not known. In an effort to identify the sources of 15-LO activity after *T. gondii* infection, Bannenberg and colleagues identified an enzymatic activity in tachyzoite forms exposed to calcium ionophore in the presence of AA in vitro.[13] Moreover, proteomics analysis of *T. gondii* tachyzoite-derived lysates revealed the presence of peptides homologous to plant-derived Type 15 Los.[13] It therefore seems probable that the induction of LX biosynthesis by *T. gondii* has been selected through the carrying of a plant-like lipoxygenase gene, which together with the actions of host-derived 5-LO results in the high-level production of LX (Fig. 1B).

The presence of high levels of LX, in turn, dampens ongoing immune responses (i.e., IL-12, IFN-γ and TNF production) so that hosts can control parasite proliferation without succumbing to the damaging consequences of excessive inflammation or tissue destruction. Suggesting that lipoxins-dependent inhibition of pro-inflammatory Type 1 responses could provide a favorable environment for transmission and propagation of the pathogen. Furthermore, it is possible to hypothesize that pathogens may take advantage of this regulatory pathway to promote host survival, or even allow a less toxic environment in which replication can occur. It is possible to conjecture that from the perspective of *T. gondii*, which is a fast-replicating pathogen, the host must be kept alive so that transmission can occur through predation. Thus, the host needs well-balanced immunity against the parasite, the number of which is kept low but not completely eliminated. To accomplish this, lipoxins are induced to keep immunity present, but not intensified.

Experimental Model of Infection with *T. gondii* and Modulation of Immune Response

Mice that receive extract derived from *T. gondii* tachyzoites (STAg) triggered endogenous production of LXA_4 that was found to inhibit microbe-induced DC migration and IL-12 production in vivo and in vitro.[12] Consistent with these findings, 5-LO-deficient animals succumbed to *T. gondii* infection at the early onset of chronic disease. Upon further investigation, it became clear that immunity against the parasite was actually increased in the absence of 5-LO, with significantly less brain cyst formation than in control animals, indicating that this was not the cause of mortality. By contrast, excessive pro-inflammatory cytokine production and massive cerebral infiltration was found, including atypical meningitis. The conclusion was that the excessive pro-inflammatory response in the brain ultimately caused the death of the 5-LO-deficient hosts.[12] *T. gondii* infection in 5-LO-deficient mice resulted in more extensive tissue pathology, mainly due to lack of LXA_4 production, as treatment of 5-LO-deficient mice with LX analogs restored the resistance to tissue pathology with no mortality associated with uncontrolled pro-inflammatory responses, in a similar manner as for wild type animals.[12]

Among the pro-inflammatory mediators, IL-12 production is critical to the development of innate and adaptive immune responses required for the control of intracellular pathogens by governing the development of IFN-γ-dependent host resistance during *T. gondii* infection. These initial steps during the interaction of *T. gondii* with immune system are critical for establishing a balanced host-parasite relationship, such that both host and parasite survive the initial phases of infection, progressing towards chronic disease. In fact, IL-12, IFN-γ and IFN-γ receptor deficient mice are extremely susceptible to *T. gondii* infection; these mice are unable to survive the early stages of acute infection.[14] DCs constitute the major IL-12-producing cell population during *T. gondii* infection in vivo and consequently, the most relevant cell population for the development of a parasite-specific Type 1 immune response.[15]

Mouse CD8α⁺ DCs produce IL-12 in response to *T. gondii* in the absence of costimulatory signals,[16] indicating that these cells can activate the immune system by recognition of parasite-derived molecules. In fact, splenic DCs stimulated with a soluble extract derived from *T. gondii* tachyzoites (STAg) produce high levels of IL-12 in vitro and in vivo in a MyD88-independent and -dependent pathways. The MyD88-independent induction of IL-12 production by DCs has been attributed to the CC-chemokine receptor (CCR5), interacting with the secreted, parasite-derived peptide, cyclophilin-18 (C-18), which mimic that signals through CCR5 inducing IL-12 production by DCs in vitro and in vivo.[17-19] CCR5 deficient mice are susceptible to *T. gondii*, with short survival times, high parasite proliferation and low detectable IL-12 and IFN-γ responses, supporting the biological significance of the unusual requirement for a chemokine receptor to participate in microbial recognition by DCs in vivo. However, the IL-12 levels observed after stimulation of DCs with C-18 are much lower than those seen after stimulation with whole parasite lysate or with a pool of tachyzoite-secreted proteins, indicating that pathways other than those initiated by CCR5/C-18 might also be important for IL-12 production by these cells.[18] In fact, studies showed that Toll Like Receptor (TLR) adaptor protein MyD88-deficient cells display severely impaired STAg-induced IL-12 responses in vivo and in vitro. Moreover, upon *T. gondii* infection, MyD88-deficient hosts had high mortality due to a lack of protective IFN-γ-mediated immunity.[19]

Recently, it was reported that TLR-11 interaction with *T. gondii*-derived prophillin is also an important signal for IL-12 induction by DCs.[20] Although the molecular basis of IL-12 gene expression is unclear, no doubt remains regarding the critical function of IL-12 in response to *T. gondii* infection. IL-12 produced in vivo triggers NK-cells to release IFN-γ. IFN-γ production is also amplified by tumor necrosis factor and IL-1β in a positive feedback loop. Together these cytokines drive the generation of Type 1 CD4⁺ and CD8⁺ T-cells,[21,22] which produce further IFN-γ, a key requirement for inducing and maintaining control of acute and chronic infection.[23]

IFN-γ is a potent microbicidal/microbiostatic factor for upregulating the expression of inducible nitric oxide synthase (iNOS) expression and the production of nitric oxide (NO) by *T. gondii*-infected macrophages. During the chronic phase, some parasites escape immune control and survive within the host for long periods, even though the immune system is continuing to survey the tissues for parasite proliferation. Some *T. gondii* virulent strains escape immune responses and eventually kill the host, despite stimulating vigorous IL-12 production and Type 1 T-cell induction. It is possible that such parasite strains represent the result of random abnormal virulence not based on evolutionary selection pressure but based on the naturally occurring host-parasite relationships. It is known that a growing list of pathogens escape immune responses through the actions of membrane receptors or cytoplasmic enzymes that inactivate or neutralize effector molecules, such a complement factors,[24] superoxide or nitric oxide.[25] Alternatively, the induction of modulatory or anti-inflammatory factors by the pathogen could inhibit the accumulation of effector cells or their activation in situ, favoring the escape of the pathogen from immune responses and the development of chronic disease.

Intracellular Mechanisms of Anti-Inflammatory Actions of LXs: SOCS Proteins

It was recently shown in an experimental ischemia model that LXA_4-analog administration prevented renal failure. Notably, such protection was associated with increased in vivo expression of several suppressors of cytokine signaling (SOCS) genes, including SOCS-1 and SOCS-2,[26] providing clues to the intracellular mechanics of LXA_4-dependent anti-inflammatory responses. SOCS are a family of intracellular proteins, typically characterized by the presence of an SH2 domain and a so-called, "SOCS Box".[27] This family of proteins was initially identified by several groups after precipitation of signaling elements of diverse cytokine receptor intracellular cascades, i.e., STATS or JAK2, to which SOCS members were associated via the SH2 domain.[28] Since then, extensive research has been reported on the first three members of this family: SOCS-1, -2 and -3.[28]

SOCS-1-deficient mice are runted and die at 3 weeks of age. At that point, the animals exhibit severe lymphopenia, fatty degeneration of the liver and macrophage infiltration of major organs.[29]

Apparently, this complex phenotype is due to uncontrolled secretion and signaling of IFN-γ.[29] These results indicate that a major role of SOCS-1 is to contain deleterious IFN-γ responses. When investigating the physiological function of SOCS-3, Marine et al, found that this protein is highly expressed in fetal liver, suggesting its participation in the regulation of fetal erythropoiesis. This role was then confirmed in SOCS-3-deficient mice, which have embryonic lethality due to massive expansion of erythroid progenitors.[30] These results indicate that the primary physiological regulatory target for SOCS-3 is the erythropoietin receptor-induced signaling cascade.

SOCS-2 has previously mainly been linked to the growth hormone (GH)/insulin-like growth factor-1 (IGF-1) signaling due the phenotype observed in SOCS-2 deficient mice. SOCS-2-deficient mice are characterized by a 40% increase in body weight,[31] a phenotype resembling that observed in GH transgenic mice,[32] acromegalic patients[33] and high-growth (hg) mice.[34]

Machado et al found that SOCS-2 is an essential intracellular mediator of the anti-inflammatory actions of LXs in vivo during toxoplasmosis.[35] Unlike WT mice, 5-LO- and AhR-deficient mice fail to up-regulate SOCS-2 expression in vivo after STAg injection, indicating that induction of SOCS-2 expression depends on LX-generated enzymes and receptors. Confirming the involvement of SOCS-2 in LXA_4-triggered immune regulatory pathways, SOCS-2-deficient and AhR-deficient mice do not show repression of IL-12 production after in vivo restimulation with STAg (DC paralysis model).[35] During infection with *T. gondii*, SOCS-2 deficient mice have significantly higher IL-12, IFN-γ and TNF production, reduced number of brain cysts, higher frequency of TNF+ and iNOS+ cells infiltrating the central nervous system and higher mortality rates—consistent with the results found in *T. gondii*-infected, 5-LO deficient mice. Additionally, *T. gondii*-infected SOCS-2 deficient mice produce significantly higher levels of chemokines and show increased monocyte and neutrophil migration into the peritoneal cavity, compared with WT counterparts.[35] Taken together, these studies demonstrate that, during *T. gondii* infection pro-inflammatory cytokine responses, leukocyte infiltration and T-cell activation are all controlled by a LX-regulatory pathway dependent on SOCS-2 in vivo.

Furthermore was also shown that SOCS-2 has a role in the intracellular pathways triggered by ATLs.[35] ATLs share the same receptors and mediate the same biological effects, as LXA_4.[36] Indeed, we have shown that WT mice, but not BOC_2 (a peptide antagonist for ALX receptor)-treated or AhR-deficient mice, upregulate SOCS-2 expression after treatment with Aspirin. Moreover, Aspirin-treated SOCS-2-deficient mice fail to inhibit neutrophil infiltration into the peritoneal cavity after thioglycollate challenge and SOCS-2 is also essential for in vivo Aspirin-dependent inhibition of lipopolysaccharide (LPS)-induced secretion of TNF, a hallmark of the anti-inflammatory actions of this drug.[35] Thus, we have shown that endogenous LXA_4 and ATL, signaling through AhR and LXAR, modulate innate and acquired immune responses through inducing SOCS-2 expression.[35]

LX-Induced SOCS2 Mediated TRAF2 and TRAF6 Proteosomal Degradation: A Major Pathway for the Anti-Inflammatory Actions of LXA_4 and ATL

SOCS family proteins are thought to mediate their actions by docking to the intracellular domains of cytokine or hormone receptors, thereby preventing binding and activation of downstream signaling elements.[28] Some SOCS proteins also facilitate proteasome-dependent degradation of transcription factors through the induction of ubiquitination.[27,28] Regardless of the target, the basic mode of action of all SOCS proteins studied to date is similar. In general, SOCS proteins associate with their targets via the SH2 domain, however this does not seem to be sufficient for inhibition. In the case of the association of JAK proteins with SOCS-1 and -3, a second domain in the amino terminus of the molecule is responsible for blocking the kinase activity.[27] In other cases, i.e., association with STATs, the inhibition occurs merely by steric hindrance of the docking site. Ultimately, the SOCS/target protein complex is subject to proteasome-mediated degradation. This occurs via poly-ubiquitination catalyzed by the complex ElonginBC/Cullin-2 bound to the SOCS-box domain in the carboxy-terminus of SOCS proteins (Fig. 3).

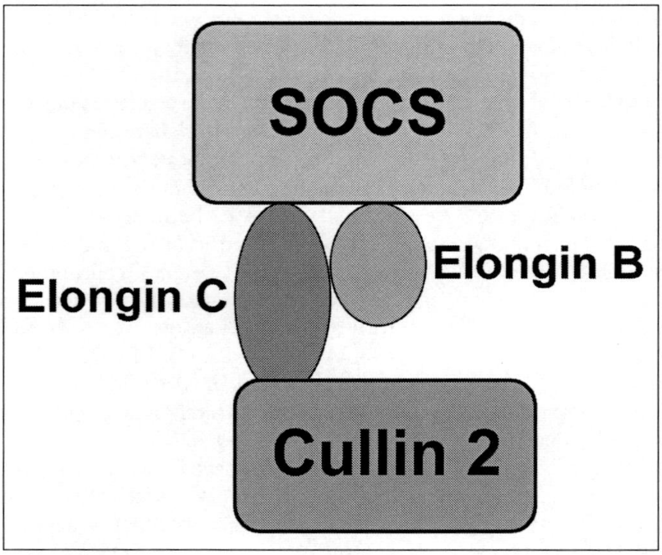

Figure 3. SOCS ubiquitin ligase complex. SOCS1 forms an E3 ubiquitin ligase complex with elongins B and C, Cullin-2. The ubiquitinated substrate is subsequently recognized and degraded by the proteasome.

Recently it was shown that LXAs-induced SOCS-2 associate with TNF receptor-associated factors (TRAFs) 2, 6.[37] TRAFs constitute a family of highly conserved adapter proteins among several species, including protozoa, nematodes, insects and mammalians. TRAFs are important in diverse biological functions, including innate and adaptive immune response, embryonic development, stress responses and bone metabolism. As adapter proteins, TRAFs elaborate receptor signal transduction as both a convergent and a divergent platform. TRAF proteins are characterized by the presence of a unique TRAF domain at the C-terminus, the TRAF-C domain. This domain plays an important role in defining biological function by mediating self-association as well as upstream interactions with receptors and other signaling molecules. The N-terminus usually contains a RING finger motif as well as several zinc finger motifs, which are important for downstream signaling events.[38]

TRAF6 is the only TRAF family member that participates in signal transduction from both TNF receptor superfamily members and IL-1/Toll-like receptor superfamily members. It is a critical element in signal transduction during innate microbial recognition (via TLRs), adaptive immune responses (via CD40) as well as TNF-mediated responses (via TNFR). TRAF6 undergoes ubiquitination when activated and can act as a junction, linking the signaling cascades initiated upon triggering of the above receptors to their downstream elements, such as NF-κB and AP-1 activation pathways.[39] It is important to point out that the ubiquitin ligation that mediates TRAF6 activation does not signal for proteasome degradation, which only occurs when ubiquitin chains are bound to its targets. Genetic deficiency of TRAF6 causes a lack of responsiveness to TLR-ligands, IL-1, IL-18 and TNF, thus leading to increased susceptibility to infection with several pathogens.[39]

Interesting that LX-induced SOCS2 association with TRAF2, 6 was described as a new molecular target and mechanism of action of LX/ATL-induced SOCS. In fact, LX-induced SOCS2 targets TRAF2 and TRAF6 for proteasomal degradation; degradation of TRAF6 renders cells exposed to LX unable to respond to stimulation through TLRs, TNF-R, IL-1R and CD40, resulting in the inhibition of pro-inflammatory cytokine production by DCs; proteasome-mediated degradation of TRAF6 plays a major role in the mechanism of action of ASA in restraining DC production of pro-inflammatory cytokines; and this pathway of targeted proteasomal

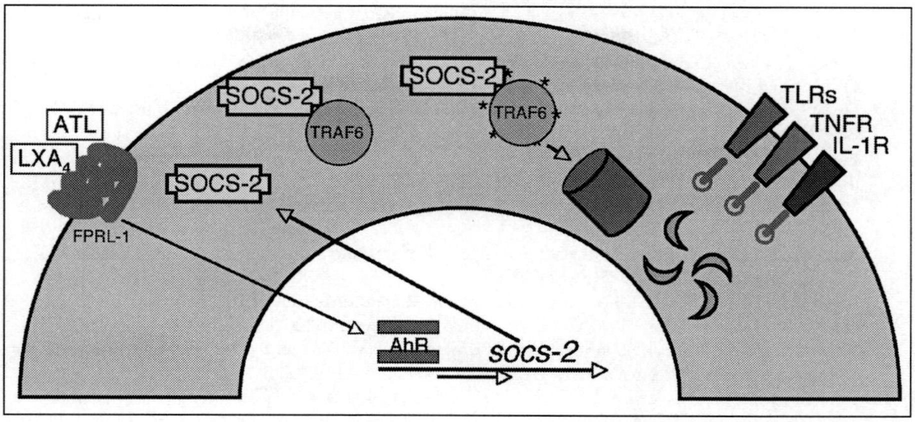

Figure 4. LXA_4-mediated TRAF6 degradation pathway. LXs and ATLs bind to two receptors, the membrane receptor ALX/FPRL-1 and the nuclear receptor, AhR. Upon ligation, AhR translocates to nucleus, dimerizes with ARNT. Transcription of SOCS-2 is induced in a AhR-dependent manner, cytoplasmic SOCS-2 proteins conjugate with TRAF6 leading to a proteasome-mediated degradation and, therefore, inhibiting TLR's, TNFR and IL-1R signaling pathways.

degradation of signaling intermediates is critical for successful modulation of inflammatory responses during infection with *T. gondii*.[37] The biological role of LX and ASA-driven TRAF2 degradation remains to be defined.

Concluding Remarks

Researches have traditionally focused on the mechanisms that underlie and control the activation of inflammatory responses. The last decade has seen growing awareness and experimental attention paid to the fact that restraint and resolution of inflammatory responses is also, perforce, under tight, active regulation. In turn, this has led to growing interest in facilitation of resolution as a therapeutic strategy for the dysregulated inflammation seen as central to the pathogenesis and expression of an increasingly wide spectrum of diseases. The mechanism(s) of action of the lipoxins, at the level of cellular signaling, are of interest. Along these lines, LXA_4 and ATL restrain innate and acquired immune responses via stimulating SOCS-2 expression in an AhR-dependent manner. LX-induced SOCS-2 targets TRAF2 and TRAF6 for proteasomal degradation that results in degradation of TRAF2, 6 inhibiting pro-inflammatory cytokine production during *T. gondii* infection and Aspirin treatment (Fig. 4). The biological role of LX and ASA-driven TRAF2 degradation remains to be defined. In addition, the emerging role for lipoxins as immune-modulatory mediators and the potential use of their inhibitory effects for pathogen survival and replication, is still a new and unclear field. The nature of the pathogen-derived signal that contributes to lipoxin generation, or whether the anti-inflammatory effects of LXA_4 have a critical role in the balance between Type 1, Type 2 and regulatory T-cell responses await to be clarified.

References

1. Martinez AJ, Sell M, Mitrovics T et al. The neuropathology and epidemiology of AIDS. A Berlin experience. A review of 200 cases. Pathol Res Pract 1995; 191(5):427-443.
2. Morisaki JH, Heuser JE, Sibley LD. Invasion of Toxoplasma gondii occurs by active penetration of the host cell. J Cell Sci 1995; 108(Pt 6):2457-2464.
3. Black MW, Boothroyd JC. Lytic cycle of Toxoplasma gondii. Microbiol Mol Biol Rev 2000; 64(3):607-623.
4. Hay J, Hutchison WM. Toxoplasma gondii—An environmental contaminant. Ecol Dis 1983; 2(1):33-43.
5. Serhan CN, Hamberg M, Samuelsson B. Lipoxins: Novel series of biologically active compounds formed from arachidonic acid in human leukocytes. Proc Natl Acad Sci USA 1984; 81(17):5335-5339.

6. Aliberti J. Host persistence: exploitation of anti-inflammatory pathways by Toxoplasma gondii. Nat Rev Immunol 2005; 5(2):162-170.
7. Funk CD, Chen XS, Johnson EN et al. Lipoxygenase genes and their targeted disruption. Prostaglandins Other Lipid Mediat 2002; 68-69:303-312.
8. Fiore S, Romano M, Reardon EM et al. Induction of functional lipoxin A4 receptors in HL-60 cells. Blood 1993; 81(12):3395-3403.
9. Fiore S, Maddox JF, Perez HD et al. Identification of a human cDNA encoding a functional high affinity lipoxin A4 receptor. J Exp Med 1994; 180(1):253-260.
10. Schaldach CM, Riby J, Bjeldanes LF. Lipoxin A4: A new class of ligand for the Ah receptor. Biochemistry 1999; 38(23):7594-7600.
11. Mandal PK. Dioxin: a review of its environmental effects and its aryl hydrocarbon receptor biology. J Comp Physiol [B] 2005; 175(4):221-230.
12. Aliberti J, Hieny S, Reis e Sousa C et al. Lipoxin-mediated inhibition of IL-12 production by DCs: A mechanism for regulation of microbial immunity. Nat Immunol 2002; 3(1):76-82.
13. Bannenberg GL, Aliberti J, Hong S et al. Exogenous pathogen and plant 15-lipoxygenase initiate endogenous lipoxin A4 biosynthesis. J Exp Med 2004; 199(4):515-523.
14. Scharton-Kersten TM, Wynn TA, Denkers EY et al. In the absence of endogenous IFN-gamma, mice develop unimpaired IL-12 responses to Toxoplasma gondii while failing to control acute infection. J Immunol 1996; 157(9):4045-4054.
15. Liu CH, Fan YT, Dias A et al. Cutting Edge: Dendritic cells are essential for in vivo IL-12 production and development of resistance against Toxoplasma gondii infection in mice. J Immunol 2006; 177(1):31-35.
16. Reis e Sousa C, Hieny S, Scharton-Kersten T et al. In vivo microbial stimulation induces rapid CD40 ligand-independent production of interleukin 12 by dendritic cells and their redistribution to T-cell areas. J Exp Med 1997; 186(11):1819-1829.
17. Aliberti J, Reis e Sousa C, Schito M et al. CCR5 provides a signal for microbial induced production of IL-12 by CD8 α+ dendritic cells. Nat Immunol 2000; 1(1):83-87.
18. Aliberti J, Valenzuela JG, Carruthers VB et al. Molecular mimicry of a CCR5 binding-domain in the microbial activation of dendritic cells. Nat Immunol 2003; 4(5):485-490.
19. Scanga CA, Aliberti J, Jankovic D et al. Cutting edge: MyD88 is required for resistance to Toxoplasma gondii infection and regulates parasite-induced IL-12 production by dendritic cells. J Immunol 2002; 168(12):5997-6001.
20. Yarovinsky F, Zhang D, Andersen JF et al. TLR11 activation of dendritic cells by a protozoan profilin-like protein. Science 2005; 308(5728):1626-1629.
21. Hunter CA, Chizzonite R, Remington JS. IL-1 beta is required for IL-12 to induce production of IFN-gamma by NK cells. A role for IL-1 beta in the T-cell-independent mechanism of resistance against intracellular pathogens. J Immunol 1995; 155(9):4347-4354.
22. Sher A, Oswald IP, Hieny S et al. Toxoplasma gondii induces a T-independent IFN-gamma response in natural killer cells that requires both adherent accessory cells and tumor necrosis factor-alpha. J Immunol 1993; 150(9):3982-3989.
23. Yap GS, Sher A. Cell-mediated immunity to Toxoplasma gondii: Initiation, regulation and effector function. Immunobiology 1999; 201(2):240-247.
24. Karp CL, Wills-Karp M. Complement and IL-12: yin and yang. Microbes Infect 2001; 3(2):109-119.
25. Son ES, Song KJ, Shin JC et al. Molecular cloning and characterization of peroxiredoxin from Toxoplasma gondii. Korean J Parasitol 2001; 39(2):133-141.
26. Leonard MO, Hannan K, Burne MJ et al. 15-Epi-16-(para-fluorophenoxy)-lipoxin A(4)-methyl ester, a synthetic analogue of 15-epi-lipoxin A(4), is protective in experimental ischemic acute renal failure. J Am Soc Nephrol 2002; 13(6):1657-1662.
27. Kile BT, Schulman BA, Alexander WS et al. The SOCS box: A tale of destruction and degradation. Trends Biochem Sci 2002; 27(5):235-241.
28. Alexander WS, Hilton DJ. The role of suppressors of cytokine signaling (SOCS) proteins in regulation of the immune response. Annu Rev Immunol 2004; 22:503-529.
29. Starr R, Metcalf D, Elefanty AG et al. Liver degeneration and lymphoid deficiencies in mice lacking suppressor of cytokine signaling-1. Proc Natl Acad Sci USA 1998; 95(24):14395-14399.
30. Marine JC, McKay C, Wang D et al. SOCS3 is essential in the regulation of fetal liver erythropoiesis. Cell 1999; 98(5):617-627.
31. Metcalf D, Greenhalgh CJ, Viney E et al. Gigantism in mice lacking suppressor of cytokine signalling-2. Nature 2000; 405(6790):1069-1073.
32. Kopchick JJ, Bellush LL, Coschigano KT. Transgenic models of growth hormone action. Annu Rev Nutr 1999; 19:437-461.
33. Colao A, Merola B, Ferone D et al. Acromegaly. J Clin Endocrinol Metab 1997; 82(9):2777-2781.

34. Bradford GE, Famula TR. Evidence for a major gene for rapid postweaning growth in mice. Genet Res 1984; 44(3):293-308.
35. Machado FS, Johndrow JE, Esper L et al. Anti-inflammatory actions of lipoxin A4 and aspirin-triggered lipoxin are SOCS-2 dependent. Nat Med 2006; 12(3):330-334.
36. Hachicha M, Pouliot M, Petasis NA et al. Lipoxin (LX)A4 and aspirin-triggered 15-epi-LXA4 inhibit tumor necrosis factor 1alpha-initiated neutrophil responses and trafficking: Regulators of a cytokine-chemokine axis. J Exp Med 1999; 189(12):1923-1930.
37. Machado FS, Esper L, Dias A et al. Native and aspirin-triggered lipoxins control innate immunity by inducing proteasomal degradation of TRAF6. J Exp Med 2008; 205(5):1077-1086.
38. Choi Y. Role of TRAF6 in the immune system. Adv Exp Med Biol 2005; 560:77-82.
39. Lomaga MA, Yeh WC, Sarosi I et al. TRAF6 deficiency results in osteopetrosis and defective interleukin-1, CD40 and LPS signaling. Genes Dev 1999; 13(8):1015-1024.

Chapter 7

Immunomodulatory Activity and Therapeutic Potential of the Filarial Nematode Secreted Product, ES-62

William Harnett* and Margaret M. Harnett

Abstract

ES-62 is a protein that is actively secreted by filarial nematodes during parasitism of the vertebrate host. The molecule is able to directly interact with a number of cells of the immune system including B-lymphocytes, dendritic cells, macrophages and mast cells. Interaction appears to be dependent on complexing with TLR4 and results in modulation of the activity of a number of signal transduction molecules including MAP kinases, PI-3 kinase and NF-κB. Immunomodulatory activity of ES-62 appears to be largely due to the presence of phosphorylcholine (PC) moieties covalently attached to N-type glycans. The net effect of ES-62's interaction with the immune system is the generation of an anti-inflammatory immunological phenotype. As a consequence of this, ES-62 demonstrates striking drug-like activity in models of disease associated with aberrant inflammation, in particular those associated with autoimmunity and allergy.

Introduction

There is currently much interest amongst scientists and clinicians in the "Hygiene Hypothesis". This hypothesis claims that the increased incidence of allergic and autoimmune diseases observed in Western countries in recent decades is due to a loss of appropriate priming of the immune system by pathogens such as parasitic worms (also called helminths) during the early years of life.[1] In support of this, it has been suspected for many years that several autoimmune disorders present with reduced incidence and severity in parts of the world with high helminth load (reviewed in ref. 2). As an example, rheumatoid arthritis (RA), a T helper 1 (Th-1)-cell mediated disease associated with high levels of pro-inflammatory cytokines such as TNF-α, is reduced in helminth-endemic areas.[3,4] In addition, but more surprisingly, given that worms tend to induce strong Th-2-Type (IL-4-, IL-5- and IL-13-dependent) immune responses, there also appears to be an inverse correlation between parasite load and atopy showing that helminths also appear to suppress Th-2-biased inflammatory disorders (reviewed in refs. 4, 5). Indeed, there is ever-increasing evidence that a number of helminth species can delay or prevent the onset of a wide range of autoimmune or allergic disorders.[4,5]

Filarial nematodes represent a group of helminths that can induce particularly severe morbidity with a significant proportion of sufferers presenting with debilitating health problems including severe skin lesions, elephantiasis and several forms of eye damage each of which may lead to blindness.[6] Infection with these organisms is long-term, with individual worms surviving for up to 10 years.[7] Parasite longevity is almost certainly dependent on suppression or modulation of the host

*Corresponding Author: William Harnett—Strathclyde Institute of Pharmacy and Biomedical Sciences, University of Strathclyde, Glasgow G4 0NR, UK. Email: w.harnett@strath.ac.uk

Pathogen-Derived Immunomodulatory Molecules, edited by Padraic G. Fallon.

immune system (reviewed in refs. 8, 9), which demonstrates impairment of lymphocyte proliferative responses and polarisation of the cytokine and antibody response. With respect to the latter, there is reduced Th-1-associated, interferon-gamma (IFN-γ) and increased Th-2-associated interleukin (IL)-4 and anti-inflammatory IL-10, cytokine production and also greatly elevated levels of IgG_4, an antibody of little value in eliminating pathogens due to an inability to activate complement or bind with high affinity to phagocytic cells (a neutralising antibody). Overall therefore, the picture is of an immune response responding to infection with a somewhat suppressed (impaired lymphocyte proliferation), anti-inflammatory (increased IL-10/IgG_4 production), Th-2- (IL-4) like, phenotype. It has been speculated that in addition to promoting parasite survival, such a phenotype is conducive to host health by limiting development of the pathological lesions referred to earlier that are suspected to result from aggressive, pro-inflammatory immune responses. There is increasing evidence that immunomodulators synthesised and secreted by the nematodes may be responsible for such immune system deviation. ES-62, a secreted protein that we discovered in the rodent filarial nematode *Acanthocheilonema viteae,*[10] and subsequently found to have well conserved orthologues in human filarial nematode parasites including *Brugia malayi* and *Onchocerca volvulus*[11,12] is one such molecule.

ES-62

ES-62 is produced in cells that subtend the oesophagus by the post-infective lifecycle stages (L4 larvae and adult worms) of *A. viteae* and can be detected in the serum of the jird host, *Meriones lybicus*.[10,11] However, work on both *A. viteae* and *B. pahangi* has shown that the ES-62 gene is transcribed throughout the helminth lifecycle, although mRNA levels are considerably higher in adult worms than L3 larvae (~5% adult levels for *A. viteae*) and microfilariae (<0.2% adult levels for *A.viteae*).[12] ES-62 mRNA is translated into a protein of 62 kDa (including posttranslational modifications) that has phosphorylcholine (PC) moieties attached via an *N*-type glycan, the latter having been trimmed to the tri-mannosyl core and then substituted with between one and four *N*-acetylglucosamine residues during oligosaccharide processing (reviewed in ref. 13). The number of PC-containing glycans present on each ES-62 molecule has not been resolved to date and the number of PC groups per glycan has been shown to be variable.

ES-62 has highest sequence homology with members of the M28 peptidase family (e.g., 38% and 37% identity with mouse and human aminopeptidases) and has been shown to possess some, albeit weak peptidase activity in vitro against synthetic substrates.[14] Interestingly, the biologically active forms of many peptidases are dimeric or tetrameric[15,16] and consistent with this, gel filtration studies and sedimentation equilibrium data demonstrated that ES-62 is a tightly bound tetramer formed from dimers.[13,17,18] Furthermore, divalent cations are known to be critical for the function of many peptidases and ES-62 has a putative metal coordination motif in its sequence; indeed, a strong magnesium (Mg^{2+}) signal was detected in its atomic emission spectrum.[18] Although a function for the peptidase component of ES-62 has not yet been convincingly demonstrated, the molecule has been shown to display a variety of immunomodulatory properties, many of which have been attributed to the presence of PC. PC is a molecular pattern associated with pathogen products from a diverse range of organisms, including bacteria, fungi and protozoa, as well as filarial and gastrointestinal nematodes (reviewed in ref. 19). It enables detection of pathogens by the host (for example via antibodies or C-reactive protein), but as revealed in detail below can also function to promote pathogen survival via modulation of the host immune response.

Immunomodulatory Properties of ES-62

ES-62 is pleiotropic, being able to exert its effects on a variety of cells of the immune system including B-lymphocytes,[20-22] antigen presenting cells (APCs) such as dendritic cells (DCs) and macrophages[23-27] and mast cells.[28] Rather than acting simply in a nonspecific immunosuppressive manner, ES-62 tends to block Th-1 responses and induce anti-inflammatory type immune responses,[29,30] although it has also been shown to block Th-2 responses in a situation in which they are associated with inflammation.[28] This latter finding was only obtained recently and has led us to conclude that our original hypothesis, namely that ES-62 is a Th-2 polarising agent is incorrect and thus to speculate that the true target of ES-62 may perhaps be a cytokine such as IL-17 (Fig. 1), which appears to act as a master regulator of both Th-1 (e.g., rheumatoid arthritis)- and Th-2 (asthma)-type inflammation.[31-33] This latter idea is currently under investigation. Specific examples of the effects of ES-62 on the mouse immune system are prevention of FcεRI-mediated mast cell degranulation,[28] inhibition of BCR-mediated B-lymphocyte proliferation,[20-22] the production of IL-10 by B1 cells,[25] reduced levels of IL-12, IFN-γ and pro-inflammatory cytokines by APCs in response to TLR ligands (BLP, LPS, CpG)[34] and the production of IgG_1 (equivalent of human IgG_4 in some aspects, e.g., inability to activate complement) rather than opsonising/complement fixing IgG_{2a} antibodies.[35]

Extensive investigation over the last fifteen years has provided a great deal of information on the mechanism of action of ES-62 and four important findings have emerged:

1. (As briefly mentioned earlier) ES-62's activity is largely dependent on its PC moiety. This belief has emerged from experiments in which normal and PC-free ES-62 were compared[35] and in which ES-62 was compared with PC-conjugated to proteins such as albumin or ovalbumin.[23,36] The take-away message is that PC-free ES-62 in general lacks ES-62's immunomodulatory activity whereas PC-conjugated proteins largely mimic it.
2. ES-62's activity is dependent on the presence of TLR4 on the target cell. This conclusion initially arose when macrophages and DCs from TLR4-KO (but not TLR2 or TLR6) mice were found to be nonsusceptible to the action of ES-62.[34] Subsequently it was found that ES-62 was largely ineffective against mouse mast cells if TLR4 levels were greatly depleted by the use of an antisense oligonucleotide approach.[28] Interestingly the interaction of ES-62 with TLR4 is distinct from that of LPS. For example, unlike LPS, ES-62 is fully active against APC derived from HeJ mice that have a mutation in the TIR domain of TLR4.[34,37]

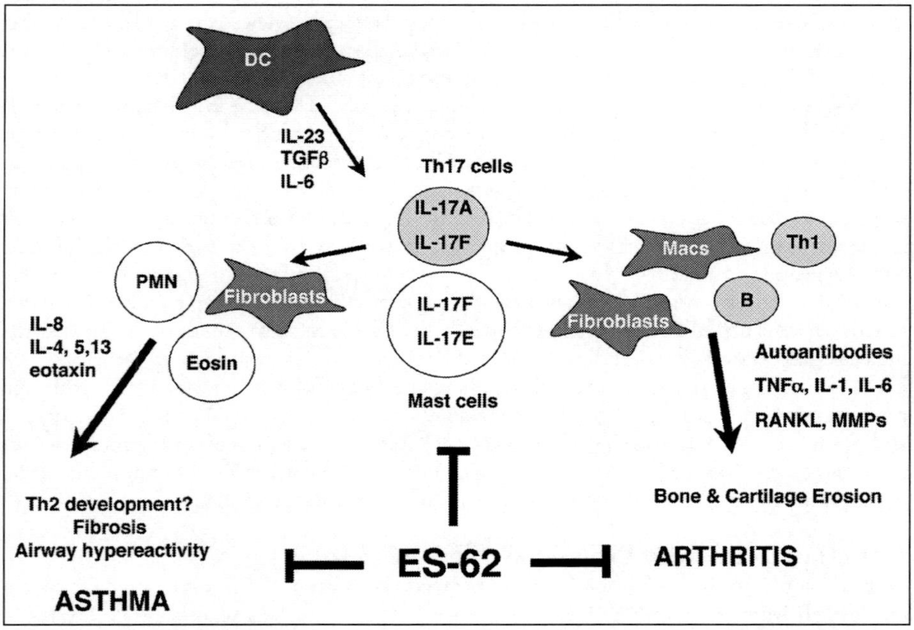

Figure 1. ES-62 may suppress inflammation by targeting IL-17 responses. IL-17 has been implicated in the pathogenesis of diseases resulting from both Th-1- and Th-2-driven inflammation. ES-62 can suppress both types of inflammation and may do this by targeting IL-17 production by mast cells and/or Th-17 cells. Macs = macrophages; MMPs = matrix metalloproteinases.

3. ES-62's activity involves modulation of signal transduction pathways. Extensive investigation has enabled us to show that exposure of cells of the immune system to ES-62 results in modulation of a number of signal transduction pathways including those dependent on PI-3 kinase, MAP kinases and NF-κB.[38,39] Modulation of such pathways can clearly explain the immunomodulatory effects of ES-62 and in some cases the mechanism of modulation has been resolved. With respect to B-lymphocytes for example, ES-62 actually activates some signalling molecules such as Erk MAP kinase resulting in the induction of negative feedback signalling such that when subsequent BCR ligation occurs, Erk activity is rapidly uncoupled by activation of the inhibitory phosphatase, PAC-1 and hence activation of the cell is prevented.[22]
4. ES-62's activity is dependent on degradation of PKC-α. This was first shown in B-lymphocytes,[20] where PKC-α is important for proliferation and hence the effect contributes to inhibition of proliferation. Subsequently it was found that ES-62 caused a similar degradation of PKC-α in mast cells. As PKC-α is necessary for PLD-coupled, sphingosine kinase-mediated, calcium mobilisation in mast cells and this pathway is necessary for degranulation, PKC-α degradation provides the molecular mechanism by which ES-62 blocks such degranulation.[28] The mechanism of ES-62-mediated PKC-α degradation has been characterised in mast cells (Fig. 2): complexing of ES-62 and TLR4 at the plasma membrane results in sequestration of PKC-α followed by caveolae/lipid raft-dependent internalisation and nonproteosomal degradation at the peri-nuclear region of the cell.[28] Sequestration of PKC-α does not take place when ES-62 is substituted by LPS, again highlighting differences between the two molecules with respect to the consequences of their interaction with TLR4.

Therapeutic Potential of ES-62

Rheumatoid arthritis (RA), as mentioned earlier, has been associated with a dysregulated Th-1 immune response in which there is excess pro-inflammatory cytokine production[40-42] within inflamed RA synovial membrane that contributes directly to cartilage/bone erosion through matrix metalloproteinase production and dysregulated chondrocyte/osteoclast function.[43,44] As our original hypothesis regarding ES-62's mechanism of action was that it inhibited Th-1-associated inflammatory responses, we investigated the prophylactic effect of ES-62 on the development of collagen-induced arthritis (CIA) in mice.[45] ES-62 significantly suppressed the severity of developing CIA and as predicted, these prophylactic effects correlated with the inhibition of collagen-specific pro-inflammatory/Th-1 cytokine production (TNF-α, IL-6 and IFN-γ) and suppression of collagen-specific IgG_{2a} antibodies. Therapeutic administration of ES-62 after CIA became clinically detectable also showed significant reduction in arthritis progression compared with vehicle-treated controls in terms of the number of subsequently recruited arthritic joints and reduced progression of articular swelling in the initially inflamed joint. Indeed, progressive articular inflammation and destruction was significantly suppressed by the nematode product.[45,46] Together these data indicate that ES-62 therapeutically and potently suppressed inflammatory CIA even when treatment was commenced after the onset of clinically detectable disease, a key predictor used in developing cytokine-targeting therapies in RA.[47] Finally, to further investigate the therapeutic potential of ES-62 in humans, we performed parallel studies on primary cultures from RA synovial fluid and membrane and compared their capacity for cytokine release in the presence or absence of ES-62. Significant suppression of LPS-induced TNF-α and IL-6 was observed in the presence of ES-62, clearly indicating that the latter can modify critical pro-inflammatory pathways ex vivo in disease relevant tissues.[45]

The idea that ES-62 activity was only directed against Th-1 type-inflammatory diseases was questioned when it was found to be active in a mouse model of allergy—ovalbumin-induced airway hypersensitivity.[28] As mentioned earlier, allergy is of course dependent on a Th-2 type of immune response and indeed since these data were obtained we have also witnessed a failure of ES-62 to modulate the activity of some Th1-inducing molecules.[48] In the airway hypersensitivity model,

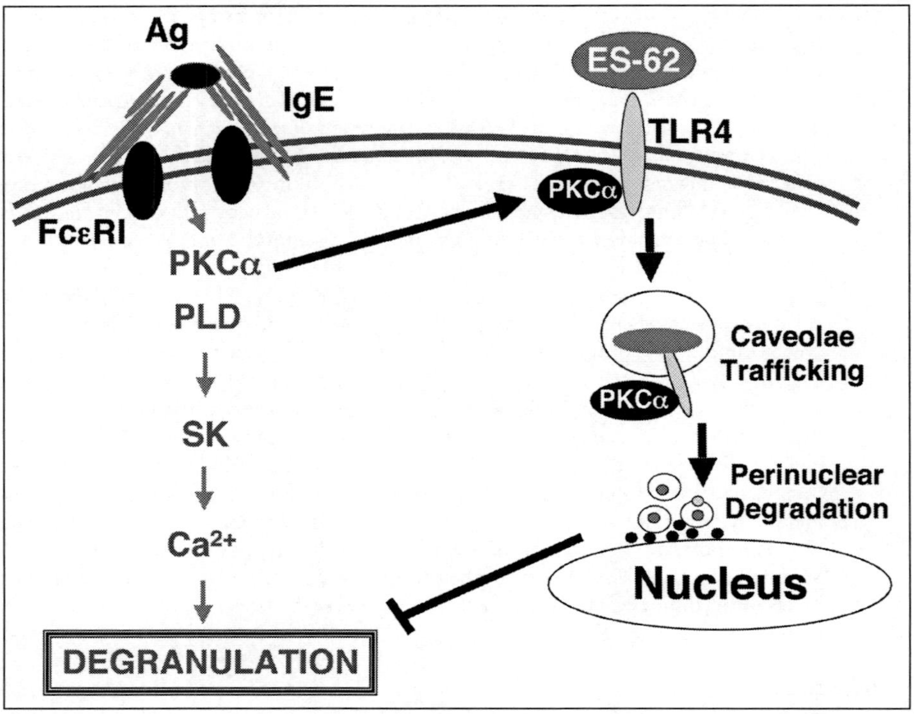

Figure 2. Filarial nematode-derived ES-62 induces hyporesponsiveness of mast cells by disrupting FcεRI signalling. Antigen (Ag)-driven ligation of IgE bound to FcεRI induces mast cell degranulation and generation of pro-inflammatory mediators and cytokines by triggering a signalling cascade involving PKCα-recruitment of the phospholipase D (PLD)-coupled, sphingosine kinase (SK)-mediated calcium mobilisation pathway. ES-62 directly blocks mast cell activation by subverting TLR4 signalling such that it results in sequestration and trafficking of PKCα for perinuclear degradation by a caveolae/lipid raft, proteosomal-independent mechanism. Such sequestration of PKCα uncouples FcεRI-mediated mast cell responses.

ES-62 was found to reduce peri-bronchial inflammation and mucosal hyperplasia, inhibit eosinophila and prevent release of the signature cytokine required for airway inflammation development, IL-4. ES-62 was also found to be active in a second model of Type I hypersensitivity, immediate-type hypersensitivity to oxazolone in the skin.[28] In this model, ES-62 targeted inflammation as shown by a reduction in ear swelling and this was correlated with inhibitory effects on mast cell activation. Subsequent in vitro analysis revealed that ES-62 directly prevented mast cell degranulation and release of mediators of allergy such as leukotrienes, prostaglandins and pro-inflammatory cytokines induced via ligation of FcεRI. It is our understanding that ES-62 is the first pathogen product described to directly inhibit mast cell effector function in this manner.

The Future

ES-62 is a large and hence immunogenic molecule whose posttranslational addition of PC is dependent on a nematode-specific pathway.[49] For these reasons, it is not suitable for use as drug. However, small PC-based derivatives offer a viable alternative. Towards this end we have shown that small PC-containing molecules (e.g., PC-glycans) can mimic some of the activities of ES-62 in vitro.[50] The next step will be to demonstrate that the same is true with respect to ES-62's anti-inflammatory activity in vivo.

References

1. Strachan DP. Hay fever, hygiene and household size. BMJ 1989; 299(6710):1259-1260.
2. Sewell DL, Reinke EK, Hogan LH et al. Immunoregulation of CNS autoimmunity by helminth and mycobacterial infections. Immunol Lett 2002; 82(1-2):101-110.
3. Beighton P, Solomon L, Valkenburg HA. Rheumatoid arthritis in a rural south african negro population. Ann Rheum Dis 1975; 34(2):136-141.
4. Dunne DW, Cooke A. A worm's eye view of the immune system: consequences for evolution of human autoimmune disease. Nat Rev Immunol 2005; 5(5):420-426.
5. Maizels RM. Infections and allergy—helminths, hygiene and host immune regulation. Curr Opin Immunol 2005; 17(6); 656-661.
6. WHO. Filariasis. Geneva: World Health Organisation, 2000.
7. Subramanian S, Stolk WA, Ramaiah KD et al. The dynamics of wuchereria bancrofti infection: a model-based analysis of longitudinal data from Pondicherry, India. Parasitology 2004; 128(Pt 5):467-482.
8. King CL. Transmission intensity and human immune responses to lymphatic filariasis. Parasite Immunol 2001; 23(7):363-371.
9. Brattig NW. Pathogenesis and host responses in human onchocerciasis: impact of onchocerca filariae and wolbachia endobacteria. Microbes Infect 2004; 6(1):113-128.
10. Harnett W, Grainger M, Kapil A et al. Origin, kinetics of circulation and fate in vivo of the major excretory-secretory product of Acanthocheilonema viteae. Parasitol Today 1989; 99:229-239.
11. Stepek G, Auchie M, Tate R et al. Expression of the filarial nematode phosphorylcholine-containing glycoprotein, ES62, is stage specific. Parasitology 2002; 125(Pt 2):155-164.
12. Stepek G, Houston KM, Goodridge HS et al. Stage-specific and species-specific differences in the production of the mRNA and protein for the filarial nematode secreted product, ES-62. Parasitol Today 2004; 128(Pt 1):91-98.
13. Harnett W, Harnett MM, Byron O. Structural/functional aspects of ES-62—a secreted immunomodulatory phosphorylcholine-containing filarial nematode glycoprotein. Curr Protein Pept Sci 2003; 4(1):59-71.
14. Harnett W, Houston KM, Tate R et al. Molecular cloning and demonstration of an aminopeptidase activity in a filarial nematode glycoprotein. Mol Biochem Parasitol 1999; 104(1):11-23.
15. Taylor A. Aminopeptidases: towards a mechanism of action. Trends in Biochemical Sciences 1993; 18(5):167-171.
16. Acosta D, Goni F, Carmona C. Characterization and partial purification of a leucine aminopeptidase from fasciola hepatica. Journal of Parasitology 1998; 84(1):1-7.
17. Harnett W, Houston KM, Amess R et al. Acanthocheilonema viteae: phosphorylcholine is attached to the major excretory-secretory product via an N-linked glycan. Exp Parasitol 1993; 77:498-502.
18. Ackerman CJ, Harnett MM, Harnett W et al. A solution structure of the filarial nematode immunomodulatory protein, ES-62. Biophys J 2003; 84(1):489-500.
19. Harnett W, Harnett MM. Phosphorylcholine: Friend or foe of the immune system? Immunology Today 1999; 20(3):125-129.
20. Deehan M, Harnett M, Harnett W. A filarial nematode secreted product differentially modulates expression and activation of protein kinase C isoforms in B-lymphocytes. J Immunol 1997; 159(12):6105-6111.
21. Deehan MR, Frame MJ, Parkhouse RM et al. A phosphorylcholine-containing filarial nematode-secreted product disrupts B-lymphocyte activation by targeting key proliferative signaling pathways. J Immunol 1998; 160(6):2692-2699.
22. Deehan MR, Harnett W, Harnett MM. A filarial nematode-secreted phosphorylcholine-containing glycoprotein uncouples the B-cell antigen receptor from extracellular signal-regulated kinase-mitogen-activated protein kinase by promoting the surface Ig-mediated recruitment of Src homology 2 domain-containing tyrosine phosphatase-1 and Pac-1 mitogen-activated kinase-phosphatase. Journal of Immunology 2001; 166(12):7462-7468.
23. Harnett W, Harnett MM. Inhibition of murine B-cell proliferation and down-regulation of protein kinase C levels by a phosphorylcholine-containing filarial excretory-secretory product. J Immunol 1993; 151:4829-4837.
24. Harnett MM, Deehan MR, Williams DM et al. Induction of signalling anergy via the T-cell receptor in cultured Jurkat T-cells by pre-exposure to a filarial nematode secreted product. Parasite Immunology 1998; 20(11):551-563.
25. Wilson EH, Katz E, Goodridge HS et al. In vivo activation of murine peritoneal B1 cells by the filarial nematode phosphorylcholine-containing glycoprotein ES-62. Parasite Immunol 2003; 25(8-9):463-466.

26. Whelan M, Harnett MM, Houston KM et al. A filarial nematode-secreted product signals dendritic cells to acquire a phenotype that drives development of Th2 cells. J Immunol 2000; 164(12):6453-6460.
27. Goodridge HS, Wilson EH, Harnett W et al. Modulation of macrophage cytokine production by ES-62, a secreted product of the filarial nematode Acanthocheilonema viteae. J Immunol 2001; 167:940-945.
28. Melendez AJ, Harnett MM, Pushparaj PN et al. Inhibition of FcepsilonRI-mediated mast cell responses by ES-62, a product of parasitic filarial nematodes. Nat Med 2007; 13(11):1375-1381.
29. Houston KM, Wilson EH, Eyres L et al. Presence of phosphorylcholine on a filarial nematode protein influences immunoglobulin G subclass response to the molecule by an interleukin-10-dependent mechanism. Infect Immun 2000; 68(9):5466-5468.
30. Marshall FA, Garside P, Harnett W et al. ES-62, an immunomodulator secreted by filarial nematodes suppresses clonal expansion and skews effector function of heterologous antigen-specific T-cells in vivo. J Immunol 2005; 175:5817-5826.
31. Afzali B, Lombardi G, Lechler RI et al. The role of T helper 17 (Th17) and regulatory T-cells (Treg) in human organ transplantation and autoimmune disease. Clin Exp Immunol 2007; 148(1):32-46.
32. Lubberts E. IL-17/Th17 targeting: on the road to prevent chronic destructive arthritis? Cytokine 2008; 41(2):84-91.
33. Linden A. Rationale for targeting interleukin 17 in the lungs. Curr Opin Investig Drugs 2003; 4(11):1304-1312.
34. Goodridge HS, Marshall FA, Else KJ et al. Immunomodulation via novel use of TLR4 by the filarial nematode phosphorylcholine-containing secreted product, ES-62. J Immunol 2005; 174(1):284-293.
35. Houston K, Wilson EH, Eyres L et al. The presence of phosphorylcholine on a filarial nematode protein influences the IgG subclass response to a molecule and by a mechanism dependent on IL-10. Infect Immunity 2000; 68(9):5466-5468.
36. Harnett MM, Kean DE, Boitelle A et al. The phosphorycholine moiety of the filarial nematode immunomodulator ES-62 is responsible for its anti-inflammatory action in arthritis. Ann Rheum Dis 2008; 67(4); 518-523.
37. Goodridge HS, McGuiness S, Houston KM et al. Phosphorylcholine mimics the effects of ES-62 on macrophages and dendritic cells. Parasite Immunol 2007; 29(3):127-137.
38. Goodridge HS, Deehan MR, Harnett W et al. Subversion of immunological signalling by a filarial nematode phosphorylcholine-containing secreted product. Cell Signal 2005; 17(1):11-16.
39. Goodridge HS, Stepek G, Harnett W et al. Signalling mechanisms underlying subversion of the immune response by the filarial nematode secreted product ES-62. Immunology 2005; 115(3):296-304.
40. Ronnelid J, Berg L, Rogberg S et al. Production of T-cell cytokines at the single-cell level in patients with inflammatory arthritides: enhanced activity in synovial fluid compared to blood. Br J Rheumatol 1998; 37(1):7-14.
41. Chabaud M, Durand JM, Buchs N et al. Human interleukin-17: A T-cell-derived proinflammatory cytokine produced by the rheumatoid synovium. Arthritis Rheum 1999; 42:963-970.
42. Gracie JA, Forsey RJ, Chan WL et al. A proinflammatory role for IL-18 in rheumatoid arthritis. J Clin Invest 1999; 104(10):1393-1401.
43. Taylor PC, Williams RO, Maini RN. Immunotherapy for rheumatoid arthritis. Curr Opin Immunol 2001; 13(5):611-616.
44. Feldmann M, Brennan FM, Maini RN. Role of cytokines in rheumatoid arthritis. Annu Rev Immunol 1996; 14:397-440.
45. McInnes IB, Leung BP, Harnett M et al. A novel therapeutic approach targeting articular inflammation using the filarial nematode-derived phosphorylcholine-containing glycoprotein ES-62. J Immunol 2003; 171(4):2127-2133.
46. McInnes IB, Gracie JA, Harnett M et al. New strategies to control inflammatory synovitis: interleukin 15 and beyond. Ann Rheum Dis 2003; 62(Suppl 2):ii51-54.
47. Williams RO, Feldmann M, Maini RN. Anti-tumor necrosis factor ameliorates joint disease in murine collagen-induced arthritis. Proc Natl Acad Sci USA 1992; 89(20):9784-9788.
48. Al-Riyami L, Egan C, Bradley JE et al. Failure of ES-62 to inhibit T-helper Type 1 responses to other filarial nematode antigens. Parasite Immunol 2008; 30(5):304-308.
49. Houston KM, Harnett W. Structure and synthesis of nematode phosphorylcholine-containing glycoconjugates. Parasitology 2004; 129(Pt 6):655-661.
50. Kean DE, Ohtsuka I, Sato K et al. Dissecting ascaris glycosphingolipids for immunomodulatory moieties—the use of synthetic structural glycosphingolipid analogues. Parasite Immunol 2006; 28(3):69-76.

Chapter 8

Helminth-Derived Immunomodulatory Molecules

Poom Adisakwattana, Sean P. Saunders, Hendrik J. Nel and Padraic G. Fallon*

Abstract

Infection of man with parasitic helminths leads to potent activation and modulation of the host immune response. This modulation of immunity by helminth infections may have bystander effects in altering, either suppressing or exacerbating, unrelated inflammatory processes. Various ongoing clinical trials are testing the therapeutic application of helminth infection of patients with inflammatory diseases, including inflammatory bowel disease and allergic disorders. Rather than the use of live helminth infection, with the potential for side effects, an alternative approach is to identify the immune modulatory molecules (IM) produced by helminths that can alter immune functions. In this review, we will focus on characterized helminth-derived IMs that may have potential to be developed as novel therapeutics for inflammatory diseases.

Introduction

Helminth parasites infect a broad range of species, including man and it is estimated that a third of the global population is infected with parasitic worms. In this chapter, we focus solely on parasitic helminths, which have a medical significance to humans. These species include the Digenean flukes of the Class Trematoda (Flatworms) and the Class Cestoda (Tapeworms), both of the phylum Platyhelminthes and the phylum Nematoda (Roundworms). Some of the major species of helminths of medical importance to humans are outlined in Table 1, presented with an arbitrary ranking based on prevalence of worldwide infections.

Helminth infections of man are often characterized by long-term chronic infections, frequently for decades, without overt morbidity and relatively few deaths. In helminth-infected individuals, the number of worms present in the host determines the levels of morbidity and pathology. Higher worm burdens result in more severe pathology and potential death. This propensity of parasitic helminths to chronically infect the human host without severe morbidity is indicative of co-evolution whereby the parasite has adapted to infect the host without killing all infected individuals. In response to infection with a foreign pathogen, such as helminths, the host will initiate a potent immune response against the parasites. Despite the attrition from the hosts immune system, parasitic helminths achieve long-term survival in the host as they have evolved multiple mechanisms to suppress immune responses and promote tolerizing regulatory responses to down-modulate inflammation. Rather than focus on the modulation of host immunity for the benefit of the parasites survival, there is a growing awareness that these processes may protect against the more deleterious aspects of pathology in unrelated inflammatory diseases. Table 2 summarizes a number of human diseases reportedly modulated by natural helminth infections. In the context of human atopy, we have postulated that the immunomodulatory

*Corresponding Author: Padraic G. Fallon—Inflammation and Immunity Group, Institute of Molecular Medicine, Trinity College Dublin, Dublin 8, Ireland. Email: pfallon@tcd.ie

Pathogen-Derived Immunomodulatory Molecules, edited by Padraic G. Fallon.

Table 1. Medical important diseases caused by parasitic helminths

Disease	Helminth
	Nematode
Ascariasis	*Ascaris lumbricoides*
Trichuriasis	*Trichuris trichiura*
Ancylostomiasis (Hookworm infection)	*Ancylostoma duodenale* *Necator americanus*
Lymphatic filariasis (Elephantiasis)	*Wuchereria bancrofti* *Brugia malayi*
Enterobiasis	*Enterobius vermicularis*
Toxocariasis (Ocular/visceral larva migrans)	*Toxocara canis*
Onchocerciasis (River Blindness)	*Onchocerca volvulus*
Strongyloidiasis	*Strongyloides spp.*
Angiostrongyliasis	*Angiostrongylus cantonensis*
Anisakiasis	*Anisakis simplex*
Trichinellosis or Trichinosis	*Trichinella spiralis*
	Trematode
Schistosomiasis (Bilharzia or snail fever)	*Schistosoma haematobium* *Schistosoma mansoni* *Schistosoma japonicum*
Paragonimiasis	*Paragonimus westermani*
Opisthorchiasis	*Opisthorchis viverrini*
	Cestode
Taeniasis	*Taenia spp.*
Cysticercosis (Neurocysticercosis)	*Taenia solium*
Alveolar Echinococcosis (Hydatid disease)	*Echinococcus multilocularis*

*The worldwide incidences of medically important helminth-induced diseases vary greatly. From 1.221 billion Ascaris infections to <1000 Trichinosis cases/year.

mechanisms employed by helminths to alter host immunity may induce regulatory responses that down-modulate inflammatory allergic processes in the infected host.[1] The potent immunosuppressive capabilities of certain parasitic helminth species has promoted research on exploiting helminths as therapeutics to treat unrelated inflammatory disease in patients. Immune modulation by helminth parasites has recently been extensively reviewed.[1-5] Therefore, the role of helminth infections in modulation of unrelated disease in humans and experimental animal models will not be covered here.[6] In this chapter we review the known helminth immunomodulatory molecules (IM), which have been shown to have a defined effects on immunity and may have therapeutic potential.

Therapeutic Use of Helminth Infections

The data in support of a protective role for helminth infections in modulating inflammation poses the clinically relevant question, "Will a worm a day keep the doctor away?" To address this question, there are a number of completed and ongoing clinical trials that involve the deliberate infection of patients with live helminth infections (Table 3). In recent clinical trials, patients with

Table 2. Studies showing that infection with helminths may modulate selected human diseases

Diseases	Helminth	Effect of Infection	Ref.
Allergic sensitization	*Schistosoma haematobium*	Lower skin-prick test positives to different allergen	7
	Schistosoma mansoni	Low skin-prick test responsiveness to inhaled allergen	8
	Ascaris lumbricoides and *Necator americanus*	Low risk of sensitization to dust mite allergen	9
Multiple sclerosis (MS)	*Hymenolepis nana* *Trichuris trichiura* *Ascaris lumbricoides* *Strongyloides stercolaris*	Improved clinical profile	10,11
Tuberculosis	*Trichuris trichiura* *Ascaris lumbricoides*	Deleterious effects on immune response to mycobacterial infection	12
	Strongyloides stercolaris	Increased pathological damage following antimycobacterial treatment with a poorer clinical outcome. Decreased immunogenicity of BCG vaccination against mycobacterial infection	

Table 3. Clinical trials involving helminth infections of patients with defined human inflammatory diseases

Disease	Helminth	Therapeutic Effect
Inflammatory bowel disease/ Irritable bowel disease		
Crohn's disease	*Trichuris suis* Ova	Decreased disease symptoms.[13,22]
Crohn's disease	*Necator amer icanus*	Decreased disease symptoms.[15]
Ulcerative colitis	*Trichuris suis* Ova	Regresses clinical symptoms.[13,22]
Autoimmunity		
Celiac disease	*Necator americanus*	Ongoing trial; recruiting Phase II[A]
Multiple Sclerosis	*Necator americanus*	Ongoing trial; not yet recruiting Phase II[B]
Allergic diseases		
Allergic Rhinoconjunctivitis	*Necator americanus*	Completed trial[C]
Asthma	*Necator americanus*	Completed trial [D] (16, 17, 23)

[A]ClinicalTrials.gov Identifier: NCT00671138; [B]ClinicalTrials.gov Identifier: NCT00630383; [C]ClinicalTrials.gov Identifier: NCT00232518; [D]ClinicalTrials.gov Identifier: NCT00469989.

inflammatory bowel disease (IBD) have been infected with two different parasitic helminth species. Repeated administration of *Trichuris suis* ova, the porcine whipworm, has been reported to be a safe and an effective treatment for IBD.[13,14] In separate studies, infection of IBD patients with the human hookworm *Necator americanus* has also been shown to have therapeutic potential.[15] Furthermore, *N. americanus* has also been utilized to treat volunteers with allergies.[16]

All studies using live helminth infections to-date have been promising, but there is a need for caution when considering such approaches with potentially virulent pathogens beyond the controlled conditions of clinical trials. Indeed, evidence in support of a protective role for helminths in unrelated inflammatory conditions is conflicting and differs between worm species. A recent meta-analysis of humans naturally infected with various helminths showed that certain species had no influence on response associated with unrelated diseases (asthma and wheeze), while infection with *Ascaris* led to potentially deleterious effects and, in contrast, hookworm infections have a strong dose-dependent protective effect.[17] Furthermore, in animal studies there is data reporting that various helminths can exacerbate the severity in models of inflammatory bowel disease[18,19] and allergic lung inflammation.[20] Alarmingly, a recent study reports that a patient with Crohn's disease, treated with *T. suis*, developed an iatrogenic infection despite the parasite being considered a nonhuman pathogen.[21] Considering the potential side effects posed by such treatments with live worms, we believe a more logical approach is to identify and investigate the IMs expressed by helminths that may be used as anti-inflammatory therapeutics. For the remainder of the chapter, we describe helminths IMs that have been identified to-date.

Helminth IMs

The development of therapeutic IMs from bacteria and viruses is a widely investigated area of research, for reasons previously discussed.[5] In comparison, though less helminth IMs have been described to-date, there are some aspects of the interplay between helminths and the human host that would suggest they are an IM source of significant promise. A fundamental feature of worm infections is that rather than being typified by transient acute infections, they modulate host immunity via the synthesis and secretion of various parasitic IMs that chronically usurp the immune response for years to decades. It is this capacity for long-term immune suppression that suggest a potential for use of helminth IMs as treatments in chronic inflammatory conditions.

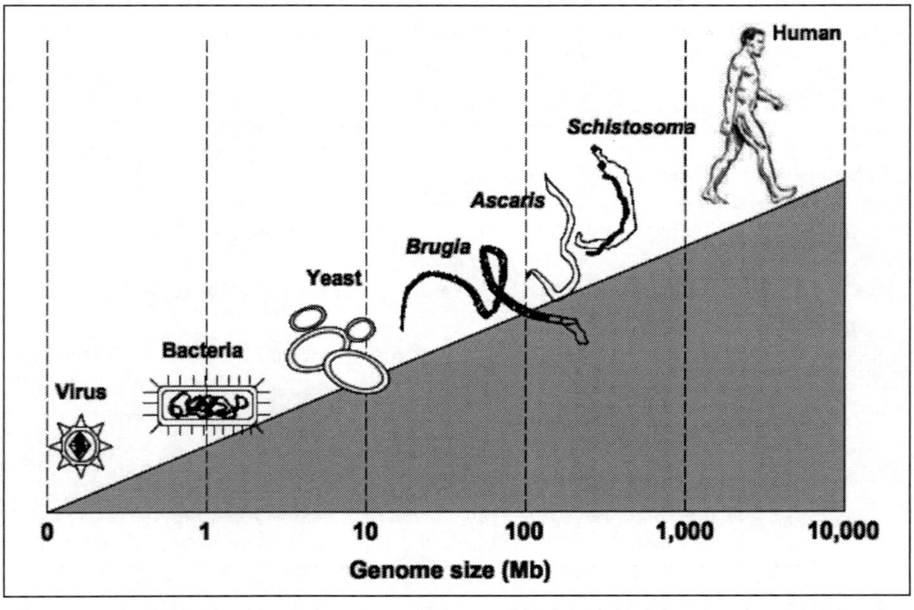

Figure 1. Comparison of genome size in different organisms. The figure illustrates the relatively large genome size (90-270 Mb) of the helminth species shown (*Brugia, Ascaris* and *Schistosoma*) relative to other organisms.

Table 4. Comparative biological differences between certain parasitic helminths and viruses

	Parasitic Helminth			
	Schistosoma	***Brugia***	***Ascaris***	**Viruses**
Genome size	270 Mb	85-95 Mb	230 Mb	2-200 kb
Expressed proteins (estimates)	15-20,000	~18,500	15-20,000	Encoding up to 200 proteins
Location	Vasculature	Lymphatics	Intestinal lumen	Obligate intracellular
Infection	Skin penetration	Insect bite	Ingesting eggs	Respiratory, alimentary or urogenital tract, skin or eye
Intermediate host/ vector	Snail	Mosquito	None	Generally none, some transmit through insects
Resident	Vasculature	Lymphatics, Vasculature	Gastro-intestinal tract	Systemic

As helminth parasites have large genomes compared with bacteria and viruses (Fig. 1), they represent a potentially greater source of diverse IMs. The large genome of helminths, with 10-20,000 expressed proteins, is possibly a prerequisite for the complexity of their life-cycles, with stage- or tissue-specific expression of genes in the multiple life-cycle stages of sexual maturation that occur in the definitive mammalian host. As many worms have intermediate hosts or vectors, this further necessitates genes that function as IMs to modulate the intermediate host's immunity (Table 4). In the mammalian host, the major interface between the immune system and the adult parasitic worms, their larvae or eggs, are molecules expressed on the surface or those that are secreted or excreted by the parasite. However, despite the known potent immunomodulatory activity of helminth parasite infections or extracts from the parasite, relatively few IMs from parasitic worms have been characterized to-date. In this chapter, we have restricted helminth IMs to only recombinant proteins or characterized native molecules with proven bioactivity (Table 5). We will describe some aspects of the modulatory activity of these helminth IMs.

ES-62

The molecule was first discovered as a 62 kDa factor in excretory/secretory products (ES-62) of *Acanthocheilonema viteae*, a filarial nematode of rodents in 1989 by Harnett et al.[40] It was also identified in other filarial nematode species that can cause severe human morbidity, such as *Wuchereria bancrofti*, *Brugia malayi* and *Onchocerca volvulus*. ES-62 is a phosphorylcholine (PC) containing glycoprotein that has been widely investigated as an immunomodulatory molecule. The active native molecule of ES-62 contains PC moieties that are modified by posttranslational processes.[40]

ES-62 exhibits immunomodulatory effects that can lead to downregulation of immunity, as well as inducing the maturation of Th2-inducing dendritic cells (DC) with a resulting Th2 cell polarized immune response.[40] In addition, it is broadly anti-inflammatory and mediates diverse effects on various cells of the immune system.[40] Functional studies of ES-62 suggested that immunomodulatory activity may rely on the PC moiety.[41] Indeed, in a recent study, PC conjugated with other carrier proteins mimics the function of native ES-62, whereas recombinant ES-62 without PC did not display immunomodulatory activity.[42] ES-62 modulates macrophages and DCs by first binding with TLR4 on their surface.[43] In addition to modulating macrophages and DCs, ES-62 forms a complex with TLR4 on the surface of human mast cells, triggering a

Table 5. Examples of IM from helminths that modulate immune responses

IM	Helminth	Modulatory Activity	Ref.
Cystatin	*Onchocerca volvulus*	1. Induces cell hyporesponsiveness	24
	Acanthocheilonema viteae	2. Induces IL-10-producing macrophages	25
ES-62	*Acanthocheilonema viteae*	1. Anti-inflammatory (prevents collagen induced arthritis)	26
		2. Protects against allergic airway hyperresponsiveness and skin hypersensitivity	27
LNFPIII	*Schistosoma mansoni*	Th2-inducing adjuvant	28,29
Lyso-PC	*Schistosoma mansoni*	Stimulates IL-10 producing regulatory T-cells	30
Sm16	*Schistosoma mansoni*	Suppresses cutaneous inflammation	31
SmCKBP	*Schistosoma mansoni*	Chemokine binding protein (suppresses acute inflammation)	32
Thioredoxin peroxidase	*Fasciola hepatica* *Schistosoma mansoni*	Induces alternatively-activated macrophages	33,34

downstream signal cascade.[27] The downstream product triggered by the ES-62/TLR4 complex directly inhibits the FcεRI-related allergic response of mast cells and blocks mast cell degranulation.[27] In terms of therapeutic potential, ES-62 reduces articular inflammation in a model of collagen-induced arthritis in the DBA-1 mouse.[26] Moreover, ES-62 efficiently protected against hypersensitivity, mediated by mast cells in the skin and lungs of a hyperresponsive allergic mouse model.[27] The data provides compelling evidence in support of further research into the therapeutic potential of ES-62, or small molecule derivatives/peptides, in a variety of immunological diseases and disorders.

Sm16

In countries where *Schistosoma mansoni* is endemic, water can contain the infective stage, the cercariae, which can attach to the skin of a human host. The cercariae penetrate the skin following the secretion of a proteolytic cocktail to gain entry into the epidermis from where they can transform into schistosomula, a larval stage, which migrate to the vasculature. Previous studies suggest that *Schistosoma mansoni* larvae utilize an immunomodulatory mechanism during penetration and migration through the skin of rodent host by suppressing inflammatory responses.[31,44] ES products from *S. mansoni* cercariae and schistosomula could activate production of IL-1 receptor antagonist from human keratinocytes in vitro[45] and cercariae also stimulate mouse skin and cultured keratinocytes to produce IL-10 that is a crucial regulatory cytokine.[46] An analysis of the bioactivity in cercariae ES products of the parasite found that a 16.8 kDa protein, designated Sm16, has partial homology with human stathmin and exerts an anti-inflammatory activity.[47] In vitro studies support a role for Sm16 in reducing expression of IL-1α in human keratinocytes, suppressing lymphocyte proliferation and down-regulating expression of ICAM-1.[48] Furthermore, consistent with a role in suppression of inflammation, intradermal delivery of Sm16 cDNA reduced cutaneous inflammation in mice.[48] Sm16 efficiently binds to the lipid bilayer presented on the cell membrane of diverse cell types that probably play a crucial role during skin penetration, which may account for its modulatory activity.[49]

S. mansoni *Chemokine Binding Protein (smCKBP)*

Production and secretion of cytokine and chemokine binding proteins (CKBP) is an immune-modulation strategy widely employed by a range of viruses.[50] SmCKBP is the first CKBP described in a parasite and is secreted solely from the eggs of *S. mansoni, S. heamatobium* and *S. japonicum*.[32] SmCKBP selectively binds to certain members of the chemokine subfamilies as well as the glycosaminoglycan heparan. Recombinant smCKBP suppresses inflammation induced in a mouse contact hypersensitivity model and blocked CXCL8-induced pulmonary inflammation.[32] The smCKBP gene has also been cloned by Hass and colleagues and is reported to be an inducer of degranulation of basophils.[51]

Thioredoxin Peroxidase (TPx)

Thioredoxin peroxidase (TPx) was first described in the yeast, *Saccharomyces cerevisiae*.[52] The enzyme is classified into the peroxiredoxin family which protects cellular components from oxidative damage by reactive oxygen species (ROS; O_2^-, H_2O_2 and $HO^\bullet$).[52] TPx is a ubiquitous enzyme that can be found in various organisms in the whole animal kingdom. In different organisms, TPx has a conserved amino acid sequence homology and a conserved TPx activity motif.[53,54] Among helminths, TPx has been identified in both trematode, nematode and cestode species, examples of which include *Fasciola hepatica, S. mansoni, Onchocerca volvulus* and *Echinococcus granulosus*.[33,53-55] Consistent with other aerobic organisms, parasites require antioxidant enzymes for protection from cell damage caused by ROS that are produced during cellular metabolic oxidation. In addition, they need antioxidant enzymes, such as TPx, for protection against ROS released from infection-stimulated host phagocytes (neutrophils, eosinophils and macrophages).[56]

Functional characterization of TPx in helminths suggests that it plays a significant role in host-parasite interaction as an antioxidant enzyme against host ROS.[57] Recombinant *F. hepatica* TPx has shown immunomodulatory activity and was demonstrated to be an inducer of alternatively activated macrophages, including the inducement and recruitment of such macrophages to the peritoneum in a mouse model.[33] Recombinant TPx directly activates a mouse macrophage cell-line to induce a polarized Th2 cell response.[33] A recent study argues that the role of TPx on alternative activation of macrophages, is not only independent of antioxidant activity, but IL-4 and IL-13 independent as well.[34] This suggests that helminth TPx activates macrophage as an initial step to developing a Th2 response.[34]

Onchocystatin

The Cystatins are ubiquitous proteinase inhibitors of cysteine proteases that regulate various biological and pathological processes by inhibition of cysteine protease activity.[58] Cystatin is secreted by human mononuclear phagocytes and during inflammatory processes its expression is down modulated which, in turn contributes to increased cytsteine protease activity.[58] In addition, the study of cystatin in DC maturation and MHC molecule processing suggested that cystatin plays a role in the intracellular control of invariant chain degradation and antigen presentation.[58]

In parasites, cystatin was first described in the filarial parasite *Onchocerca volvulus*, hence termed onchocystatin, which causes river blindness.[59] The cellular hyporeactivity induced by secreted onchocystatin when added to human peripheral blood mononuclear cells (PBMC) in vitro, is comparable to some aspects of the state of immune hyporesponsiveness seen in patients infected with *O. volvulus*.[24] In addition, onchocystatin stimulated human PBMCs to produce significant levels of IL-10 compared to unstimulated PBMCs and down-regulated the expression of HLA-DR proteins and costimulatory molecules on human monocytes.[60] Interestingly, an in vivo study with the administration of filarial cystatin (*Acanthocheilonema viteae*) to an allergic airway hyperresponsive mouse model supported the role of cystatin in suppression of allergic hyperreactivity by induction of IL-10-derived macrophages.[25] Furthermore, in a chemically-induced colitis mouse model, cystatin regressed the macrophage-mediated inflammatory infiltration and epithelial damage in the colon.[25] Immunomodulatory cystatins are present in other parasitic nematodes, which contrast with the absence of such activity in cystatins from free-living nematode *Caenorhabditis elegans*.[58,61] Cystatin has been also identified in the blood fluke *S. mansoni*, with a potential function as a cysteine protease inhibitor.[62] However, the immunomodulatory potential of schistosome cystatin has yet to be fully addressed.

Helminth Glycans and Glycolipids as IMs

Recent studies demonstrate that carbohydrate and lipid determinants within cell surface and secreted glyco- or lipo-conjugates are capable of eliciting an immunomodulatory potential.[12,63-65] In parasitic helminths, several different carbohydrate and lipid molecules from the parasitic surface or ES product are potent modulators of the host immune response. The most widely investigated helminth carbohydrate is Lacto-N-fucopentaose III (LNFPIII)[64] and, while with respect to helminth lipid IMs, schistosomal lysophosphatidylserine (lyso-PS) is the best characterized.[30]

LNFPIII

Previous studies on the role of carbohydrates in schistosome eggs demonstrated that glycans in soluble egg antigen promoted an enhanced Th2 response.[28] LNFPIII is one such glycan that is present in schistosome soluble egg antigens with potent modulatory activity. Administration of LNFPIII conjugated with carrier molecule into a mouse model demonstrated strong Th2 responses, thus functioning as a potent Th2-inducing adjuvant.[28] In addition, LNFPIII induces in mice expansion of IL-10-producing B-1 cells[66,67] a peritoneal macrophage population that can suppress $CD4^+$ T-cell,[68,69] evoke alternative activation of macrophages in vivo[70] and also induces DC to develop into a DC2 phenotype.[29,71] With its potent in vivo modulatory activity, it will be of interest to further evaluate the effects of LNFPIII administration in animal models of inflammatory diseases.

Lyso-PS

Lyso-PS was identified in a screen for innate interactions between human DCs and different classes of lipids from schistosome worms or eggs.[30,72] Human DCs treated with schistosome phosphatidylserine (PS) were modulated such that they primed naive T-cells to suppress proliferation of autologous T-cells, via the production of IL-10 in a TLR2-dependent manner.[30,72] The specific factor that induced IL-10-producing regulatory cells was shown to be lyso-PS.[30]

Table 6. Helminth-derived cytokine and chemokine homologues

Cytokine and Chemokine Homoloques	Helminth	Modulatory Activity	Ref.
IFN-γ	*Trichuris muris*	Polarizes Th-1 immune response (inappropriate antihelminthic immune response)	35
HMGB1	*Schistosoma mansoni*	Probably modulates host immune responses to promote helmith egg dissemination	36
MIF	*Trichuris muris*	Probably controls migration of immune cells to the infection site	37
TCTP	*Brugia malayi*	Antioxidant (prevent oxidative damage)	38
TGF-β	*Brugia malayi*	Binds to human TGF-β receptor (Possibility of immunomodulation)	39

Helminth Cytokine and Chemokine Homologues as IM

The host immune response to helminths is partially controlled by cytokines as immune cell signaling molecules, and chemokines to induce cell-specific chemotaxis, in order to elicit the appropriate antihelminthic response. Prolonged co-existence of helminths and their hosts have resulted in parasite developing homologues to human cytokines and chemokines that manipulates the host immune response in such a way as to prolong parasite infestation. Examples of helminth cytokine and chemokine homologues are shown in Table 6 and described herein.

Interferon (IFN)-γ

The generation of a Type 2 immune response by the hosts expels infections by many helminths, thus it would be advantageous for the helminth to evoke a counter regulating Type 1 response. Research on *Trichuris muris*, a mouse model of human infecting nematode *T. trichiura*, a homologue to a Type 1 immune response driving cytokine, IFN-γ, was uncovered.[35] This protein had the capacity to bind to the mouse IFN-γ receptor with similar downstream effects as the host cytokine, including Th1 lymphocyte differentiation and function.[35] The secretion of a IFN-γ homologue by a helminth could alter the local cytokine milieu and facilitate its survival in the host.

Macrophage Migration Inhibition Factor (MIF)

Many other helminths, such as *Trichinella spiralis*, *Brugia pahangi*, *Ancylostoma ceylanicum*, have an orthologue MIF.[73] MIF was one of the first cytokines discovered and has been shown to play a role in promoting inflammation through macrophage and T-cell activation as well as IgE synthesis.[74] It has also been shown to play a central role in host protection against some helminths, specifically *Taenia crassiceps*.[75] Using recombinant helminth MIFs, it has been shown that many of these orthologs have the ability to be chemoattractant to human monocytes in vitro and thus might have the ability to control the influx of immune cells and the egress of primed antigen presenting cells from the infection site.[37,76] Some heminth MIFs also have the ability to interact with the human MIF receptor (CD74), resulting in monocyte activation and their subsequent cytokine production, capable of affecting appropriate antihelminth immune responses.[37]

Transforming Growth Factor-β (TGF-β)

TGF-β homologues, or TGF-β-like protein or TGF-β-like ligand, have been observed in all subgroups of helminths. The filarial worm, *B. malayi*, secretes a TGF-β-like protein that is capable of binding to the mammalian TGF-β receptor[39] with possible immune modulation. Several in vivo and in vitro studies have demonstrated the interaction of *S. mansoni*-derived TGF-β receptors, SmTβRI and SmTβRII, with the human TGFβ1 ligand that is ultimately capable of resulting in gene specific transcription.[77-81] This provides evidence for a possible involvement in the regulation of early schistosome development in response to host-derived factors. In the cestode *Echinococcus multilocularis*, 2 Smad protein family members have been identified that are crucially involved in intracellular TGF-β signaling[82] providing further evidence for a TGF-β-like signaling cascade within helminths as a mechanism of host-parasite interaction.

Translationally Controlled Tumor Protein (TCTP)

TCTP, also known as tumor protein, translational-controlled 1 (TPT1), is expressed in various organisms, including helminths and has broad biological activities.[83] TCTP homologues from the human filarial parasites *Wuchereria bancrofti* and *Brugia malayi*, have been shown to possess both histamine-releasing function as well as calcium-binding properties in vitro.[84] Their precise function in helminths do not seem clear, but in vivo injection of isolated helminth TCTP into mice induced the accumulation of eosinophils implying a role in the induction of host pathology as a possible survival strategy.[84] A recent study on the Bm-TCTP homologue has implicated a role as an antioxidant protein that could provide protection against oxidative damage brought on by a host antihelminthic immune response.[38]

High-Mobility Group Box 1 (HMGB1)

HMGB1 is an abundant nuclear protein released from dying mammalian cells, as well as being secreted from activated macrophages, that can induce the production of TNF-α and lead to inflammation and tissue repair (reviewed in ref. 85). Homologues of HMGB1 have been identified among the *Schistosoma* species.[36] SmHMGB1 is expressed in large quantities in adult female, or in schistosoma eggs and egg-secreted antigens, prompting speculation that the molecule plays a role in modulating host immune responses to promote helminth egg dissemination.[36] Recombinant *S. m* HMGB1 has been shown to induce the release of pro-inflammatory cytokines from macrophages in vivo implying a significant role for this cytokine-like molecule in eliciting host immune responses.[36]

Conclusions

A rational approach to exploit the immune modulatory activity of helminths is the characterization of the IMs from helminths that can mimic the desirable anti-inflammatory effects of infection. In this article we have outlined various helminth IM described to-date. In the future, the identification of new helminth-derived IM has the potential to lead to the development of novel therapeutic strategies for the treatment of a range of human immunological and nonimmunological disorders.

Acknowledgement

The authors are supported by Science Foundation Ireland. We are grateful to the Fallon group for reading and commenting on the manuscript.

References

1. Fallon PG, Mangan NE. Suppression of TH2-type allergic reactions by helminth infection. Nat Rev Immunol 2007; 7:220-230.
2. Yazdanbakhsh M, Kremsner PG, van Ree R. Allergy, parasites and the hygiene hypothesis. Science 2002; 296:490-494.
3. Maizels RM, Yazdanbakhsh M. Immune regulation by helminth parasites: cellular and molecular mechanisms. Nat Rev Immunol 2003; 3:733-744.
4. Harnett W, Goodridge HS, Harnett MM. Subversion of immune cell signal transduction pathways by the secreted filarial nematode product, ES-62. Parasitology 2005; 130(Suppl):S63-68.
5. Fallon PG, Alcami A. Pathogen-derived immunomodulatory molecules: future immunotherapeutics? Trends Immunol 2006; 27:470-476.
6. Dunne DW, Cooke A. A worm's eye view of the immune system: consequences for evolution of human autoimmune disease. Nat Rev Immunol 2005; 5:420-426.
7. van den Biggelaar AHJ, van Ree R, Rodrigues LC et al. Decreased atopy in children infected with Schistosoma haematobium: a role for parasite-induced interleukin-10. The Lancet 2000; 356:1723-1727.
8. Araujo MI, Lopes AA, Medeiros M et al. Inverse association between skin response to aeroallergens and Schistosoma mansoni infection. Int Arch Allergy Immunol 2000; 123:145-148.
9. Flohr C, Tuyen LN, Lewis S et al. Poor sanitation and helminth infection protect against skin sensitization in Vietnamese children: A cross-sectional study. J Allergy Clin Immunol 2006; 118:1305-1311.
10. Jorge Correale MF. Association between parasite infection and immune responses in multiple sclerosis. Annals of Neurology 2007; 61:97-108.
11. Jorge Correale MFGR. Helminth infections associated with multiple sclerosis induce regulatory B-cells. Annals of Neurology 2008; 64:187-199.
12. Resende Co T, HCS Toossi Z et al. Intestinal helminth co-infection has a negative impact on both anti-Mycobacterium tuberculosis immunity and clinical response to tuberculosis therapy. Clin Exp Immunol 2007; 147:45-52.
13. Summers RW, Elliott DE, Urban JF et al. Trichuris suis therapy in Crohn's disease. Gut 2005; 54:87-90.
14. Summers RW, Elliott DE, Urban JJF et al. Trichuris suis therapy for active ulcerative colitis: A randomized controlled trial. Gastroenterology 2005; 128:825-832.
15. Croese J, O'Neil J, Masson J et al. A proof of concept study establishing Necator americanus in Crohn's patients and reservoir donors. Gut 2006; 55:136-137.
16. Falcone FH, DI Pritchard. Parasite role reversal: worms on trial. Trends Parasitol 2005; 21:157-160.

17. Leonardi-Bee J, Pritchard D, Britton J. The parasites in asthma collaboration. Asthma and current intestinal parasite infection: systematic review and meta-analysis. Am J Respir Crit Care Med 2006; 174:514-523.
18. Smith P, Mangan NE, Walsh CM et al. Infection with a helminth parasite prevents experimental colitis via a macrophage-mediated mechanism. J Immunol 2007; 178:4557-4566.
19. Hunter MM, Wang A, McKay DM. Helminth infection enhances disease in a murine TH2 model of colitis. Gastroenterology 2007; 132:1320-1330.
20. Mangan NE, van Rooijen N, McKenzie AN et al. Helminth-modified pulmonary immune response protects mice from allergen-induced airway hyperresponsiveness. J Immunol 2006; 176:138-147.
21. Kradin RL, Badizadegan K, Auluck P et al. Iatrogenic Trichuris suis infection in a patient with Crohn disease. Arch Pathol Lab Med 2006; 130:718-720.
22. Summers RW, Elliott DE, Qadir K et al. Trichuris suis seems to be safe and possibly effective in the treatment of inflammatory bowel disease. Am J Gastroenterol 2003; 98:2034-2041.
23. Mortimer K, Brown A, Feary J et al. Dose-ranging study for trials of therapeutic infection with Necator Americanus in humans. Am J Trop Med Hyg 2006;75:914-920.
24. Doetze A, Satoguina J, Burchard G et al. Antigen-specific cellular hyporesponsiveness in a chronic human helminth infection is mediated by Th3/Tr1-type cytokines IL-10 and transforming growth factor-β but not by a Th1 to Th2 shift Int Immunol 2000; 12:623-630.
25. Schnoeller C, Rausch S, Pillai S et al. A helminth immunomodulator reduces allergic and inflammatory responses by induction of IL-10-producing macrophages. J Immunol 2008; 180:4265-4272.
26. McInnes IB, Leung BP, Harnett M et al. A Novel therapeutic approach targeting articular inflammation using the filarial nematode-derived phosphorylcholine-containing glycoprotein ES-62. J Immunol 2003; 171:2127-2133.
27. Melendez AJ, Harnett MM, Pushparaj PN et al. Inhibition of FcεRI-mediated mast cell responses by ES-62, a product of parasitic filarial nematodes. Nat Med 2007; 13:1375-1381.
28. Okano M, Satoskar AR, Nishizaki K et al. Lacto-N-fucopentaose III Found on Schitosoma mansoni egg antigens functions as adjuvant for proteins by inducing Th2-type response. J Immunol 2001; 167:442-450.
29. Thomas PG, Carter MR, Da'dara AA et al. A Helminth glycan induces APC maturation via alternative NF-κB activation independent of IκBα degradation. J Immunol 2005; 175:2082-2090.
30. van der Kleij D, Latz E, Brouwers JFHM et al. A Novel Host-Parasite Lipid Cross-talk. Schistosomal lyso-phosphatidylserine activates Toll-like receptor 2 and affects immune polarization. J Biol Chem 2002; 277:48122-48129.
31. Jenkins SJ, Hewitson JP, Jenkins GR. Modulation of the host's immune response by schistosome larvae. Parasite Immunol 2005; 27:385-393.
32. Smith P, Fallon RE, Mangan NE et al. Schistosoma mansoni secretes a chemokine binding protein with antiinflammatory activity. J Exp Med 2005; 202:1319-1325.
33. Donnelly S, O'Neill SM, Sekiya M. Thioredoxin peroxidase secreted by Fasciola hepatica induces the alternative activation of macrophages. Infect Immun 2005; 73:166-173.
34. Donnelly S, Stack CM, O'Neill SM et al. Helminth 2-Cys peroxiredoxin drives Th2 responses through a mechanism involving alternatively activated macrophages. FASEB J 2008; 22:4022-4032.
35. Grencis RK, Entwistle GM. Production of an interferon-gamma homologue by an intestinal nematode: functionally significant or interesting artefact? Parasitology 1997; 115(Suppl):S101-106.
36. Gnanasekar M, Velusamy R, He Y-X. Cloning and characterization of a high mobility group box 1 (HMGB1) homologue protein from Schistosoma mansoni. Mol Biochem Parasitol 2006; 145:137-146.
37. Zang X, Taylor P, Wang JM et al. Homologues of human macrophage migration inhibitory factor from a parasitic nematode. Gene cloning, protein activity and crystal structure. J Biol Chem 2002; 277:44261-44267.
38. Gnanasekar M, K Ramaswamy. Translationally controlled tumor protein of Brugia malayi functions as an antioxidant protein. Parasitol Res 2007; 101:1533-1540.
39. Gomez-Escobar N, Lewis E, Maizels RM. A Novel member of the transforming growth factor-β (TGF-β) superfamily from the filarial nematodes Brugia malayi and B. pahangi. Exp Parasitol 1998; 88:200-209.
40. Harnett W, IB McInnes, MM Harnett. ES-62, a filarial nematode-derived immunomodulator with anti-inflammatory potential. Immunol Lett 2004; 94:27-33.
41. Houston KM, Wilson EH, Eyres L et al. Presence of phosphorylcholine on a filarial nematode protein influences immunoglobulin G subclass response to the molecule by an interleukin-10-dependent mechanism. Infect Immun 2000; 68:5466-5468.
42. Goodridge HS, McGuiness S, Houston KM et al. Phosphorylcholine mimics the effects of ES-62 on macrophages and dendritic cells. Parasite Immunol 2007; 29:127-137.

43. Goodridge HS, Marshall FA, Else KJ et al. Immunomodulation via novel use of TLR4 by the filarial nematode Phosphorylcholine-Containing secreted product, ES-62. J Immunol 2005; 174:284-293.
44. Mountford AP, Trottein F. Schistosomes in the skin: a balance between immune priming and regulation. Trends Parasitol 2004; 20:221-226.
45. Ramaswamy K, He YX, Salafsky B. ICAM-1 and iNOS expression increased in the skin of mice after vaccination with gamma-irradiated cercariae of Schistosoma mansoni. Exp Parasitol 1997; 86:118-132.
46. Trottein F, Descamps L, Nutten S et al. Schistosoma mansoni activates host microvascular endothelial cells to acquire an anti-inflammatory phenotype. Infect Immun 1999; 67:3403-3409.
47. Valle C, Festucci A, Calogero A et al. Stage-specific expression of a Schistosoma mansoni polypeptide similar to the vertebrate regulatory protein Stathmin. J Biol Chem 1999; 274:33869-33874.
48. Rao KV, He YX, Ramaswamy K. Suppression of cutaneous inflammation by intradermal gene delivery. Gene Ther 2002; 9:38-45.
49. Holmfeldt P, Brännström K, Sellin ME et al. The Schistosoma mansoni protein Sm16/SmSLP/SmSPO-1 is a membrane-binding protein that lacks the proposed microtubule-regulatory activity. Mol Biochem Parasitol 2007; 156:225-234.
50. Alcami A. Viral mimicry of cytokines, chemokines and their receptors. Nat Rev Immunol 2003; 3:36-50.
51. Schramm G, Falcone FH, Gronow A et al. Molecular Characterization of an Interleukin-4-inducing factor from Schistosoma mansoni eggs. J Biol Chem 2003; 278:18384-18392.
52. Kim K, Kim IH, Lee KY et al. The isolation and purification of a specific "protector" protein which inhibits enzyme inactivation by a thiol/Fe(III)/O2 mixed-function oxidation system. J Biol Chem 1988; 263:4704-4711.
53. Chandrashekar R, Curtis KC, Lu W et al. Molecular cloning of an enzymatically active thioredoxin peroxidase from Onchocerca volvulus. Mol Biochem Parasitol 1998; 93:309-312.
54. Kwatia MA, Botkin DJ, Williams DL. Molecular and enzymatic characterization of Schistosoma mansoni thioredoxin peroxidase. J Parasitol 2000; 86:908-915.
55. Margutti P, Ortona E, Delunardo F et al. Thioredoxin peroxidase from Echinococcus granulosus: a candidate to extend the antigenic panel for the immunodiagnosis of human cystic echinococcosis. Diagn Microbiol Infect Dis 2008; 60:279-285.
56. Dzik JM. Molecules released by helminth parasites involved in host colonization. Acta Biochim Pol 2006; 53:33-64.
57. Henkle-Dührsen K, Kampkötter A. Antioxidant enzyme families in parasitic nematodes. Mol Biochem Parasitol 2001; 114:129-142.
58. Hartmann S, Lucius R. Modulation of host immune responses by nematode cystatins. Int J Parasitol 2003; 33:1291-1302.
59. Lustigman S, Brotman B, Huima T et al. Molecular cloning and characterization of onchocystatin, a cysteine proteinase inhibitor of Onchocerca volvulus. J Biol Chem 1992;267:17339-17346.
60. Schonemeyer A, Lucius R, Sonnenburg B et al. Modulation of human T-cell responses and macrophage functions by Onchocystatin, a secreted protein of the filarial nematode Onchocerca volvulus. J Immunol 2001; 167:3207-3215.
61. Murray J, Manoury B, Balic A et al. Bm-CPI-2, a cystatin from Brugia malayi nematode parasites, differs from Caenorhabditis elegans cystatins in a specific site mediating inhibition of the antigen-processing enzyme AEP. Mol Biochem Parasitol 2005; 139:197-203.
62. Morales FC, Furtado DR, Rumjanek FD. The N-terminus moiety of the cystatin SmCys from Schistosoma mansoni regulates its inhibitory activity in vitro and in vivo. Mol Biochem Parasitol 2004; 134:65-73.
63. van der Kleij D, Yazdanbakhsh M. Control of inflammatory diseases by pathogens: lipids and the immune system. Eur J Immunol 2003; 33:2953-2963.
64. Thomas PG, Harn DA. Immune biasing by helminth glycans. Cellular Microbiology 2004; 6:13-22.
65. Capron A, Riveau G, Capron M. Schistosomes: the road from host-parasite interactions to vaccines in clinical trials. Trends Parasitol 2005; 21:143-149.
66. Velupillai P, Harn DA. Oligosaccharide-specific induction of interleukin 10 production by B220+ cells from schistosome-infected mice: a mechanism for regulation of CD4+ T-cell subsets. Proc Natl Acad Sci USA 1994; 91:18-22.
67. Velupillai P, Secor WE, Horauf Harn AM. B-1 cell (CD5+B220+) outgrowth in murine schistosomiasis is genetically restricted and is largely due to activation by polylactosamine sugars. J Immunol 1997; 158:338-344.
68. Terrazas LI, Walsh KL, Piskorska D et al. The schistosome oligosaccharide lacto-N-neotetraose expands Gr1+ cells that secrete anti-inflammatory cytokines and inhibit proliferation of naive CD4+ cells: A potential mechanism for immune polarization in helminth infections. J Immunol 2001; 167:5294-5303.

69. Atochina O, Daly-Engel T, Piskorska D et al. A Schistosome-expressed immunomodulatory glycoconjugate expands peritoneal Gr1+ macrophages that suppress naive CD4+ T cell proliferation via an IFN-γ and nitric oxide-dependent mechanism. J Immunol 2001; 167:4293-4302.
70. Atochina O, Da'dara AA, Walker M. The immunomodulatory glycan LNFPIII initiates alternative activation of murine macrophages in vivo. Immunology 2008; 125:111-121.
71. Thomas PG, Carter MR, Atochina O et al. Maturation of dendritic cell 2 phenotype by a helminth glycan uses a Toll-Like Receptor 4-dependent mechanism. J Immunol 2003; 171:5837-5841.
72. van der Kleij D, van Remoortere A, Schuitemaker JHN et al. Triggering of innate immune responses by schistosome egg glycolipids and their carbohydrate epitope GalNAcβ1-4(Fucα1-2Fucα1-3)GlcNAc. J Infect Dis 2002; 185:531-539.
73. Vermeire JJ, Cho Y, Lolis E et al. Orthologs of macrophage migration inhibitory factor from parasitic nematodes. Trends Parasitol 2008; 24:355-363.
74. Senter PD, Al-Abed Y, Metz CN et al. Inhibition of macrophage migration inhibitory factor (MIF) tautomerase and biological activities by acetaminophen metabolites. Proc Natl Acad Sci USA 2002; 99:144-149.
75. Rodriguez-Sosa M, Rosas LE, David JR et al. Macrophage migration inhibitory factor plays a critical role in mediating protection against the helminth parasite Taenia crassiceps. Infect Immun 2003; 71:1247-1254.
76. Cho Y, Jones BF, Vermeire JJ et al. Structural and functional characterization of a secreted hookworm macrophage migration inhibitory factor (MIF) that interacts with the human MIF receptor CD74. J Biol Chem 2007; 282:23447-23456.
77. Beall MJ, McGonigle S, Pearce EJ. Functional conservation of Schistosoma mansoni Smads in TGF-β signaling. Mol Biochem Parasitol 2000; 111:131-142.
78. Beall MJ, Pearce EJ. Human transforming growth factor-beta activates a receptor serine/threonine kinase from the intravascular parasite Schistosoma mansoni. J Biol Chem 2001; 276:31613-31619.
79. Osman A, Niles EG, LoVerde PT. Identification and characterization of a Smad2 homologue from Schistosoma mansoni, a transforming growth factor-beta signal transducer. J Biol Chem 2001; 276:10072-10082.
80. Osman A, Niles EG, LoVerde PT. Expression of functional Schistosoma mansoni Smad4: role in Erk-mediated transforming growth factor beta (TGF-beta) down-regulation. J Biol Chem 2004; 279:6474-6486.
81. Osman A, Niles EG, Verjovski-Almeida S et al. Schistosoma mansoni TGF-beta receptor II: role in host ligand-induced regulation of a schistosome target gene. PLoS Pathog 2006; 2:e54.
82. Zavala-Góngora R, Kroner A, Wittek B et al. Identification and characterisation of two distinct Smad proteins from the fox-tapeworm Echinococcus multilocularis. Int J Parasitol 2003; 33:1665-1677.
83. Bommer UA, Thiele BJ. The translationally controlled tumour protein (TCTP). Int J Biochem Cell Biol 2004; 36:379-385.
84. Gnanasekar M, Rao KVN, Chen L et al. Molecular characterization of a calcium binding translationally controlled tumor protein homologue from the filarial parasites Brugia malayi and Wuchereria bancrofti. Mol Biochem Parasitol 2002; 121:107-118.
85. Bianchi ME, Manfredi AA. High-mobility group box 1 (HMGB1) protein at the crossroads between innate and adaptive immunity. Immunol Rev 2007; 220:35-46.

Chapter 9

Fungal-Derived Immune Modulating Molecules

Tania C. Sorrell* and Sharon C.A. Chen

Abstract

Invasive fungal infections are an increasing clinical problem for which new therapeutic approaches are needed. Understanding the initial interaction between fungi and the host offers potential for development of new drugs or vaccines. It has recently been recognized that like other pathogens, fungi initially interact with the innate immune system via binding between fungus-specific chemical signatures (pattern-associated molecular patterns or PAMPs) and pattern recognition receptors (PRRs) on mononuclear phagocytes. Fungal PAMPs are restricted to complex carbohydrates in the cell wall, including mannoproteins, phospholipomannan, β-glucans and possibly chitin. These PAMPs bind specifically to two classes of PRR in phagocyte membranes, toll-like receptors and C-lectin-like receptors, through which they initiate signaling responses that culminate in release of pro- and anti-inflammatory cytokines, link the innate immune response with the adaptive immune response and initiate phagocytosis and intracellular killing. Isolated PAMPs have been used to dissect phagocyte responses in vitro and have revealed mechanisms by which host cells can tailor innate immune responses to individual pathogens. The interactions are complex and are yet to be translated into a clear understanding of the roles of the respective PAMPs and PRRs in vivo. Recent advances in this area in relation to the pathogenesis of fungal infections are summarized in this chapter.

Introduction

Recent increases in invasive fungal infections, with their attendant high morbidity and mortality, their predilection for immunocompromised hosts and the relatively limited armamentarium of antifungal therapies, have focused attention on research into fungal diseases and their pathogenesis. In fact, humans are relatively resistant to fungal infections since few of the estimated thousands of environmental fungi are pathogenic, even in the presence of immunocompromise. Apart from dimorphic fungi, such as *Histoplasma capsulatum, and Coccidioides immitis*, which affect apparently healthy hosts and are limited in geographic distribution, the major life-threatening pathogens affect primarily immunocompromised hosts and have a global distribution. These pathogens include the yeasts *Candida* and *Cryptococcus* and the moulds, *Aspergillus* and, to a lesser extent, Zygomycetes, *Fusarium* and *Scedosporium*.

The outcome of fungus-host interactions is determined by the invasive strategies of the pathogen balanced against the relative efficacy of the host response. Containment relies on distinct, but interlinked immune recognition systems, whereas fungal strategies for invasion include tissue destruction, immune evasion and/or suppression of host responses. The initial recognition of pathogens by the host innate immune system is achieved via cellular pattern-recognition recep-

*Corresponding Author: Professor Tania C. Sorrell—Rm 3114, Level 3, ICPMR Bldg, Westmead Hospital, Darcy Rd., Westmead NSW 2145, Australia. Email: t.sorrell@usyd.edu.au

Pathogen-Derived Immunomodulatory Molecules, edited by Padraic G. Fallon.

tors (PRRs) on mononuclear phagocytes and neutrophils and soluble (opsonic) proteins such as complement and the collectins[1] which recognize conserved, pathogen-specific chemical signatures known as PAMPs (pathogen-associated molecular patterns).[2] Engagement of these PRRs initiates ingestion and killing of the pathogen and activation of inflammatory responses. Of the four major classes of phagocyte-associated PRRs (Toll-like receptors (TLRs), C-type lectin receptors (CLRs), nucleotide-binding oligomerization domain (NOD)-like receptors (NLRs) and retinoic acid-inducible gene I (RIGI) receptors) that initiate cellular signaling responses to pathogens, to date, only TLRs and CLRs have been implicated in binding of fungal PAMPs.

The complex and sometimes redundant phagocyte signaling pathways that lead to production of proinflammatory, regulatory and suppressive cytokines and/or chemokines, have been the subject of several recent reviews.[3-8] In vertebrates, a second line of defense, (adaptive immunity) is initiated and modulated by the innate immune system through the interactions of antigen-presenting cells especially dendritic cells, with specialized B-cells and T-cells, which recognize a wide spectrum of microbial antigens.[9]

This review will focus on the interaction between PAMPs and phagocyte PRRs in the pathogenesis of fungal infections.

Morphotypes of Yeasts, Moulds and Dimorphic Fungi Influence Pathogenicity and Innate Immune Responses

In contrast to bacterial pathogens, morphotype switching is essential for the pathogenicity of yeasts, moulds and dimorphic fungi as exemplified by *Candida albicans, Aspergillus fumigatus* and *H. capsulatum,* but the relationship between fungal PAMPs and pathogenicity has been investigated only recently. *C. albicans* morphotypes include yeasts, which are formed by separation of mother and daughter cells during growth and two filamentous forms, pseudohyphae and hyphae. Filamentation results from lack of separation of a growing hyphal form from the yeast cell. The yeast form of *C. albicans* is a normal commensal on certain human mucosal surfaces, where it survives and multiplies without eliciting a host response. The filamentous form is required for tissue invasion and the yeast form, for haematogenous dissemination of infection.[10] The mould, *A. fumigatus,* has three morphotypes, resting conidia, swollen conidia and hyphae. Spores of *A. fumigatus* (resting conidia) are unbiquitous in the environment, are poorly pathogenic and elicit a minimal inflammatory response. Following inhalation, most conidia are phagocytosed and killed by alveolar macrophages. Those which escape phagocytosis lose their hydrophobic coat and swell as they begin to germinate, then form germ tubes and ultimately, mature hyphae, which are the invasive form of the fungus. Swollen conidia initiate rapid recruitment of neutrophils[11] and elicit much higher levels of pro-inflammatory cytokines (TNFα, CCL3/MIP-1α and CXCL2/MIP2) than resting conidia or hyphae.[12] In the case of the dimorphic fungus, *H. capsulatum,* resting conidia do not elicit PRR-mediated acute inflammatory responses. Following inhalation they germinate into the parasitic yeast form at physiological temperature; this conversion is absolutely required for pathogenicity.[13]

Fungal PAMPs—The Fungal Cell Wall

The major distinguishing feature between fungi and mammalian cells, both of which are eukaryotes, is the rigid fungal cell wall, which maintains cellular shape and integrity, is continually remodeled during growth and is the organ by which the mammalian host first distinguishes a potential pathogen as "nonself". The majority of fungus-specific components (PAMPs) that are recognized by a mammalian host are glucose-containing macromolecules such as mannoproteins, β-glucans and chitin. These make up the bulk of the fungal cell wall and interact with the host when they are exposed at the fungal cell surface. The structure of the fungal cell wall and the pattern of PAMPs exposed at the cell surface is genus-, species- and morphotype-dependent.

Cell walls of *C. albicans* and *A. fumigatus* have been studied in detail and exemplify structural differences between yeasts and moulds. *Cryptococcus neoformans* is unique in that the cell wall is surrounded by a loosely woven polysaccharide capsule comprised predominantly of glucuronylxylomannan (GXM) (for review, see),[14] which is itself a PAMP.

The *C. albicans* cell wall is comprised predominantly of mannoproteins, β (1,3)-glucans, β (1,6)-glucans and chitin (a β-(1,4)-linked polymer of *N*-acetylglucosamine (GlcNAc), the relative amounts and arrangement of which vary with morphotype.[15] The architecture of the inner cell wall incorporates skeletal and matrix components that provide strength and a degree of plasticity and are arranged in general in three layers, an inner, chitin-rich layer, a central scaffold of β-1,3-linked glucans, linked to the chitin inner layer and a β-1,6-linked glucan outer layer (shown diagrammatically in Fig. 1). The β-1,6-linked glucans are critical for attachment of a series of highly glycosylated proteins, the major class being glycosylphosphatidylinositol (GPI)-anchored mannoproteins, which make up most of the outer cell wall, contribute to its integrity and are displayed at the cell surface. A second group of secretory cell wall proteins (Pir CWPs) reside predominantly in the inner "skeletal" layer rather than outer protein layer and are attached directly to β-1,3-linked glucans.

Recent evidence indicates that in addition to mannoproteins, some β-glucan is exposed on the surface of *C. albicans* yeasts,[16] at sites that presumptively represent bud scars.[17] During cell division, B-glucans and chitin become exposed at the site of separation of the daughter from the mother

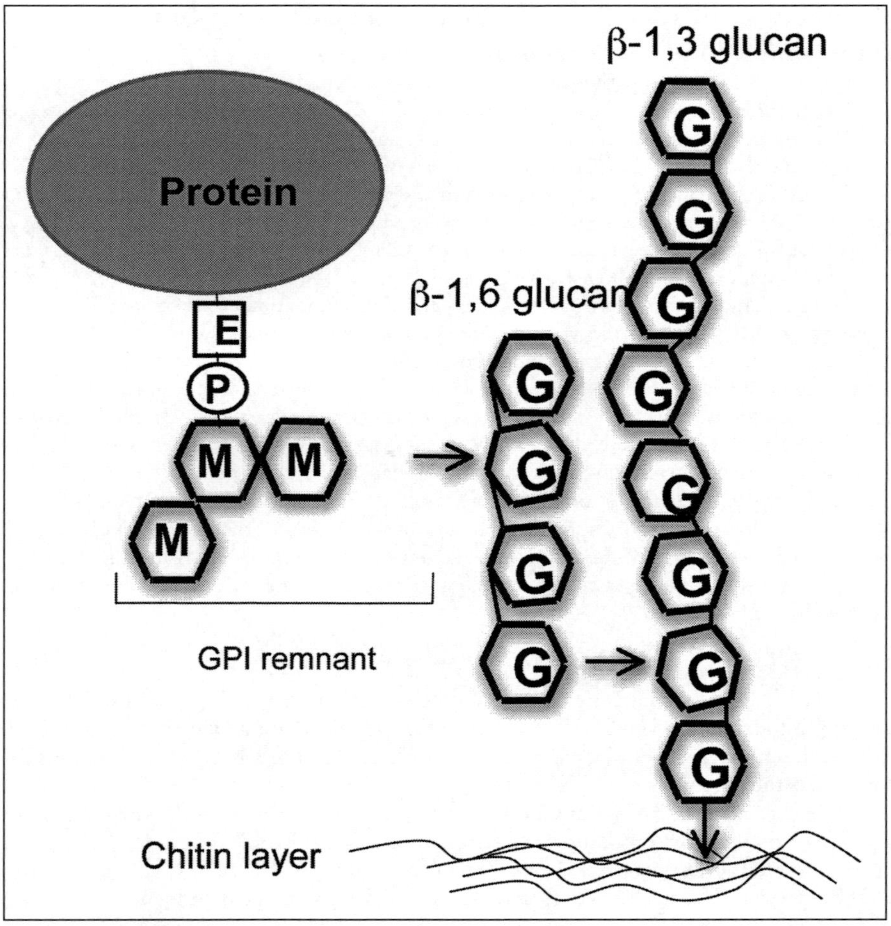

Figure 1. Model structure of the yeast cell wall based on studies in *S. cerevisiae* and *C. albicans*.

cell.[17] Neither β-glucans nor chitin are exposed on the surfaces of pseudohyphae or hyphae.[17] Heat-killing of yeasts results in exposure of high levels of B-glucans.[18]

The mycelium of *A. fumigatus* is comprised of an alkali-soluble fraction (mainly α-(1,3)-glucans with some galactomannan[19] and an alkali-insoluble fraction responsible for maintaining rigidity of the cell wall. Its major carbohydrate polymers are galactomannan (consisting of a core chain of α-mannose residues with a short side-chain of β-1,5- galactofuranose residues,[20] chitin, branched β-1,3/β1,6 glucans and linear β(1,3)/(1,4)-glucans.[21] These are arranged in a three dimensional structural core comprised of chitin, galactomannan and linear β(1,3)/(1,4)-glucans, linked covalently to the nonreducing end of β-1,3- glucan side chains.[19] The structure of the conidial cell wall differs from that of the mycelium though little is known about the chemical organization of its polysaccharides. The outermost layer is comprised of interwoven rodlet fascicles made up of hydrophobic proteins (hydrophobins) which confer hydrophobic properties to the conidia. During swelling (early germination) the inner layer of the cell wall becomes exposed.[22]

Phagocyte-Induced Exposure of PAMPS on the Fungal Cell Surface

Differential exposure of fungal PAMPs also occurs in response to interactions with host cells[23] and other external factors. For example, the hydrophobic coat of *Aspergillus* resting conidia is stripped off as germination commences in macrophage phagolysosomes.[24] β-glucans are exposed at the cell surface by treatment with the antifungal drug, caspofungin, which inhibits synthesis of cell wall glucans[25] and following death of fungal cells at sites of infection.

In addition, certain PAMPs such as β-glucan, galactomannan or cryptococcal capsular material are released into tissues and biological fluids during infection and exert immunomodulatory effects.[3,26-28]

PAMP-PRR Interactions and Down-Stream Effects

Fungal PAMPs, their receptors (PRRs) and the fungal species on which they have been found are summarized in Table 1. PAMP-PRR binding initiates characteristic signaling events in phagocytes. These events have generally been identified by stimulation of receptors with specific ligands or zymosan (a cell wall preparation of *S. cerevisiae,* which is comprised of β-glucans and mannan) and confirmed in many, but not all, cases using intact fungal cells (see Fig. 1). The definitive roles of each PAMP in phagocytosis, killing and activation of chemokine and cytokine signaling pathways have not all been established and some are controversial (for review also see).[1]

Toll-Like Receptors and Effects of Binding by Fungal PAMPs

TLRs are a family of Type I transmembrane glycoprotein receptors which are characterized by an extracellular leucine-rich repeat domain and a cytoplasmic signaling domain homologous to that of the interleukin-1 receptor (IL-R), called the Toll/IL-1 receptor (TIR) domain.[6] Using a set of isogenic glycosylation mutants of *C. albicans,* Netea et al showed that mannoproteins of the outer layer of the cell wall bind to TLR4 via the short linear chains of their O-linked mannosyl residues.[29] Another PAMP, the glycolipid phospholipomannan, has been identified on the surface of *C. albicans.* Phospholipomannan is composed of a hydroxyl fatty acid linked to phytosphingosine, with a hydrophilic polysaccharide domain consisting of a linear chain of β-1,2-linked mannose residues.[30] It is shed by yeasts on contact with macrophages and binds to them, inducing TNF-α secretion and protein tyrosine kinase-dependent signal transduction, similar to that induced by whole yeast cells.[31]

Following ligand binding to TLR2 and TLR4 the cytoplasmic adaptor molecule myeloid differentiation factor 88 (MyD88), associates to the TLR cytoplasmic tail and initiates a signaling cascade that leads to activation of nuclear factor kappa B (NFκB) and mitogen-activated protein (MAP) kinases.[6] Another adaptor TIRAP/Mal, is required to recruit MyD88 to the receptor.[6] Activation of these pathways in general leads to production of proinflammatory chemokines and cytokines (Fig. 2).

Table 1. Fungal PAMPs and their receptors

Pamp	PRR	Fungal Species Studied	References
Mannans			
	TLR4	*C. albicans, A. fumigatus, C. neoformans, Pneumocystis carinii*	3,8,24,29,55,56
	Dectin2	*C. albicans*	40
	DC-SIGN	*C. albicans, A. fumigatus* conidia, *Chrysosporium tropicum*	55,57,58
	Mannose receptor	*C. albicans*	29
	MBL (serum protein)	*C. albicans, A. fumigatus*	1
	SP-A, SP-D (lung surfactant)	*C. albicans, A. fumigatus, C. neoformans, P. carinii, H. capsulatum.*	1
Phospholipomannan	TLR2	*C. albicans, A. fumigatus, Coccidiodes posadasii*	48,24,59,60
Galactomannan	*PTX3 (pentraxin-3)	*A. fumigatus*	61
Glucuronoxylomannan	CD14/TLR 4 CD14 (human cells)	*C. neoformans* *A. fumigatus*	56
β-glucans	Dectin-1,	*C. albicans, A. fumigatus, S. cerevisiae,*	1,3,24,47,62
	CR3 (opsonin)	*P. carinii, Coccidiodes, Histoplasma capsulatum*	1
	SP-D (opsonin)	*C. albicans, A. fumigatus C. albicans, A. fumigatus, P. carinii*	1
β-1,2 mannosides	Galectin-3	*Candida* species	45
DNA	TLR9	*C. neoformans*	63

*PTX3 is a soluble pattern recognition receptor produced by macrophages and other cells in response to inflammatory mediators which binds to *Aspergillus* conidia but not hyphae and facilitates their ingestion by macrophages. PTX3-/- mice (genetic backgrounds 129/Sv- C57Bl/6 mixed and 129/Sv) exhibited markedly increased mortality in a model of IPA (invasive pulmonary aspergillus) and survival was significantly increased by systemic plus local administration of PTX3.[61]

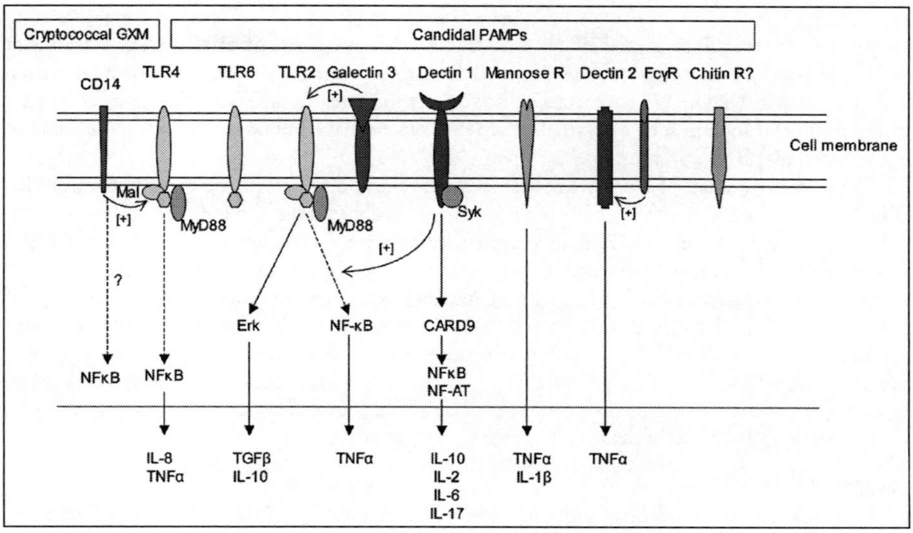

Figure 2. Model of PAMP-PRR interactions of *C. albicans* yeast cells and cryptococcal capsular GXM with monocytes/macrophages based on studies using individual PAMPs, zymosan, GXM and intact *Candida* cells. Adapted in part from reference 8.

C. Neoformans Glucuronoxylomannan (GXM)

Studies in CHO cells transfected with TLR2, TLR4 and/or CD14 revealed that GXM, the major component of the cryptococcal capsule, binds to these PRRs and that activation of NF-κB requires cooperation between TLR4 and CD14. In human PBMCs however, GXM-induced nuclear translation of NF-κB resulted in neither TNF-α secretion nor activation of MAP kinase pathways.[26] Human monocyte-derived macrophages were shown by another group to bind GXM via TLR4, CD14, FcγR11 and CD18 receptors with resultant upregulation of TLR4 and CD14 protein synthesis but no stimulation of NF-κB. As in the previous study, there was no increase in production of TNF-α though the anti-inflammatory cytokine, IL-10 was induced.[32] The reason for these discrepancies is unexplained but could be due to differences in experimental conditions.[32]

C-Type Lectin-Like Receptors and Effects of Binding by Fungal PAMPs

C-type lectin receptors (CLRs) are a family of proteins containing one or more structurally-related C-type, lectin-like domains. Transmembrane CLRs which mediate fungal binding include the mannose receptor (MR), Dectin 1, Dectin 2 and the dendritic cell receptor, DC-SIGN (reviewed in).[1]

Mannose Receptor

Oligosaccharides that terminate in mannose, fucose and GlcNAc[33] are recognized by and bind to, carbohydrate recognition domains (CRDs) 4-8 in the extracellular region of the MR.[34] Terminal residues of N-bound mannans from *C. albicans* are preferentially recognized by MR on monocytes and macrophages,[35] consistent with recognition of α-linked oligomannoses with branched rather than linear structures.[29,36] Although MR can induce NFκB activation and production of cytokines such as IL-12, IL-8, Il-1β, IL-6 and granulocyte-macrophage colony-stimulating factor in response to fungi, the mechanism for this is not known as the receptor lacks the classical signaling motifs in its cytoplasmic tail. It appears to mediate binding of fungi to phagocytes.

Dectin 1

By constructing lipid-linked oligosaccharide micro-arrays, Palma et al demonstrated that only 1,3-linked glucose oligomers (at least 10-11 mer in length) of microbial, not mammalian origin, bound to Dectin 1.[37] Furthermore, 11-13 mer glucose oligomers displayed on liposomes triggered TNFα secretion by a dectin 1-expressing macrophage cell line, mimicking the effect of macromolecular B-glucan binding.[37]

The cytoplasmic tail of Dectin 1 contains an immunoreceptor tyrosine-based activation-like (ITAM-like) motif, through which it mediates intracellular signaling (see Fig. 2) either independently or in collaboration with TLRs. Dectin 1-mediated responses include production of TNFα, macrophage inflammatory protein 2, IL-2, IL-10, IL-6 and IL-23, stimulation of the phagocyte respiratory burst, activation of phospholipase A2, and cyclooxygenase 2 and phagocytosis.[1,8,12,38,39] Some responses, such as the production of IL-10, IL-6 and IL-2 are mediated directly by Dectin 1, whereas the production of proinflammatory chemokines and cytokines requires collaborative signaling with TLR2.[8] TLR2 ligation appears to produce a weak proinflammatory cytokine response without production of IL-12 or IFNγ release. The complexity of these signaling responses in *C. albicans* is discussed in detail in a recent review by Netea et al.[8]

Dectin 2

Terminal mannose residues on high-mannose structures ($Man_9GLCNAc_2$) in fungi (*C. albicans,* zymosan), predominantly hyphal forms, bind specifically but with low affinity to the extracellular domain of Dectin 2.[1,35] The cytoplasmic tail of Dectin 2 appears to associate with the Fc receptor γ chain, a signaling adaptor molecule associated with several other transmembrane receptors. This interaction does not play a significant role in tethering yeasts to the cell surface[35] though hyphal forms of *C. albicans* have been reported to upregulate secretion of TNFα and IL-1R antagonist by RAW macrophages.[40] The role of Dectin 2 in antifungal immunity is not known.

DC-SIGN

In contrast to binding to Dectin 2, an internal sequence of high-mannose oligocaccharides binds with high affinity to the dendritic cell-specific receptor, DC-SIGN,[41] which forms tetramers at the cell surface.[42] DC-SIGN is proposed to mediate fungal uptake by dendritic cells. Through induction of the Raf-kinase pathway, it may also increase production of the immunosuppressive cytokine IL-10 by modulating TLR-mediated responses.[43]

S-Type Lectin—Galectin 3

Galectin 3 (gal 3) is an S-lectin with specificity for β-galactose and is expressed on a large number of host cells. Through their unusual β-1,2-linkages, mannoproteins and phospholipomannan bind to gal 3 on macrophages.[44] Gal 3 binding appears essential for phagocytosis[23] and may directly mediate extracellular killing of *C. albicans*.[45]

Cooperativity between Receptors

The integrated response to a pathogen depends on the mosaic of PAMP-PRR complexes that are engaged. Costimulation via multiple PAMP-PRR interactions can increase both sensitivity and specificity of immune recognition, for example, activation of phagocyte responses via β-glucan-Dectin1/TLR2 receptor complexes.[29,46] The sequence of molecular events in cooperative responses is not established but it is possible that β-glucan/Dectin 1 cooperativity occurs because β-glucan interacts solely with Dectin 1 and induces an amplification loop on cytokine release which is induced by the interaction of a different PAMP such as phospholipomannan with TLR2.[46-48] It has also been reported that CD14 engagement is needed for recognition of mannan by TLR4.[49]

PRR Arrays on Mononuclear Phagocytes

The differential expression of PRRs on different mononuclear phagocytes results in a cell-type specific response and hence further increases the specificity of innate immune responses to

Table 2. Role of PRR deficiencies in susceptibility to fungal infection in mouse models

Receptor Deficiency	Pathogen	Morphotype	Route of Infection	*Susceptibility to Infection	Reference
TLR4	*C. albicans*	Yeast	Intravenous	Increased	64
TLR4	*C. albicans*	Yeast	Intravenous	Not increased	4
TLR4	*C. albicans*	Hyphae	Intravenous	Not increased (survived longer than WT)	4
TLR2	*C. albicans*	Yeast	Intravenous	More resistant	65
TLR2	*C. albicans*	Yeast	Intravenous	Increased	66
TLR2	*C. albicans*	Yeast	Intravenous	Not increased	4
MyD88	*C. albicans*	Yeast	Intravenous	Increased	4
MyD88	*C. albicans*	Yeast	Intravenous	Increased	67
TLR6	*C. albicans*	Yeast	Intravenous	Not increased	68
TLR9	*C. albicans*	Yeast	Intravenous	Not increased	4
Dectin 1	*C. albicans*	Yeast	Intravenous	Increased	69
Dectin 1	*C. albicans*	Yeast	Intravenous	Not increased	70
CARD9	*C. albicans*	Yeast	Intravenous	Increased	71
IL-17	*C. albicans*	Yeast	Intravenous	Increased	72
TLR4	*A. fumigatus*	Conidia	Intranasal	Not increased	73
TLR4	*A. fumigatus*	Conidia	Intranasal	Increased	4
TLR2	*A. fumigatus*	Conidia	Intranasal	Not increased	73
TLR2	*A. fumigatus*	Conidia	Intranasal	More resistant	4
MyD88	*A. fumigatus*	Conidia	Intranasal	Not increased	73

continued on next page

Table 2. Continued

Receptor Deficiency	Pathogen	Morphotype	Route of Infection	*Susceptibility to Infection	Reference
MyD88	*A. fumigatus*	Conidia	Inhalation of aerosol	Increased early in infection- (Fungal burden, tissue necrosis increased at <72h)	74
WT plus local instillation of s-Dectin 1-fusion protein (with murine IgG1 Fc or mutant human IgG1 Fc fragment)	*A. fumigatus*	Conidia	Caudal oropharynx	Increased (lung fungal burden and broncho-alveolar fluid cell infiltrate)	12
TLR4 (C3H/HeJ mice**)	*C. neoformans*	Yeast	Intranasal	Not increased	75
TLR4 (C3H/HeJ mice)	*C. neoformans*	Yeast	Intravenous	Not increased	75
TLR2	*C. neoformans*	Yeast	Intranasal	Increased	75
TLR2	*C. neoformans*	Yeast	Intravenous	Not increased	75
MyD88	*C. neoformans*	Yeast	Intranasal	Increased	75
MyD88	*C. neoformans*	Yeast	Intravenous	Increased	75

*Measured as one or more of survival, organ fungal burdens, histopathology (fungi and host response) **C3H/HeJ mice have a naturally occurring point mutation in the TLR4 gene which renders the receptor unable to signal.[76]

individual pathogens. Monocytes display high levels of TLRs including membrane-associated TLRs 2, 4 and 6 and the cytosolic TLR9 and moderate levels of the CLRs Dectin 1 and MR. Macrophages display high levels of TLRs 2, 4, 6 and 9 and the CLR Dectin 1, as well as galectin 3, MR, CR3 and FcγR. Dendritic cells display TLRs 2, 4 and 9, Dectin 1, MR, DC-SIGN, CR3 and FcγR and neutrophils display moderate levels of the TLRs TLR2 and 4, Dectin 1 and high levels of CR3 and FcγR.[8] Dectin 2 is expressed at low level on dendritic cells and monocytes and at high level on inflammatory monocytes and tissue macrophages.[50]

In vivo Studies

Translating the roles of signaling pathways and innate immune functions subserved by PAMP-PRR interactions in vitro to the pathogenesis of fungal infections in the mammalian host has proven difficult, as studies in transgenic animals deficient in genes for various PRRs have yielded conflicting results (see Table 2). The most consistent findings have been in MyD88-deficient mice, which are more susceptible than wild-type animals to systemic candidiasis and cryptococcosis and to pulmonary aspergillosis, at least in the early stages of *Aspergillus* infection (for references, see Table 2).

Associations between PRR Polymorphisms and Fungal Diseases in Humans

The relationship between polymorphisms in PRRs and susceptibility to fungal infections has been studied infrequently in humans. The TLR4 Asp299Gly polymorphism has been associated with an increased incidence of chronic cavitary aspergillosis and candidemia[51,52] but not with increased susceptibility to, or severity of, urogenital candidiasis.[53] Another TLR4 polymorphism at Thr399Ile was also associated with candidemia.[52] In a small study of chronic mucocutaneous candidiasis there was no association with TLR2 or TLR4 polymorphisms.[54] Patients with allergic bronchopulmonary aspergillosis were noted to have an increased incidence of a TLR9 polymorphism (allele G on T 1237c).[51]

Conclusions and the Future

Much progress has been made in elucidating the role of fungal PAMPs and PRRs in the pathogenesis of fungal infections, primarily based on studies using individual PAMPs and phagocytic cells in vitro. Inconsistent results from studies using intact fungal cells and in animal models underline the complexity of the process and invite much further investigation before our understanding of the specificity of the innate immune response in vivo can be advanced.

Acknowledgements

The authors thank Dr Julianne Djordjevic for producing Figure 1. TCS' work is funded by a Centre of Clinical Research Excellence grant (# 264625) from the National Health and Medical Research Council of Australia.

References

1. Willment JA, Brown GD, Willment JA et al. C-type lectin receptors in antifungal immunity. Trends Microbiol 2008; 16:27-32.
2. Janeway CA Jr. The immune system evolved to discriminate infectious nonself from noninfectious self. Immunol Today 1992; 13:11-16.
3. Levitz SM. Interactions of toll-like receptors with fungi. Microbes and Infection 2004; 6:1351-1355.
4. Bellocchio S, Montagnoli C, Bozza S et al. The contribution of the toll-like/IL-1 receptor superfamily to innate and adaptive immunity to fungal pathogens in vivo. J Immunol 2004; 172:3059-3069.
5. Phadke AP, Mehrad B. Cytokines in host defense against Aspergillus: recent advances. Med Mycol 2005; 43(Suppl 1):S173-176.
6. Akira S, Uematsu S, Takeuchi O et al. Pathogen recognition and innate immunity. Cell 2006; 124:783-801.
7. Filler SG. Candida-host cell receptor-ligand interactions. Curr Opin Microbiol 2006; 9:333-339.
8. Netea MG, Brown GD, Kullberg BJ et al. An integrated model of the recognition of Candida albicans by the innate immune system. Nat Revs Microbiol 2008; 6:67-78.

9. Hoebe K, Janssen E, Beutler B et al. The interface between innate and adaptive immunity. Nat Immunol 2004; 5:971-974.
10. Saville SP, Lazzell AL, Monteagudo C et al. Engineered control of cell morphology in vivo reveals distinct roles for yeast and filamentous forms of Candida albicans during infection. Eukaryot Cell 2003; 2:1053-1060.
11. Schaffner A, Douglas H, Braude A. Selective protection against conidia by mononuclear and against mycelia by polymorphonuclear phagocytes in resistance to Aspergillus. Observations on these two lines of defense in vivo and in vitro with human and mouse phagocytes. J Clin Invest 1982; 69:617-631.
12. Steele C, Rapaka RR, Metz A et al. The beta-glucan receptor dectin-1 recognizes specific morphologies of Aspergillus fumigatus. PLoS Pathogens 2005; 1:e42.
13. Nemecek JC, Wuthrich M, Klein BS et al. Global control of dimorphism and virulence in fungi. Science 2006; 312:583-588.
14. Casadevall A, Perfect, JR. Cryptococcus neoformans. Washington DC: ASM Press, 1998.
15. Klis FM, de Groot P, Hellingwerf K. Molecular organization of the cell wall of Candida albicans. Med Mycol 2001; 39(Suppl 1):1-8.
16. Torosantucci A, Bromuro C, Chiani P et al. A novel glycoconjugate vaccine against fungal pathogens. J Exp Med 2005; 202:597-606.
17. Gantner BN, Simmons RM, Underhill DM et al. Dectin-1 mediates macrophage recognition of Candida albicans yeast but not filaments. EMBO J 2005; 24:1277-1286.
18. Netea MG, Van der Graaf C, Van der Meer JW et al. Recognition of fungal pathogens by Toll-like receptors. Eur J Clin Microbiol Infect Dis 2004; 23:672-676.
19. Bernard M, Latge JP. Aspergillus fumigatus cell wall: composition and biosynthesis. Med Mycol 2001; 39(Suppl 1):9-17.
20. Latge JP, Kobayashi H, Debeaupuis JP et al. Chemical and immunological characterization of the extracellular galactomannan of Aspergillus fumigatus. Infect Immun 1994; 62:5424-5433.
21. Fontaine T, Simenel C, Dubreucq G et al. Molecular organization of the alkali-insoluble fraction of Aspergillus fumigatus cell wall. J Biol Chem 2000; 275:27594-27607.
22. Tronchin G, Bouchara JP, Ferron M et al. Cell surface properties of Aspergillus fumigatus conidia: correlation between adherence, agglutination and rearrangements of the cell wall. Can J Microbiol 1995; 41:714-721.
23. Poulain D, Jouault T, Poulain D et al. Candida albicans cell wall glycans, host receptors and responses: elements for a decisive crosstalk. Curr Opin Microbiol 2004; 7:342-349.
24. Gersuk GM, Underhill DM, Zhu L et al. Dectin-1 and TLRs permit macrophages to distinguish between different Aspergillus fumigatus cellular states. J Immunol 2006; 176:3717-3724.
25. Wheeler RT, Fink GR, et al. A drug-sensitive genetic network masks fungi from the immune system. PLoS Pathogens 2006; 2:e35.
26. Shoham S, Huang C, Chen JM et al. Toll-like receptor 4 mediates intracellular signaling without TNF-α release in response to Cryptococcus neoformans polysaccharide capsule. J Immunol 2001; 166:4620-4626.
27. Obayashi T, Yoshida M, Mori T et al. Plasma (1→3)-beta-D-glucan measurement in diagnosis of invasive deep mycosis and fungal febrile episodes. Lancet 1995; 345:17-20.
28. Pirofski LA. Of mice and men, revisited: new insights into an ancient molecule from studies of complement activation by Cryptococcus neoformans. Infect Immun 2006; 74:3079-3084.
29. Netea MG, Gow NA, Munro CA et al. Immune sensing of Candida albicans requires cooperative recognition of mannans and glucans by lectin and Toll-like receptors. J Clin Invest 2006; 116:1642-1650.
30. Trinel PA, Borg-von-Zepelin M, Lepage G et al. Isolation and preliminary characterization of the 14- to 18-kilodalton Candida albicans antigen as a phospholipomannan containing beta-1,2-linked oligomannosides. Infect Immun 1993; 61:4398-4405.
31. Jouault T, Fradin C, Trinel PA et al. Early signal transduction induced by Candida albicans in macrophages through shedding of a glycolipid. J Infect Dis 1998; 178:792-802.
32. Monari C, Bistoni F, Casadevall A et al. Glucuronoxylomannan, a microbial compound, regulates expression of costimulatory molecules and production of cytokines in macrophages. J Infect Dis 2005; 191:127-137.
33. Stahl PD, Rodman JS, Miller MJ et al. Evidence for receptor-mediated binding of glycoproteins, glycoconjugates and lysosomal glycosidases by alveolar macrophages. Proc Natl Acad Sci U S A 1978; 75:1399-1403.
34. Linehan SA, Martinez-Pomares L, Gordon S. Macrophage lectins in host defence. Microbes Infect 2000; 2:279-288.
35. McGreal EP, Rosas M, Brown GD et al. The carbohydrate-recognition domain of Dectin-2 is a C-type lectin with specificity for high mannose. Glycobiology 2006; 16:422-430.

36. Kery V, Krepinsky JJ, Warren CD et al. Ligand recognition by purified human mannose receptor. Arch Biochem Biophys 1992; 298:49-55.
37. Palma AS, Feizi T, Zhang Y et al. Ligands for the beta-glucan receptor, Dectin-1, assigned using "designer" microarrays of oligosaccharide probes (neoglycolipids) generated from glucan polysaccharides. (erratum appears in J Biol Chem 2006 Aug 25; 281(34):24999). J Biol Chem 2006; 281:5771-5779.
38. Brown GD, Brown GD. Dectin-1: a signalling nonTLR pattern-recognition receptor. Nature Revs Immunol 2006; 6:33-43.
39. Herre J, Marshall AS, Caron E et al. Dectin-1 uses novel mechanisms for yeast phagocytosis in macrophages. Blood 2004; 104:4038-4045.
40. Sato K, Yang XL, Yudate T et al. Dectin-2 is a pattern recognition receptor for fungi that couples with the Fc receptor gamma chain to induce innate immune responses. J Biol Chem 2006; 281:38854-38866.
41. Feinberg H, Mitchell DA, Drickamer K et al. Structural basis for selective recognition of oligosaccharides by DC-SIGN and DC-SIGNR. Science 2001; 294:2163-2166.
42. Mitchell DA, Fadden AJ, Drickamer K. A novel mechanism of carbohydrate recognition by the C-type lectins DC-SIGN and DC-SIGNR. Subunit organization and binding to multivalent ligands. J Biol Chem 2001; 276:28939-28945.
43. Gringhuis SI, den Dunnen J, Litjens M et al. C-type lectin DC-SIGN modulates Toll-like receptor signaling via Raf-1 kinase-dependent acetylation of transcription factor NF-kappaB. Immunity 2007; 26:605-616.
44. Fradin C, Poulain D, Jouault T. beta-1,2-linked oligomannosides from Candida albicans bind to a 32-kilodalton macrophage membrane protein homologous to the mammalian lectin galectin-3. Infect Immun 2000; 68:4391-4398.
45. Kohatsu L, Hsu DK, Jegalian AG et al. Galectin-3 induces death of Candida species expressing specific beta-1,2-linked mannans. J Immunol 2006; 177:4718-4726.
46. Gantner BN, Simmons RM, Canavera SJ et al. Collaborative induction of inflammatory responses by Dectin-1 and Toll-like receptor 2. J Exp Med 2003; 197:1107-1117.
47. Brown GD, Herre J, Williams DL et al. Dectin-1 mediates the biological effects of beta-glucans. J Exp Med 2003; 197:1119-1124.
48. Jouault T, Ibata-Ombetta S, Takeuchi O et al. Candida albicans phospholipomannan is sensed through toll-like receptors. J Infect Dis 2003; 188:165-172.
49. Tada H, Nemoto E, Shimauchi H et al. Saccharomyces cerevisiae- and Candida albicans-derived mannan induced production of tumor necrosis factor alpha by human monocytes in a CD14- and Toll-like receptor 4-dependent manner. Microbiol Immunol 2002; 46:503-512.
50. Taylor PR, Reid DM, Heinsbroek SE et al. Dectin-2 is predominantly myeloid restricted and exhibits unique activation-dependent expression on maturing inflammatory monocytes elicited in vivo. Eur J Immunol 2005; 35:2163-2174.
51. Carvalho APA, Pitzurra L, Romani L et al. Polymorphisms in toll-like receptor genes and susceptibility to pulmonary aspergillosis. J Infect Dis 2008; 197:618-621.
52. Van der Graaf CA, Netea MG, Morre SA et al. Toll-like receptor 4 Asp299Gly/Thr399Ile polymorphisms are a risk factor for Candida bloodstream infection. Eur Cytokine Netw 2006; 17:29-34.
53. Morre SA, Murillo LS, Spaargaren J et al. Role of the toll-like receptor 4 Asp299Gly polymorphism in susceptibility to Candida albicans infection. J Infect Dis 2002; 186:1377-1379; author reply 1379.
54. van der Graaf CA, Netea MG, Drenth IP et al. Candida-specific interferon-gamma deficiency and toll-like receptor polymorphisms in patients with chronic mucocutaneous candidiasis. Neth J Med 2003; 61:365-369.
55. Cambi A, Gijzen K, de Vries JM et al. The C-type lectin DC-SIGN (CD209) is an antigen-uptake receptor for Candida albicans on dendritic cells. Eur J Immunol 2003; 33:532-538.
56. Steele C, Marrero L, Swain S et al. Alveolar macrophage-mediated killing of Pneumocystis carinii f. sp. muris involves molecular recognition by the Dectin-1 beta-glucan receptor. J Exp Med 2003; 198:1677-1688.
57. Serrano-Gomez D, Leal JA, Corbi AL et al. DC-SIGN mediates the binding of Aspergillus fumigatus and keratinophylic fungi by human dendritic cells. Immunobiology 2005; 210:175-183.
58. Serrano-Gomez D, Dominguez-Soto A, Ancochea J et al. Dendritic cell-specific intercellular adhesion molecule 3-grabbing nonintegrin mediates binding and internalization of Aspergillus fumigatus conidia by dendritic cells and macrophages. J Immunol 2004; 173:5635-5643.
59. Mambula SS, Sau K, Henneke P et al. Toll-like receptor (TLR) signaling in response to Aspergillus fumigatus. J Biol Chem 2002; 277:39320-39326.
60. Viriyakosol S, Fierer J, Brown GD et al. Innate immunity to the pathogenic fungus Coccidioides posadasii is dependent on Toll-like receptor 2 and Dectin-1. Infect Immun 2005; 73:1553-1560.
61. Garlanda C, Hirsch E, Bozza S et al. Nonredundant role of the long pentraxin PTX3 in antifungal innate immune response. Nature 2002; 420:182-186.

62. Rappleye CA, Eissenberg LG, Goldman WE et al. Histoplasma capsulatum alpha-(1,3)-glucan blocks innate immune recognition by the beta-glucan receptor. Proc Natl Acad Sci USA 2007; 104:1366-1370.
63. Nakamura KMA, Xiao G, Hatta M et al. Deoxynucleic acids from Cryptococcus neoformans activate myeloid dendritic cells via TLR9-dependent pathway. J Immunol 2008; 180:4067-4074.
64. Netea MG, Van Der Graaf CA, Vonk AG et al. The role of toll-like receptor (TLR) 2 and TLR4 in the host defense against disseminated candidiasis. J Infect Dis 2002; 185:1483-1489.
65. Netea MG, Sutmuller R, Hermann C et al. Toll-like receptor 2 suppresses immunity against Candida albicans through induction of IL-10 and regulatory T-cells. J Immunol 2004; 172:3712-3718.
66. Villamon E, Gozalbo D, Roig P et al. Toll-like receptor-2 is essential in murine defenses against Candida albicans infections. Microbes Infect 2004; 6:1-7.
67. Villamon E, Gozalbo D, Roig P et al. Myeloid differentiation factor 88 (MyD88) is required for murine resistance to Candida albicans and is critically involved in Candida-induced production of cytokines. Eur Cytokine Netw 2004; 15:263-271.
68. Netea MG, van de Veerdonk F, Verschueren I et al. Role of TLR1 and TLR6 in the host defense against disseminated candidiasis. FEMS Immunol Med Microbiol 2008; 52:118-123.
69. Taylor PR, Tsoni SV, Willment JA et al. Dectin-1 is required for beta-glucan recognition and control of fungal infection. Nat Immunol 2007; 8:31-38.
70. Saijo S, Fujikado N, Furuta T et al. Dectin-1 is required for host defense against Pneumocystis carinii but not against Candida albicans. Nat Immunol 2007; 8:39-46.
71. Gross O, Gewies A, Finger K et al. Card9 controls a nonTLR signalling pathway for innate antifungal immunity. Nature 2006; 442:651-656.
72. Huang W, Na L, Fidel PL et al. Requirement of interleukin-17A for systemic anti-Candida albicans host defense in mice. J Infect Dis 2004; 190:624-631.
73. Dubourdeau M, Athman R, Balloy V et al. Aspergillus fumigatus induces innate immune responses in alveolar macrophages through the MAPK pathway independently of TLR2 and TLR4. J Immunol 2006; 177:3994-4001.
74. Bretz C, Gersuk G, Knoblaugh S et al. MyD88 signaling contributes to early pulmonary responses to Aspergillus fumigatus. Infect Immun 2008; 76:952-958.
75. Yauch LE, Mansour MK, Shoham S et al. Involvement of CD14, toll-like receptors 2 and 4 and MyD88 in the host response to the fungal pathogen Cryptococcus neoformans in vivo. Infect Immun 2004; 72:5373-5382.
76. Hoshino K, Takeuchi O, Kawai T et al. Cutting edge: Toll-like receptor 4 (TLR4)-deficient mice are hyporesponsive to lipopolysaccharide: evidence for TLR4 as the Lps gene product. J Immunol 1999; 162:3749-3752.

Chapter 10

The Immunosuppresive Tick Salivary Protein, Salp15

Ignacio J. Juncadella and Juan Anguita*

Abstract

The interaction between Ixodid ticks and their mammalian hosts is a complex relationship. While the mammalian host tries to avoid the completion of the feeding process, the tick has devised strategies to counteract these attempts. Tick saliva contains a vast array of pharmacological activities that presumably aid the tick to evade host responses, including anti-complement, oxidative and innate and adaptive immune responses. The characterization of these activities has gained momentum in the last several years. One of the best studied activities present in tick saliva corresponds to the antigen known as Salp15, which binds specifically to the T-cell coreceptor CD4 resulting in the specific repression of $CD4^+$ T-cell activation. We discuss here the current state of our knowledge of the mode of action of this salivary protein.

Introduction

Ixodid ticks transmit a diverse array of infectious agents of medical and veterinary importance,[1] including the causative agents of Lyme disease and human granulocytic anaplasmosis.[2,3] Ticks are considered to be second only to mosquitoes worldwide as vectors of human pathogens. Ticks are obligate blood-feeding ectoparasites that belong to the family *Ixodidae*. The geographical distribution of *Ixodes* spp. ticks is varied: *I. scapularis* is located in eastern US, while *I. pacificus* is found in the Western part of the country. In Europe, *I. ricinus* is the main vector for the bacteria that cause Lyme disease.

The tick life cycle is comprised of three developmental stages: larvae, nymph and adult. Tick larvae can acquire bacteria from infected hosts and subsequently transmit them to mammals once they molt to nymphs.[4] During the nymph stage, *I. scapularis* preferentially feeds on small wild rodents, where transmission of vector-borne pathogens to the mammalian host is most efficient. The adult tick attaches and feeds on medium to large mammals including humans and can also contribute to pathogen transmission.[4]

Control of tick populations and pathogen transmission are primary research targets both in veterinary and human medicine. Several strategies are currently used to try to achieve this goal, although none has been demonstrated as particularly efficient. The use of acaricides is prevalent and is used in combination with the only approved vaccine formulation that exists: a recombinant Bm86 from the cattle tick *Boophilus microplus*.[5,6] This vaccine uses an antigen of unknown function in the tick. Anti-Bm86 antibodies are detrimental to adult tick survival. The antibodies target the antigen in the tick midgut cells provoking cell damage and blood leakage, which results

*Corresponding Author: Juan Anguita—Department of Veterinary and Animal Sciences. University of Massachusetts Amherst. Paige Lab. 161 Holdsworth Way. Amherst, MA 01003, USA. Email: janguita@vasci.umass.edu

Pathogen-Derived Immunomodulatory Molecules, edited by Padraic G. Fallon.

in tick death or reduced fecundity. Thus, the efficacy of the vaccine is most evident at the next generation of ticks.[5]

Unfortunately, the research aimed at the discovery of novel vaccine candidates in other tick species has not been as successful. Serine protease inhibitors (Serpins) have been shown to promote protection in cattle against *Rhipicephalus appendiculatus* ticks[7,8] and have been proposed as targets for other ticks, including *B. microplus* and *Haemaphysalis longicornis*.[9-11] Other targets include potential physiologically important proteins, such as those with enzymatic activity required for tick survival (reviewed by Willadsen[6]). A combination vaccine with different antigens has been proposed as a likely candidate to elicit strong antitick immunity. Tick saliva antigens are first exposed to the mammalian host during feeding and thus can be used as vaccine candidates. The presence of several immunomodulators in tick saliva however, can hinder the generation of efficient vaccine targets. Here, we will briefly discuss the interaction between ticks, their mammalian hosts and the pathogens they transmit. We will also discuss progress made in the understanding of the function of the immunomodulator, Salp15, with a brief discussion on its potential applications.

Interaction of Tick Saliva with the Mammalian Host

In contrast to other hematophagous arthropods, hard ticks feed for extended periods of time. *I. scapularis* requires a 5 to 7 day attachment period to the mammalian host in order to feed to repletion.[12] The long period of tick engorgement could allow the host to generate immune responses directed towards secreted components present in saliva. In turn, *I. scapularis* ticks attempt to modulate these responses in order to effectively attach to and feed upon its host.[12-16] Tick saliva is abundantly secreted into the host during feeding.[17] Feeding involves sucking blood from a hematoma formed by mandible laceration of the host tissue. During tick attachment to its host, a glue-like anchoring substance is secreted from their salivary glands to prevent detachment from the feeding site.[18] The injected saliva contains a cocktail of potent pharmacologically bioactive components able to maintain the liquid state of host blood.[19] Ixolaris, an inhibitor of the factor X-tissue factor VIIIa complex[20] and Salp14, an inhibitor of factor Xa,[21] represent two prototypical anticoagulants that have been identified in *I. scapularis* saliva. Bioactive compounds such as these support tick feeding and consequently may also facilitate the transmission of human pathogens.

Immune Response to Tick Feeding and Tick Immunity

Many tick-host relationships are particularly characterized by the acquisition of resistance to tick feeding which develops as a result of repeated tick infestations.[22] In many animal species, acquired resistance to ticks is characterized by reduced engorgement, increased duration of feeding, blocked molting and death of engorging ticks.[23] This phenomenon of resistance is known as acquired tick immunity. Acquired resistance to ticks was first described by Trager in 1939, who showed that guinea pigs in which *Dermacentor variabilis* (the common dog tick) had fed became resistant to future tick bites.[24] Many authors have extended these findings over the last 70 years. Repeated exposure of rabbits, cattle, dogs and guinea pigs to ticks have been shown to interfere with vector feeding, molting and fecundity.[24-26] Tick immunity can interfere with the transmission of several pathogens, including *Borrelia burgdorferi* to guinea pigs,[27] *Babesia bovis* to cattle[28] or *Franciscella tularensis* to rabbits.[29]

Tick rejection is characterized by cutaneous reactions marked by edema with inflammatory infiltrates.[23] Following the entry of the tick's hypostome into the host skin, a localized inflammatory response can be initiated. Activation of the complement system in the vicinity of the tick bite site can also result in the generation of mediators of inflammation, chemotaxis, opsonins followed by the chemoattraction of basophils and other cells linked to host resistance.[30] Besides local inflammation, tick feeding can also induce a complex array of host immune responses involving antigen presenting cells, T and B lymphocytes, antibodies and pro-inflammatory cytokines (reviewed in refs. 23 and 31). T-cells are key elements in the generation of an appropriate adaptive immune response including antibody production and cell mediated immunity.

Tick immunity can be partially transferred to guinea pigs never exposed to ticks by using serum from tick-immune animals[32-34] and antibodies, complement and cutaneous basophil hypersensitivity have been implicated in resistance.[22,23,31,35-38] Tick feeding induces the generation of homocytotropic antibodies that bind to Fc receptors on the surface of basophils and mast cells. These antibodies complex with tick saliva components during tick feeding, resulting in the release of biologically active molecules during degranulation.[39-44] A delayed-type hypersensitivity reaction has also been associated with the rejection of ticks by sensitized hosts.[45] The immunization of guinea pigs with tick salivary gland extracts is sufficient for the development of tick immunity and cutaneous basophil hypersensitivity responses.[45-47]

Tick Modulation of the Host Immune Response

Ticks have evolved several mechanisms to counteract host responses during feeding, especially in hosts that are most relevant for their life cycle, such as rodents. Thus, tick salivary proteins exert pleiotropic effects that include the inhibition of the complement cascade,[19] the impairment of NK cell function,[48] the reduction in homocytotrophic and circulating antibody titers,[1] the repressed production of cytokines such as IL-2 and IFNγ[49-51] and the inhibition of T-lymphocyte proliferation.[16,49,50]

A role for immunosuppressants as modulators or potential targets of vaccine development has been proposed.[5] The use of inactive forms of immunodulators in vaccine formulations could trigger the generation of antibodies with blocking activity with the potential to counteract the immunodulatory action of tick saliva during feeding. Willadsen suggested that antigenic tick immunomodulators could induce an immune response to the suppressors, "thereby allowing the development of resistance to repeated parasite infestation or a more effective defense against vector-borne disease".[5]

Saliva Proteins and the Transmission of Human Pathogens

Activities present in tick saliva may also contribute to the pathogenesis of tick-transmitted micro-organisms by facilitating their transmission from the tick to the host and/or from the host to the tick.[52,53] As vector-borne pathogens have evolved with the arthropods, they have developed an intimate relationship that facilitates their survival and transmission.[54] At least two examples exist of co-evolutionary strategies devised by pathogens to persist in the mammalian or the arthropod hosts. Both *B. burgdorferi* and *Anaplasma phagocytophilum* have devised mechanisms to take advantage of the secreted products during tick feeding. *B. burgdorferi* transit through the salivary glands results in the induction of Salp15.[55] This secreted protein binds to the outer surface protein C of the spirochete providing protection against antibody-mediated killing of the micro-organism.[55] More recently, Salp16, another salivary antigen produced during tick feeding,[56] has been demonstrated to provide survival advantages to the causative agent of Human Granulocytic Anaplasmosis, *A. phagocytophilum*.[54] The acquisition of mammalian viruses may also be facilitated by tick saliva,[7] although the process has not been assigned to specific components.

The acquisition of pathogens by ticks is also facilitated by salivary proteins. For example, Salp25D, an antioxidant secreted during tick feeding allows *B. burgdorferi* to counteract the production of reactive oxygen species in their traslocation from the host to the tick.[53] This allows the spirochete to colonize the arthropod during tick feeding and complete their natural infectious cycle.

These examples provide evidence of co-evolutionary mechanisms that result in the maintenance of a beneficial relationship between ticks and pathogens. However, while the presence of tick saliva proteins that are beneficial for the completion of the infectious cycle of these micro-organisms has been described, no information is available that provides evidence of advantages provided to the vector by the presence of the micro-organism. Thus, the relationship between both may be merely of a parasitic nature, although it is possible that unknown mechanisms would result in advantages to the tick due to the presence of these bacteria. Further research should clarify this point.

Identification and Cloning of the Salivary Protein, Salp15

As discussed before, tick saliva contains many activities, which aid the arthropod in the evasion of host defenses and allow the completion of the blood meal. A few species of mammals have been reported to develop immunity to tick bites after repeated exposures, such as guinea pigs and rabbits, while the consensus is that others are not able to mount efficient immune responses to tick antigens. Therefore, the presence of specific antibodies to salivary proteins in species that develop tick immunity may permit the identification of genes that may be critical for vector survival during its interaction with the mammalian host. The specific antigenic and functional components of *I. scapularis* saliva has been characterized and reports have described the proteomic component of tick saliva of different species[57-61] as well as the antibody response during tick attachment.[62]

The salivary protein, Salp15 was identified by screening an *I. scapularis* cDNA expression library with sera from rabbits that had been sensitized to ticks by three infestations of 100 nymph ticks per animal.[56] The screening of 100,000 cDNA plaques using the tick immune sera yielded 47 clones that were reactive. These 47 clones represented 14 genes coding *I. scapularis* saliva proteins in various degrees of repetition.[56] One of the genes corresponded to a 14.7 kDa antigen that was named Salp15, following the nomenclature Salivary protein (Salp) followed by its molecular weight.[56] The expression of the 14 identified genes, including Salp15, was confirmed by RT-PCR and showed that in several cases the expression is induced upon tick feeding.[56] Salp15 could be readily detected in skin biopsies of mice infested with *I. scapularis* two days after tick attachment demonstrating that it is deposited at the site of feeding.[49]

Efforts to discover homologues of Salp15 in different species of *Ixodes* ticks have been recently successful. Hovius and collaborators have recently found three Salp15 homologues in *Ixodes ricinus*.[63] The homologues have been named Salp15 Iric-1, 2 and 3 due to their similarity to the *I. scapularis* Salp15. Salp15 Iric-1 was found to have the most similarity to *I. scapularis* Salp15 (80% homology at the amino acid level), while Iric-2 and Iric-3 showed lower degree of homology (62 and 63%, respectively). The homology between the three *I. ricinus* Salp15 is highest at the carboxyl terminus domain of all three homologues, which is responsible for the immunosuppressive activity of the protein (see below) (Fig. 1). This high homology suggests that these proteins may be part of a family in which members are conserved between different tick species.[63] Salp15 Iric-1 can be considered an orthologue of *I. scapulari* Salp15 resulting from a speciation event while Iric-2 and Iric-3 may be paralogues of Iric-1 arising from duplication events within a particular lineage. The biological functions of the *I. ricinus* homologues need to be further investigated but one can speculate to be comparable to the function of *I. scapularis* Salp15, given the high homology within the carboxyl terminal domain which mediates *I. scapularis* Salp15 immunosuppressive function.[64] The activity of these homologues regarding their capacity to help *B. burgdorferi* sensu lato (*B. burgdorferi* sensu strictu, *B. afzelii* and *B. garinii*, which together are responsible for Lyme borreliosis in Europe) during their colonization of the mammalian host by interacting with the different OspC variants in these European species is also unknown. Since intuitively, evolutionary pressure to maintain a parasitic micro-organism such as *B. burgdorferi* is not expected, it would be extremely interesting to determine whether this activity is also conserved in *I. ricinus* Salp15 homologues related to the different genospecies. This in turn, would suggest that the presence of *B. burgdorferi* and perhaps other micro-organisms in the tick is beneficial for the latter.

The presence of multi-copy genes in the salivary glands of different species of ticks has been well documented. The presence of several copies of Salp15 homologues has been described in *I. pacificus*.[58] However, to date, no other homologue of Salp15 has been described in *I. scapularis*.

Salp15 Inhibits IL-2 Production and CD4$^+$ T-Cell Proliferation

The initial analysis of Salp15 revealed a weak homology between the saliva protein and active motifs present in inhibin A (motif 4 and 7), a member of the tumor growth factor-β (TGFβ) super-family.[49] TGFβ signaling has a nonredundant essential role in the immune system, which includes regulating natural killer (NK)-cell responses, facilitating IgA class-switching, promoting the development of certain lymphocyte subsets and limiting T-cell mediated autoimmunity.[65]

```
Salp15   MESFVAMKVVCILFLVGVVAANESGPTKADASTADKDTKKNNVQLRFPNYISNHQKLALK  60
Iric-1   MESFVAMKVVCITVLFVIVAVNESATSEARTSSAAKETKKKNVTLHFPSYIRNPQKLALE  60
Iric-2   MESFVAMKVVCIVLLFVIAAEAAS-TGKNPVGDAPNGKKNN-ITFNFPPYVPNHHAFASS  58
Iric-3   MESFVAMKVVCIILLFVIAAEAES-INEKSDVEPSKGKNNSGLQFKFPPYVPNHKAFALR  59
         ************ .*. :.*   *   :     . :.::.  : :.** *: * : :*

Salp15   LLKICKDSKSSHNSLS----SRSSDVINDKYVDFKNCTFLCKHGNDVNVTLNLPEDTPC  115
Iric-1   LLEICKNNKS-RNSLP----STNYSAINDKYVDFKNCTFLCKHAEDRNVTLDLPPNTLC  114
Iric-2   LWKLCEESTPKPEMKIVDSRTTYTRRINDLQVNFKDCTFLCKRRFD-NVTLDLPKNTPC  116
Iric-3   LLSLCEQG-------------IYGTKINDLKVDFKNCTFLCIRKYE-NLTLPLPEDTPC  104
         * .:*::.                  ***  *:**:***** :  : *:** ** :* *

Salp15   GPNGQTCAEKNKCVGHIPGC                                         135
Iric-1   GPNGETCAEKSKCVGHIPGC                                         134
Iric-2   GPKNQTCENKDQCVPHIPGC                                         136
Iric-3   GPNNQTCHKKDECVGYIPGC                                         124
         **:.:** :*.:** :****
```

Figure 1. Amino acid sequence alignment of Salp15 and the 3 homologues present in *I. ricinus*. Identical amino acid residues is indicated by an asterisk (*) in blue, while conserved substitutions are presented as (.) or (:), in red. The amino acids corresponding to the sequence of P11 in Salp15 are delimited by the box.

TFGβ is also important for the generation of regulatory T-cells and the differentiation of T helper 17 cells.[65] Furthermore, TGFβ inhibits the production of cytokines, such as IL-2 and the proliferation of T-cells,[66] which suggested that Salp15 may have immunomodulatory properties.[49]

As a result of the weak similarities to the TGFβ super-family member, the effect of Salp15 on the proliferation of T-cells was examined. The proliferative response to TCR-triggered ligation was impaired by Salp15 treatment. This effect was not the result of a potential cytotoxic effect of the saliva protein.[49] The levels of interleukin-2 (IL-2) produced by $CD4^+$ T-cells stimulated in the presence of Salp15 were lowered in a dose dependent manner.[49] Furthermore, the effect of Salp15 on IL-2 production was completely abolished by the addition of Salp15 anti-sera suggesting a specific effect of the suppressive protein. Salp15 also diminished the expression levels of CD25 (the high affinity subunit of the IL-2 receptor) as soon as 12 h that persisted throughout the activation period.[49] The inhibition of T-cell activation by Salp15 was specifically the result of diminished production IL-2 because the addition of exogenous recombinant IL-2 reversed the effect of Salp15 during the activation period. This reversal was evident by increased T-cell proliferation. Furthermore, the expression of CD25 augmented to similar levels even in the presence of Salp15.[49] Interestingly, a stronger inhibitory effect of Salp15 on IL-2 production was observed at lower concentrations of αCD3 or when the cells are activated in the absence of costimulatory signals,[49] strongly suggesting that Salp15 may interfere with proximal TCR-mediated activation.

Salp15 Specifically Interacts with the T-Cell Coreceptor CD4

Binding of Salp15 to membrane preparations purified from $CD4^+$ T-cells indicated that Salp15 associates with a protein component on the surface of T-cells.[49] Confocal microscopy showed Salp15 staining on the surface of purified $CD4^+$ but not $CD8^+$ T-cells, suggesting that it binds to a component specific to $CD4^+$ T-cells.[67] Furthermore, Salp15 colocalized with CD4 on the surface of both unstimulated and αCD3/CD28 stimulated $CD4^+$ T-cells.[67] Since Salp15 may interfere with TCR-mediated signaling, key surface molecules including CD4, CD3, CD28 and components of the TCR complex were immunoprecipitated from lysates prepared from $CD4^+$ T-cells that have been pretreated with Salp15. Only the CD4 molecule immunoprecipitated with Salp15 suggesting CD4 as the receptor of Salp15.[67] This association was confirmed using nonlymphocyte cell lines expressing CD4, such as HeLa-CD4 or 3T3.CD4 cells (Fig. 2). In correlation, the effect of Salp15 on T-cell activation was negligible in CD4-deficient T-cells.[64] Thus, the immunosuppressive activity of Salp15 is likely to depend on the specific interaction between Salp15 and CD4.

Classical microtiter assays using purified soluble CD4 (sCD4) containing either all four (D1-D4) or the first two (D1-D2) most extracellular domains revealed saturable binding kinetics with Salp15 with a dissociation constant (K_d) in the nanomolar range.[67] These results also suggested that the salivary protein interacts with domains 1 and 2 of CD4. In correlation, the αCD4 monoclonal antibody MT310, which is D1-specific[68] competed the binding of Salp15 to sCD4, while the D3D4-specific mAb OKT4[69] did not.[67]

The identification of the residues of Salp15 responsible for its interaction with CD4 was initially performed with the use of 20 amino acid-long peptides spanning the entire amino acid sequence of Salp15 (P1-P11). These peptides were analyzed for their capacity to interact with CD4. Only P11, the carboxyl terminal region of Salp15 interacted with soluble CD4 (D1D2 or D1D4).[67] Interestingly, the P11 peptide was found to be sufficient to inhibit IL-2 production and prevent T-cell activation, recapitulating the immunosuppressive activity of the salivary protein.[64] Furthermore, a mutant version of Salp15 (Salp15ΔP11) that lacks the carboxyl terminal region corresponding to the peptide P11 failed to bind sCD4 and to recapitulate the inhibitory effect of Salp15 (unpublished observations).

Salp15 Causes Conformational Rearrangements in CD4

In order to investigate the mechanism by which the interaction of Salp15 with CD4 results in such dramatic downstream effects, we performed structural analysis using small angle X-ray scattered (SAXS) assays with proteins in solution. The SAXS-based models of CD4, Salp15 and

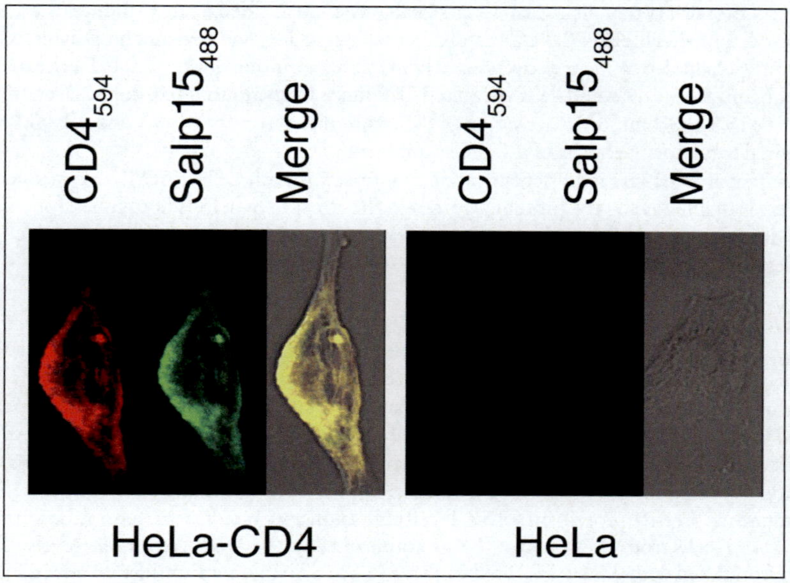

Figure 2. Colocalization of CD4 and Salp15 on the surface of HeLa cells ectopically expressing CD4 (HeLa-CD4). Confocal micrograph corresponding to staining of surface CD4 with a specific antibody labeled with Alexa $Fluor_{594}$ and Salp15 labeled with Alexa $Fluor_{488}$.

CD4-Salp15 complexes revealed differences in their globular structures when comparing the unliganded proteins with their complexes.[70] Ab initio structure restoration of the measured SAXS data representing the average of five independent solutions for Salp15 resulted in three-dimensional models showing two asymmetrical globular domains with 75 Å and 25 Å in their long axis, with an overall length of ~101 Å.[70] The analysis of sCD4 (both D1D2 and D1D4 versions) correlated extremely precisely with X-ray crystallographic data, validating the results obtained.[70]

SAXS data obtained from complexes of Salp15 and sCD4 (both D1D2 and D1D4) generated models that when compared to the published structures of unligated sCD4, showed additional volumes that suggested that Salp15 binds to one side of the apex of domain 1.[70] Furthermore, binding of Salp15 to sCD4 induces a structural rearrangement throughout the long axis of the coreceptor that "straightens" the bent Z-shape of the sCD4 molecule and affects the ability of the src-family tyrosine kinase Lck to interact with CD4.[70] Lck interacts with cytoplasmic domains of CD4 forming disulfide bridges.[71] The significance of this association is accentuated by the deficient activation seen in Jurkat cells that contain mutations in the cysteines residues involved in this interaction.[71] From this experiments, it was concluded that the immunosuppressive effect of Salp15 on T-cell activation is the product of an inappropriate interaction between CD4 and Lck which could translate into insufficient activation of proximal substrates of the kinase that are required for signal propagation.

T-Cell Signaling Pathways Inhibited by Salp15

T-cells receptor engagement causes the phosphorylation of Lck, which is coupled to the downstream activation of PLCγ1.[72] During TCR signaling, PLCγ1 hydrolyzes phosphatidyl-inositol[4,5]-bisphosphate (PIP_2) to inositol[1,4,5]-trisphosphate (IP_3) and diacylglycerol (DAG), leading to calcium mobilization and protein kinase-C (PKC) activation, respectively.[73] Lck is considered almost the first tyrosine kinase to get activated during TCR signaling.[74] Furthermore, the Zeta-chain associated protein, Zap-70, is an immediate substrate of Lck and its phosphorylation is a requisite for PLCγ1 hydrolysis.[75] Since, Salp15 interaction with CD4 affects further Lck association with the coreceptor, the effect of the salivary protein in early T-cell signals was studied. Indeed, Salp15 induces the defective phosphorylation of Lck, Zap-70, PLCγ1 and other early signaling intermediate components of T-cell activation.[64,67] Salp15 treatment also repressed lipid raft organization during $CD4^+$ T-cell activation, actin reorganization events and TCR-triggered Ca^{+2} fluxes from intracellular stores. Overall, the effect of Salp15 is the diminished activation of the transcription factors NF-AT and NF-$_k$B,[49] which are essential regulatory elements of *il-2* gene expression.

The effect of Salp15 is highly dependent on its interaction with CD4, since IL-2-dependent Lck activation is unaffected by the protein. Moreover, the effect of Salp15 is not global: for example, Salp15 does not affect AP-1 DNA binding activation[49] or other signaling pathways activated for example, by IFNγ (unpublished observations).

In Vivo Function of Salp15

The physiological importance of Salp15 would reside in its ability to modulate the adaptive immune response in vivo to tick proteins that become in contact with the mammalian host during feeding. Previous experiments have shown that T-lymphocytes purified from mice repeatedly exposed to *I. scapularis* nymph ticks do not produce high levels of IFNγ after ex vivo restimulation.[51] It is plausible to speculate that Salp15 may be at least partially responsible for this effect. The initial characterization of the effect of Salp15 in vivo was confirmed by assessing the effect of the immunosuppressive protein on $CD4^+$ T-cell activation and T-cell-mediated antibody production. $CD4^+$ T-cells from thioredoxin (TR)-immunized animals produced similar levels of IFNγ when restimulated with TR or TR fused to Salp15 (TR-Salp15), confirming that effector T-cells are not affected by Salp15.[49] In contrast, $CD4^+$ T-cells purified from TR-Salp15-immunized animals produced lower IFNγ in response to TR or TR-Salp15 indicating that Salp15 had inhibited the activation of antigen-specific $CD4^+$ T-cells.[49] This also resulted in reduced levels of antigen-specific IgG, while the levels of IgM, which do not require T-cell help were not affected.[49] Finally, delayed type hypersensitivity reactions (DTH), which have been associated with the tick immunity, were impaired in mice immunized with keyhole limpet hemocyanine (KLH) in the presence of Salp15.[49]

The role of Salp15 in the generation, or lack thereof, of tick immunity remains to be elucidated. It is tempting to speculate that the ability of certain mammalian species but not others to mount tick immune responses reside at least partially in the ability of Salp15 to differentially interact with CD4 from those species in which tick immunity fails to materialize, such as mice. While ticks that are made deficient in salp15 expression by shRNA injection do not show signs of deficient feeding,[55] this is not a surprising finding, since the development of tick immunity requires several rounds of tick infestation in order to allow the generation of acquired immune responses. Thus, the role of Salp15 in the prevention of the development of tick immunity would be evident in mice only after several rounds of infestations, as it is required for the generation of tick immunity in rabbits and guinea pigs.

Therapies Based on Salp15

Salp15 is one of the best-described salivary proteins. The specificity of the protein for CD4 and the downstream effects that it exerts on $CD4^+$ T-cells make it a candidate for potential therapies in conditions in which the repression of T-cell activation is required. For example, the capacity of Salp15 to prevent the development of experimental asthma has been recently described.[76] When Balb/c mice were sensitized with ovalbumin in aluminum hydroxide in the presence of Salp15, they displayed significantly reduced symptoms of allergic asthma, including airway hyperresponsiveness, eosinophilia, antigen-specific IgG1 and IgE, mucus cell metaplasia and Th2 cytokine secretion, compared to control-treated mice.[76] These experiments show that there is promise in the potential use of this protein for the inhibition of T-cell activation and potential pathologies. Much work needs to be done, however, to identify targets during ongoing pathologies, since the preventative treatment of conditions such as asthma is not a plausible course of action, with the exception perhaps of transplants.

Concluding Remarks

Much research has been completed regarding the activity that the tick salivary protein, Salp15 exerts on naïve T-cells during their activation and its potential therapeutic uses. Further work needs to be done to refine the uses of this salivary antigen during tick feeding, pathogen transmission and the treatment of conditions that course with the activation of naïve T-cells.

Acknowledgements

Supported by the National Institutes of Health and a Hatch project.

We thank Tonya C Bates for critically reading the manuscript.

References

1. Wikel SK, Bergman D. Tick-host immunology: Significant advances and challenging opportunities. Parasitol Today 1997; 13(10):383-389.
2. Burgdorfer W, Barbour AG, Hayes SF et al. Lyme disease-a tick-borne spirochetosis? Science 1982; 216(4552):1317-1319.
3. Chen SM, Dumler JS, Bakken JS et al. Identification of a granulocytotropic Ehrlichia species as the etiologic agent of human disease. J Clin Microbiol 1994; 32(3):589-595.
4. Kurtenbach K, Kampen H, Dizij A et al. Infestation of rodents with larval Ixodes ricinus (Acari: Ixodidae) is an important factor in the transmission cycle of borrelia burgdorferi s.l. in german woodlands. J Med Entomol 1995; 32(6):807-817.
5. Willadsen P. Tick control: thoughts on a research agenda. Vet Parasitol 2006; 138(1-2):161-168.
6. Willadsen P. The molecular revolution in the development of vaccines against ectoparasites. Vet Parasitol 2001; 101(3-4):353-368.
7. Muleng A, Sugino M, Nakajim M et al. Tick-Encoded serine proteinase inhibitors (serpins); potential target antigens for tick vaccine development. J Vet Med Sci 2001; 63(10):1063-1069.
8. Imamura S, Namangala B, Tajima T et al. Two serine protease inhibitors (serpins) that induce a bovine protective immune response against Rhipicephalus appendiculatus ticks. Vaccine 2006; 24(13):2230-2237.
9. Andreotti R, Gomes A, Malavazi-Piza KC et al. BmTI antigens induce a bovine protective immune response against Boophilus microplus tick. Int Immunopharmacol 2002; 2(4):557-563.
10. Sugino M, Imamura S, Mulenga A et al. A serine proteinase inhibitor (serpin) from Ixodid tick Haemaphysalis longicornis; cloning and preliminary assessment of its suitability as a candidate for a tick vaccine. Vaccine 2003; 21(21-22):2844-2851.
11. Imamura S, da Silva Vaz Junior I, Sugino M et al. A serine protease inhibitor (serpin) from haemaphysalis longicornis as an antitick vaccine. Vaccine 2005; 23(10):1301-1311.
12. Binnington KC, Kemp DH. Role of tick salivary glands in feeding and disease transmission. Adv Parasitol 1980; 18:315-339.
13. Ramachandra RN, Wikel SK. Modulation of host-immune responses by ticks (Acari: Ixodidae): effect of salivary gland extracts on host macrophages and lymphocyte cytokine production. J Med Entomol 1992; 29(5):818-826.
14. Kopecky J, Kuthejlova M, Pechova J. Salivary gland extract from Ixodes ricinus ticks inhibits production of interferon-gamma by the upregulation of interleukin-10. Parasite Immunol 1999; 21(7):351-356.
15. Wikel SK. Tick modulation of host immunity: an important factor in pathogen transmission. Int J Parasitol 1999; 29(6):851-859.
16. Urioste S, Hall LR, Telford SR 3rd et al. Saliva of the lyme disease vector, Ixodes dammini, blocks cell activation by a nonprostaglandin E2-dependent mechanism. J Exp Med 1994; 180(3):1077-1085.
17. Wang H, Nuttall PA. Immunosglublin binding proteins in ticks: new targets for vaccine development against a blood-feeding parasite. Cell Mol Life Sci 1999; 56:286-295.
18. Kaufman WR. Tick-host interaction: a synthesis of current concepts. Parasitol Today 1989; 5:47-56.
19. Ribeiro JM, Schneider M, Guimaraes JA. Purification and characterization of prolixin S (nitrophorin 2), the salivary anticoagulant of the blood-sucking bug rhodnius prolixus. Biochem J 1995; 308(Pt 1):243-249.
20. Francischetti IM, Valenzuela JG, Andersen JF et al. Ixolaris, a novel recombinant tissue factor pathway inhibitor (TFPI) from the salivary gland of the tick, Ixodes scapularis: identification of factor X and factor Xa as scaffolds for the inhibition of factor VIIa/tissue factor complex. Blood 2002; 99:3602-3612.
21. Narasimhan S, Koski RA, Beaulieu B et al. A novel family of anticoagulants from the saliva of Ixodes scapularis. Insect Mol Biol 2002; 11:641-650.
22. Willadsen P. Immunity to ticks. Adv Parasitol 1980; 18:293-311.
23. Wikel SK. Host immunity to ticks. Annu Rev Entomol 1996; 41:1-22.
24. Trager W. Acquired immunity to ticks. J Parasitol 1939; 25:57.

25. Kay BH, Kemp DH. Vaccines against arthropods. Am J Trop Med Hyg 1994; 50(6 Suppl):87-96.
26. Mulenga A, Sugimoto C, Sako Y et al. Molecular characterization of a haemaphysalis longicornis tick salivary gland-associated 29-kilodalton protein and its effect as a vaccine against tick infestation in rabbits. Infect Immun 1999; 67(4):1652-1658.
27. Nazario S, Das S, de Silva AM et al. Prevention of borrelia burgdorferi transmission in guinea pigs by tick immunity. Am J Trop Med Hyg 1998; 58(6):780-785.
28. Francis J, Little DG. Resistance of droughtmaster cattle to tick infestation and babesiosis. Aust Vet J 1964; 40:247-253.
29. Bell JF, Stewart SJ, Wikel SK. Resistance to tick-borne Francisella tularensis by tick-sensitized rabbits: allergic klendusity. Am J Trop Med Hyg 1979; 28(5):876-880.
30. Brossard M, Wikel SK. Tick immunobiology. Parasitology 2004; 129:S161-S176.
31. Brossard M, Wikel SK. Immunology of interactions between ticks and hosts. Med Vet Entomol 1997; 11(3):270-276.
32. Shapiro SZ, Voigt WP, Fujisaki K. Tick antigens recognized by serum from a guinea pig resistant to infestation with the tick Rhipicephalus appendiculatus. J Parasitol 1986; 72(3):454-463.
33. Brossard M, Girardin P. Passive transfer of resistance in rabbits infested with adult Ixodes ricinus L: humoral factors influence feeding and egg laying. Experientia 1979; 35(10):1395-1397.
34. Wikel SK, Allen JR. Acquired resistance to ticks. I. passive transfer of resistance. Immunology 1976; 30(3):311-316.
35. Ganapamo F, Rutti B, Brossard M. Immunosuppression and cytokine production in mice infested with Ixodes ricinus ticks: a possible role of laminin and interleukin-10 on the in vitro responsiveness of lymphocytes to mitogens. Immunology 1996; 87(2):259-263.
36. Ganapamo F, Rutti B, Brossard M. In vitro production of interleukin-4 and interferon-gamma by lymph node cells from BALB/c mice infested with nymphal Ixodes ricinus ticks. Immunology 1995; 85(1):120-124.
37. Nithiuthai S, Allen JR. Significant changes in epidermal Langerhans cells of guinea-pigs infested with ticks (Dermacentor andersoni). Immunology 1984; 51(1):133-141.
38. Mbow ML, Rutti B, Brossard M. IFN-gamma IL-2 and IL-4 mRNA expression in the skin and draining lymph nodes of BALB/c mice repeatedly infested with nymphal Ixodes ricinus ticks. Cell Immunol 1994; 156(1):254-261.
39. Allen JR. Tick resistance: basophils in skin reactions of resistant guinea pigs. Int J Parasitol 1973; 3(2):195-200.
40. Brown SJ, Graziano FM, Askenase PW. Immune serum transfer of cutaneous basophil-associated resistance to ticks: mediation by 7SIgG1 antibodies. J Immunol 1982; 129(6):2407-2412.
41. Benveniste J, Egido J, Gutierrez Millet V et al. Detection of immediate hypersensitivity in rabbits by direct basophil degranulation. J Allergy Clin Immunol 1977; 59(4):271-279.
42. Brossard M, Fivaz V. Ixodes ricinus L: mast cells, basophils and eosinophils in the sequence of cellular events in the skin of infested or re-infested rabbits. Parasitology 1982; 85(Pt 3):583-592.
43. Inokuma H, Kemp DH, Willadsen P. Comparison of prostaglandin E2 (PGE2) in salivary gland of boophilus microplus, Haemaphysalis longicornis and Ixodes holocyclus and quantification of PGE2 in saliva, hemolymph, ovary and gut of B. microplus. J Vet Med Sci 1994; 56(6):1217-1218.
44. Wikel SK. Histamine content of tick attachment sites and the effects of H1 and H2 histamine antagonists on the expression of resistance. Ann Trop Med Parasitol 1982; 76(2):179-185.
45. Brossard M, Monneron JP, Papatheodorou V. Progressive sensitization of circulating basophils against Ixodes ricinus L. antigens during repeated infestations of rabbits. Parasite Immunol 1982; 4(5):335-361.
46. Askenase PW, Bagnall BG, Worms MJ. Cutaneous basophil-associated resistance to ectoparasites (ticks). I. transfer with immune serum or immune cells. Immunology 1982; 45(3):501-511.
47. Brown SJ, Bagnall BG, Askenase PW. Ixodes holocyclus: kinetics of cutaneous basophil responses in naive and actively and passively sensitized guinea pigs. Exp Parasitol 1984; 57(1):40-47.
48. Kopecky J, Kuthejlova M. Suppressive effect of Ixodes ricinus salivary gland extract on mechanisms of natural immunity in vitro. Parasite Immunol 1998; 20(4):169-174.
49. Anguita J, Ramamoorthi N, Hovius JW et al. Salp15, an Ixodes scapularis salivary protein, inhibits $CD4^+$ T-cell activation. Immunity 2002; 16(6):849-859.
50. Ferreira BR, Silva JS. Saliva of Rhipicephalus sanguineus tick impairs T-cell proliferation and IFN-gamma-induced macrophage microbicidal activity. Vet Immunol Immunopathol 1998; 64(3):279-293.
51. Schoeler GB, Manweiler SA, Wikel SK. Ixodes scapularis: effects of repeated infestations with pathogen-free nymphs on macrophage and T-lymphocyte cytokine responses of BALB/c and C3H/HeN mice. Exp Parasitol 1999; 92(4):239-248.
52. Wikel SK. Tick modulation of host immunity: an important factor in pathogen transmission. Int J Parasitol 1999; 29(6):851-859.

53. Narasimhan S, Sukumaran B, Bozdogan U et al. A tick antioxidant facilitates the Lyme disease agent's succesful migration from the mammalian host to the arthropod vector. Cell Host Microbe 2007; 2:7-18.
54. Sukumaran B, Narasimhan S, Anderson JF et al. An Ixodes scapularis protein required for survival of Anaplasma phagocytophilum in tick salivary glands. J Exp Med 2006; 203(6):1507-1517.
55. Ramamoorthi N, Narasimhan S, Pal U et al. The Lyme disease agent exploits a tick protein to infect the mammalian host. Nature 2005; 436(7050):573-577.
56. Das S, Banerjee G, DePonte K et al. Salp25D, an Ixodes scapularis Antioxidant, Is 1 of 14 immunodominant antigens in engorged tick salivary glands. J Infect Dis 2001; 184(8):1056-1064.
57. Valenzuela JG. Exploring tick saliva: from biochemistry to 'sialomes' and functional genomics. Parasitology 2004; 129:S83-S94.
58. Francischetti IM, My Pham V, Mans BJ et al. The transcriptome of the salivary glands of the female western black-legged tick Ixodes pacificus (Acari: Ixodidae). Insect Biochem Molec Biol 2005; 35:1142-1161.
59. Oleaga A, Escudero-Poblacion A, Camafeita E et al. A proteomic approach to the identification of salivary proteins from the argasid ticks Ornithodoros moubata and Ornithodoros erraticus. Insect Biochem Mol Biol 2007; 37:1149-1159.
60. Valenzuela JG, Francischetti IM, My Pham V et al. Exploring the sialome of the tick Ixodes scapularis. J Exp Biol 2002; 205:2843-2864.
61. Mans BJ, Andersen JF, Francischetti IM et al. Comparative sialomics between hard and soft ticks: implications for the evolution of blood-feeding behavior. Insect Biochem Mol Biol 2008; 38:42-58.
62. Narasimhan S, DePonte K, Marcantonio N et al. Immunity against Ixodes scapularis salivary proteins expressed within 24 hours of attachment thwarts tick feeding and impairs Borrelia transmission. PLoS ONE 2007; 2:e451.
63. Hovius JW, Ramamoorthi N, Van't Veer C et al. Identification of salp15 homologues in Ixodes ricinus ticks. Vector Borne Zoonotic Dis 2007; 7:296-303.
64. Juncadella IJ, Garg R, Ananthanarayanan SK et al. T-cell signaling pathways inhibited by the tick saliva immunosuppressor, salp15. FEMS Immunol Med Microbiol 2007; 49(3):433-438.
65. Rubtsov YP, Rudensky AY. TGFbeta signalling in control of T-cell-mediated self-reactivity. Nat Rev Immunol 2007; 7:443-453.
66. McCartney-Francis NL, Wahl SM. Transforming growth factor beta: a matter of life and death. J Leukoc Biol 1994; 55:401-409.
67. Garg R, Juncadella IJ, Ramamoorthi N et al. Cutting edge: CD4 is the receptor for the tick saliva immunosuppressor, salp15. J Immunol 2006; 177(10):6579-6583.
68. Sutor GC, Dreikhausen U, Vahning U et al. Neutralization of HIV-1 by anti-idiotypes to monoclonal anti-CD4. potential for idiotype immunization against HIV. J Immunol 1992; 149(4):1452-1461.
69. Moore JP, Sattentau QJ, Klasse PJ et al. A monoclonal antibody to CD4 domain 2 blocks soluble CD4-induced conformational changes in the envelope glycoproteins of human immunodeficiency virus Type 1 (HIV-1) and HIV-1 infection of $CD4^+$ cells. J Virol 1992; 66(8):4784-4793.
70. Ashish, Juncadella IJ, Garg R et al. Conformational rearrangement within the soluble domains of the CD4 receptor is ligand specific. J Biol Chem 2007; PMID:18045872.
71. Glaichenhaus N, Shastri N, Littman DR et al. Requirement for association of p56lck with CD4 in antigen-specific signal transduction in T-cells. Cell 1991; 64:511-520.
72. Yang H, Shen F, Herenyiova M et al. Phospholipase C (EC 3.1.4.11): a malignancy linked signal transduction enzyme. Anticancer Res 1998; 18:1399-1404.
73. Bae YS, Cantley LG, Chen CS et al. Activation of phospholipase C-gamma by phosphatidylinotisol 3,4,5-trisphosphate. J Biol Chem 1998; 273:4465-4469.
74. Horejsi V, Zhang W, Schraven B. Transmembrane adaptor proteins: organizers of immunoreceptor signalling. Nat Rev Immunol 2004; 4(8):603-616.
75. Williams BL, Irvin BJ, Sutor SL et al. Phosphorylation of Tyr319 in ZAP-70 is required for T-cell antigen receptor-dependent phospholipase C-gamma1 and raas activation. EMBO J 1999; 18:1832-1844.
76. Paveglio SA, Allard J, Mayette J et al. The tick salivary protein, salp15, inhibits the development of experimental asthma. J Immunol 2007; 178:7064-7071.

Chapter 11

The Serpin Saga; Development of a New Class of Virus Derived Anti-Inflammatory Protein Immunotherapeutics

Alexandra Lucas,* Liying Liu, Erbin Dai, Ilze Bot, Kasinath Viswanathan, Ganesh Munuswamy-Ramunujam, Jennifer A. Davids, Mee Y. Bartee, Jakob Richardson, Alexander Christov, Hao Wang, Colin Macaulay, Mark Poznansky, Robert Zhong, Leslie Miller, Erik Biessen, Mary Richardson, Collin Sullivan, Richard Moyer, Mark Hatton, David A. Lomas and Grant McFadden

Wisdom in infinite thought,
Expanding time, endlessly sought,
An ancient star journeying here
So far, so near, bright seer
This unseen dance
A brilliant chance
—Anonymous

Abstract

Serine proteinase inhibitors, also called serpins, are an ancient grouping of proteins found in primitive organisms from bacteria, protozoa and horseshoe crabs and thus likely present at the time of the dinosaurs, up to all mammals living today. The innate or inflammatory immune system is also an ancient metazoan regulatory system, providing the first line of defense against infection or injury. The innate inflammatory defense response evolved long before acquired, antibody dependent immunity. Viruses have developed highly effective stratagems that undermine and block a wide variety of host inflammatory and immune responses. Some of the most potent of these immune modifying strategies utilize serpins that have also been developed over millions of years, including the hijacking by some viruses for defense against host immune attacks. Serpins represent up to 2-10 percent of circulating plasma proteins, regulating actions as wide ranging as thrombosis, inflammation, blood pressure control and even hormone transport. Targeting serpin-regulated immune or inflammatory pathways makes evolutionary sense for viral defense and many of these virus-derived inhibitory proteins have proven to be highly effective,

*Corresponding Author: Alexandra Lucas—University of Florida, 1600 SW Archer Road, PO Box 100277, Gainesville, FL 32610-0277, USA.
Email: alexandra.lucas@medicine.ufl.edu, alexluc1@ufl.edu

Pathogen-Derived Immunomodulatory Molecules, edited by Padraic G. Fallon.

working at very low concentrations—even down to the femptomolar to picomolar range. We are studying these viral anti-inflammatory proteins as a new class of immunomodulatory therapeutic agents derived from their native viral source. One such viral serpin, Serp-1 is now in clinical trial (conducted by VIRON Therapeutics, Inc.) for acute unstable coronary syndromes (unstable angina and small heart attacks), representing a 'first in class' therapeutic study. Several other viral serpins are also currently under investigation as anti-inflammatory or anti-immune therapeutics. This chapter describes these original studies and the ongoing analysis of viral serpins as a new class of virus-derived immunotherapeutic.

Innate Immunity

Many investigators have studied the antigen-dependent, antibody-mediated immune response, which is only found in vertebrates. Over a century ago, however, Ilya Ilyich Mechnikov described a more ancient and yet extraordinarily powerful immune response, known as the innate immune system. Mechnikov studied the response of a transparent starfish (bipinanaria) to wood splinters and recorded the early massing of cells around these splinters inside this organism. Mechnikov thus provided the first description of the cell-based innate immune response that forms the first defense response to injury or infection.[1] This 'inflammatory', cell-based immune system recognizes and then eradicates or blocks pathogen and parasite infection, invasion and dissemination long before antibodies are formed and the acquired, antibody dependent immune response is activated.[2-6] The innate immune response also orchestrates the first stages of tissue repair after other forms of injury produced by physical or chemical insults.

The vascular endothelium, together with the circulating inflammatory blood cells, monocytes/macrophages, T-lymphocytes and polymorphonuclear leukocytes (also called neutrophils), recognize patterns of microbial molecular expression through pattern recognition receptors (PRR) forming the prelude to this innate response. The PRRs now recognized include toll-like receptors (TLRs), nucleotide binding and oligomerization domain-like receptors (NLRs), C-type lectin-like receptors (CLRs), cytoplasmic double stranded RNA (dsRNA) helicase-like receptors and cytoplasmic dsDNA receptors.[7,8] These receptors comprise an alarm system that alerts inflammatory cells to danger or infection, signaling through MyD88, NFκB and MAPK signal-transduction pathways. The endothelial cell layer is the innermost layer of cells in the arterial tree and is composed of miles of interconnected cells, a living carpet of cells that encompasses the vasculature, the cardiac valves and the inner chambers of the heart. This endothelial cell layer is in constant contact with the circulating blood. Injury or infection of the endothelium causes loss and/or activation of endothelial cells with increasing expression of selectins on the activated cells. These selectins, when expressed, slow down circulating leukocytes that pass by in the blood stream. Once slowed, mononuclear cells (leukocytes composed of neutrophils, monocytes and lymphocytes) can then recognize cell adhesion molecules and adhere to the endothelium and in turn become activated.[9] Circulating and activated inflammatory cells also can recognize connective tissue and lipids exposed under areas of damaged endothelium. The activated endothelium expresses increased amounts of selectins and adhesion molecules that further stimulate cell adherence and activation. Once activated, inflammatory mononuclear cells, together with endothelial cells, begin to release chemoattractant proteins, particularly chemokines, which bind to surface glycosaminoglycans (GAGs). Also induced are pro-inflammatory immune signaling molecules, cytokines and growth factors that signal cells to migrate through the vessel wall and into the surrounding tissue, become further activated, proliferate and then release more inflammatory cytokines. Damaged cells also can become apoptotic and act as small cytokine release factories, further stimulating this inflammatory response.

Platelets, small clotting cell fragments derived from megakaryocytes, are also activated at sites of damaged or apoptotic endothelium. Platelets carry reserves of proteins in storage granules that are released upon platelet activation. Initially, platelets adhere to areas of arterial or tissue damage, secreting pro-inflammatory proteins from storage granules into this mix of cells and proteins. This activated and inflamed milieu then further stimulates cell invasion and activation. The clot forming (thrombotic) and clot dissolving (thrombolytic) cascades are made up of sequentially activated serine protease enzymes. These protease cascades are initiated by intrinsic and extrinsic clotting factors (factor VII and tissue factor complex and thrombin) that cleave fibrinogen to form fibrin (Fig. 1). Thrombosis and fibrin formation occurs on the surface of platelets and the fibrin mesh that is deposited on and around the activated platelets, endothelial cells and leukocytes. Clot deposition will then stimulate fibrinolysis, the serine protease cascade that acts to break up the forming clot and maintain a natural homeostasis in the arterial wall, the balance between the clot forming (coagulation) and clot dissolving (fibrinolytic) pathways. When cells become apoptotic and engage in cell suicide, there is again a change in activation and in some cases, such as apoptotic endothelial cells and monocytes/macrophages, there is an increased release of cytokines and clot activating serine proteases into the local tissues which leads to a crescendo in the inflammatory responses.[10,11] Extrinsic apoptotic pathways can also be driven by serine proteases, particularly granzyme B.

At one further level of complexity, the connective tissue components, specifically collagen, elastin and glycosaminoglycans (GAGs) have many roles only now being uncovered. Connective tissues surrounding cells provide storage sites for release of growth factors and cytokines. Thus breakdown of these cellular-embedding and tensile-building materials can lead to a local increase in released inflammatory factors. The connective tissue layers also form adhesive platforms on which chemokines and cells adhere and through which cells migrate and invade tissues. Additionally GAGs directly regulate serine protease and serpin activation. The serpin antithrombin III is well known to have a 1000 fold or greater increase in activity when exposed to infusions of the GAG heparin, a drug commonly used in vascular patients.[12,13] When cells lose adhesion to connective tissue or basement membranes, an apoptotic state can be induced further activating these pathways. Serine proteases activate the proteases then that break down collagen, elastin and GAGs, the matrix metalloproteinases (MMPs). The fibrinolytic serine proteases, tissue- and urokinase-type plasminogen activators (tPA and uPA, respectively), plasmin and the thrombotic protease thrombin also directly cleave collagen and elastin. Thus the serine protease pathways in the coagulation and fibrinolytic pathways interact on many levels with the inflammatory and apoptotic responses.[11,12]

In summary, inflammatory cells and activating factors perform in concert to initiate host innate immune/inflammatory responses that heal sites of infection and injury but can also damage tissues when present in excess. Serine protease pathways are regulated by serpins, and many viruses have acquired and developed their own serpins over many millions of years of evolution that are designed to modulate host immune responses. These viral serpins probably target a variety of innate sensors, particularly PRR signals such as the extracellular TLRs and intracellular NLRs that trigger innate immune responses.

Serine Protease Inhibitors/Serpins

Serpin Structure and Function

The symphony of interactive responses between cells, cytokines, serine proteases, connective tissue and growth factors acts as a composite, and it is unclear if there is a single, controlling conductor for the inflammatory pathways. Several of these pathways, however, are known to be regulated by serpins (serine protease inhibitors), guiding these factors to play in concert. The term serpin was first introduced by Carrell and Travis in 1985[14] and describes a family of proteins with up to 30% sequence identity (ranging up to 70% when limited to hydrophobic sequences) that is believed to have arisen from countless gene duplications of an ancestral gene. Serpins are large complex proteins that exist in strained, latent and cleaved (inactive) conditions (Fig. 1A). The serpin basic protein secondary structure is fairly well conserved consisting of a 350 amino acid core with of 3 β-sheets (A, B and C) and 7 to 9 α-helices, labeled hA-hI.[15] The reactive site loop and the β-sheet A are labile and have key roles in serpin inhibitory function. The reactive site loop (RSL) sits exposed above the serpin folding framework, thus presenting the P1-P1′ site as bait. The β-sheet opens to incorporate the RSL loop after cleavage by a target protease.[15] Of course, each of the serpins has a variable inhibitory activity for a range of proteases and thus can

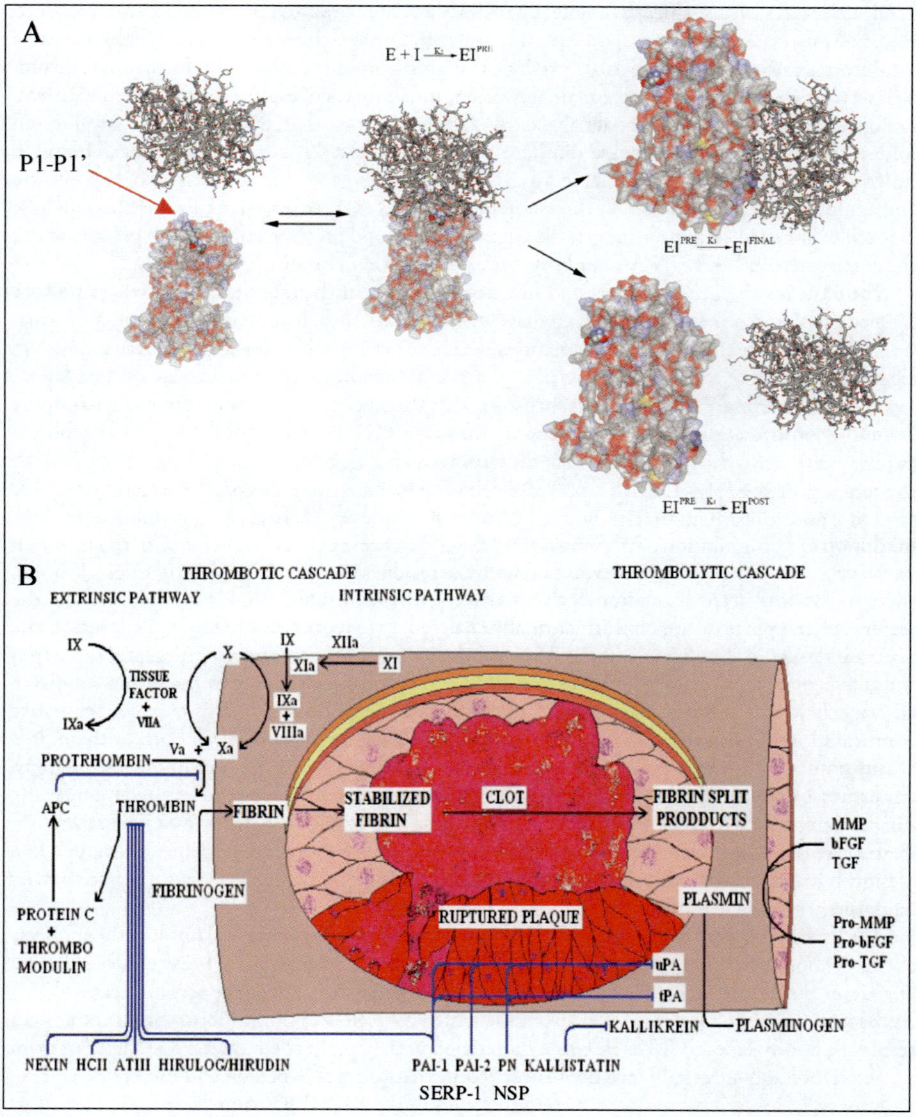

Figure 1. A) Serpin binding to target proteases occurs via interaction of the protease with the RSL P1-P1′ scissile bond. This interaction can result in either (1) a form of suicide inhibition wherein the serpin RSL is cleaved, but the protease remains bound to the serpin and is dragged across the face of the serpin to remain stuck to the opposite pole of the serpin (top serpin/protease interactive pathway) or (2) the serpin RSL is cleaved and the serpin rendered inactive (the bottom serpin/protease interactive pathway). B) Diagram of thrombotic and thrombolytic pathways as well as potential targeted pathways for the mammalian, PAI-1 and viral, Serp-1, serpins in an injured arterial wall.

be a true inhibitor for selected proteases or can be a target for protease cleavage (Fig. 1A). The earliest evolved serpins are found in simple arachaea and bacteria. Other serpin-related proteins are found in the more complex horseshoe crab, unchanged since the time of the dinosaurs are still used for clot lysis assays.[16] Serpins represent a large proportion of circulating proteins in the blood of all vertebrates, with estimates ranging up to 2-10% of human serum proteins. Serpins have been classified into clades A-I based upon structure and function and these complex and highly effective inhibitors are now known to regulate processes ranging from thrombosis (clot formation), thrombolysis (clot dissolution), complement activation, inflammatory responses, sperm development, to hormone transport.[17,18] There are also serpin family members that, while having classical serpin folding structure, lack functional serpin-like inhibitory activity.[17] These serpins have been found to function in a noninhibitory capacity in diverse roles such as hormone transport, tumor suppression and as molecular chaperones[15,17,18] Three notable examples of these serpin family members include chicken ovalbumin, angiotensinogen the angiotensin (blood pressure controlling) precursor and the corticosteroid (SERPIN A6) and thyroxin (SERPIN A7) binding globulin.[15,18]

The basis for the inhibitory activity of true serpins lies in the fact that the active, inhibitory molecules exist in a metastable intermediate state, instead of their more stable latent conformation.[19] Of the more than 80 crystal structures that have been resolved for serpins to date, there is a clear indication that the inhibitory serpins prefer a metastable native state instead of their latent, lower energy folding.[15] Although it is surprising that the serpins avoid a more thermodynamically favorable conformation, this state is necessary for the inhibitory activity of the serpin. The folding pathway that yields the most stable state for serpins can be triggered by the cleavage of the RSL by the target protease (Fig. 1A). This protease-response mechanism relies on the serpin metastable structure and conformational mobility.[20] Within this pathway, there are two possible outcomes, inhibition or noninhibition of the substrate protease.[21] However, they both begin with the approach of the protease to the RSL and interaction with the residues surrounding the P1-P1′ scissile bond, thereby resulting in the formation of a noncovalent Michaelis-like complex in which neither the serpin nor the protease are conformationally changed.[15,22] Hydrolysis of the scissile bond by the protease results in the cleavage of the RSL and the formation of an acyl intermediate; the serpin thus efficiently traps the targeted protease and drags it across to the face of the serpin following cleavage at its P1′ position with the aid of the covalent acyl bond.[15,20] This changes the native conformation of both the serpin and the protease, causing loss of activity in both of them. It is at this point that the two potential pathways diverge. In the case of the noninhibitory pathway, the protease is able to deacylate the acyl intermediate complex before loop insertion and trapping can occur, resulting in the release of the active protease along with the inactive cleaved serpin.[15,23] In the case of the inhibitory pathway, both the serpin and the protease remaines entrapped in a 1:1 inhibitory complex exhibiting an extremely long half-life that effectively removes both parties from further biological roles.[15]

The RSL is also a key factor in serpin function consisting of a region of ~20 residues that projects above the body of the serpin, presenting the P-P1′ bond as a protease bait.[24] Any conformational change in the overall structure of the RSL will also significantly affect the serpin's activity. This has been observed in the case of antithrombin, which by itself is a poor inhibitor of thrombin, as a result of a poorly exposed RSL.[18,25] Upon interaction with heparin cofactor, the RSL flips out from its native partially inserted (into beta-sheet) confirmation thus exposing and greatly enhancing its activity.[26] Similarly alteration of the amino acids in the P1-P1′ position also results in the loss of activity as observed in the case of viral serpin Serp-1 mutant where replacing the P1-P1′ R-N sequence with A-A results in a total loss of serpin protease inhibitory activity and anti-inflammatory activity.[27,28] Interestingly, mutation of the Serp-1 P2-P7 arm of the RSL to a series of Ala (A) residues in the Serp-1 (Ala_6) mutant, leads to a complete reversal of anti-inflammatory actions and instead creates a now highly pro-inflammatory and pro-thrombotic protein.[27]

As noted above, the structural scaffold of the serpin has some advantages, but the thermodynamically unfavorable state means the active serpin can transition to spontaneous conformational changes as observed for PAI-1.[24] The large energetic barrier between the serpin's native state and its latent state prevents the inactive protein from spontaneously attaining this activated state.[24] This makes serpins less forgiving to mutations which can cause them to fold into their more stable, noninhibitory or latent state, in which the RSL is inserted into β-sheet A.[29] Such mutations can lead to disorders known as serpinopathies that are characterized by the misfolding of the serpins and formation of serpin oligomers, i.e., polymeric forms of the protein.[13] The polymers are formed by insertion of the RSL of one serpin into the β-sheet of another serpin, known as the "trans" or foreign insertion.[24,29] Thus, misfolding of serpins can lead to an inactive conformation of the serpin whose thermodynamic stability is comparable with that of the cleaved serpin.

Thrombotic and Thrombolytic Serpins

At sites of injury to any tissue in the body, damaged blood vessels respond to limit blood loss through vasoconstriction and clot formation (thrombosis). With endothelial cell damage there is a loss of anticoagulant signals and the exposed inner connective tissue layers of arterial and venous walls stimulate platelet activation. These activated platelets form a surface on which serine proteases in the clot thrombotic pathways create a clot. Such clots are driven by factor VII, factor IX and tissue factor complexes in the extrinsic cascade and factors IX and VIII in the intrinsic cascade all of which lead to the activation of factor X and thrombin formation (Fig. 1B). Thrombin in turn activates fibrinogen to form fibrin and factor XIII leads to cross linking to form a fibrin and platelet mesh. This clot forms when there is damage, whether caused by outside physical trauma or internal damage such as high cholesterol, diabetes or smoking. The sudden formation of a thrombus on the damaged inner arterial surface occludes the arterial lumen and blocks blood flow. This causes heart attacks (myocardial infarctions) and strokes (cerebrovascular accidents) and/or gangrene with peripheral vascular occlusions. Antithrombin III (AT III, SERPIN C1) and heparin cofactor II (HC II, SERPIN D1) are the main inhibitors of the thrombotic protease pathway.

The thrombolytic pathway is best known for its role in clot breakdown or dissolution. 'Clot busting' or thrombolytic therapies are used to treat acute heart attacks and strokes where thrombosis in coronary and cerebrovascular arteries occludes blood flow to the heart or brain. These thrombolytic agents include streptokinase and tPA or uPA (urokinase). The clot dissolving drugs are mammalian or bacterium derived serine proteases, plasminogen activators that cleave the pro-form of plasminogen to form active plasmin. The clot dissolving thrombolytic cascade also has a central role in acute inflammation. tPA, uPA, plasmin, the uPA receptor (uPAR) and the mammalian serpin that inhibits these serine proteases, plasminogen activator inhibitor-1 (PAI-1) are all up-regulated at sites of tissue injury (Fig. 1B). Thus the thrombolytic pathway is part of an acute phase, inflammatory response to tissue injury. The uPA/uPAR complex also regulates this pathway, as described by Blasi.[30] Cellular invasion, whether inflammatory cells responding to tissue damage or invasive tumor cells (Fig. 1B), can be initiated by the uPA/uPAR complex. S Collen and P Carmeliet's groups have utilized mouse genetic 'knock out' models to demonstrate the key roles of uPA, tPA, uPAR and PAI-1 in arterial responses to injury in vascular disease.[31]

tPA and uPA have chemoattractant activities attracting cells into areas of damage. The uPA/uPAR complex sits at the leading edge of invading cells where uPA activates plasminogen to form plasmin, and the plasminogen activators together with plasmin also activate matrix degrading pro-enzymes, the pro form of matrix metalloproteinases (pro-MMPs), to form active MMPs that degrade local connective tissue, collagen and elastin. These actions are believed to allow cells to invade damaged or infected tissues by creating a path in the connective tissue layers through which cells migrate. The plasminogen activators (PAs) also activate growth factors and can release growth factors from connective tissue stores.

Native mammalian serpins regulate these pathways, with PAI-1 (SERPIN E1) functioning as the major serpin regulator for the tPA and uPA pathways. Other serpins that target and regulate this system include PAI-2 (SERPIN B2), PAI-3 (SERPIN A5), α-2 antiplasmin (SERPIN F2), protease nexin-1 (PN-1, SERPIN E2) and neuroserpin (SERPIN I1). These serpins exhibit differing degrees of inhibitory activity where PAI-1 has a 20 fold faster interaction with tPA than neuroserpin, whereas neuroserpin has a 20 fold faster inhibition rate than protease PN-1.[32,33] PAI-1 can also alter its regulatory patterns in the presence of vitronectin to become a stronger inhibitor of thrombin. PAI-1 is up-regulated increased amounts in inflammatory disorders such as unstable arterial plaque (unstable angina and impending heart attacks). PAI-1 forms a tripartite complex with uPA and uPAR which is then internalized, effectively blocking uPA/uPAR complex activity.

PAI-1 can also act on circulating serine proteases and under varied conditions, as with the presence of the GAG heparan sulfate, PAI-1 can inhibit thrombin. uPAR is a GPI linked nontransmembrane protein and relies upon a large array of associated membrane proteins that exist associated with other proteins in a large and complex lipid raft. Proteins associated with uPAR in the lipid raft complex include integrins, lipoprotein related proteins (LRP or alpha 2 macroglobulin), chemokine receptors, as well as many other proteins that can provide relays for signaling into the cell. The uPA/uPAR complex has been reported to modify cell signals through the intracellular signaling pathways (Fig. 1B).

PAI-1 binding blocks uPA/uPAR mediated actions, and the inhibition or knock out of PAI-1 has been shown to reduce plaque growth in animal models of vascular injury.[33,34] Similarly, excessive expression of uPA in the rabbit carotid after angioplasty led to increased plaque growth.[35] This observation has however been variable, with PAI-1 demonstrating exacerbation of vascular plaque growth in other models.[36] Similar variability has been seen in studies assessing PAI-1 up-regulation. For example, one study reported that elevated PAI-1 levels were associated with reduced restenosis after angioplasty[37] whereas many other reports have associated elevated PAI-1 with increased inflammation and risk of vascular disease.[38] Neuroserpin is up-regulated at sites of cerebral injury and is reported to reduce cerebral ischemic scarring in mouse stroke models. In preliminary work our lab has detected reductions in plaque growth in rodent models after neuroserpin infusions.[39] Thus, overall, native mammalian serpins have had varied and sometimes contradictory effects in both animal models and in patient clinical trials on inflammatory responses and vascular disease.

Serine Proteases and Serpins in Apoptotic Pathways

Apoptosis, or cellular suicide, can be activated by proteases. Aspartate specific cysteine proteases, named caspases, are considered the predominate mediators of classical cellular apoptosis (Fig. 2). The proteolytic cleavage and activation of a series of cysteine proteases is considered a key pathway to apoptosis, wherein caspases 8, 9 and 10 are believed to initiate the activities of caspases 3, 6, 9 that execute the apoptotic command through cleavage of a large number of cellular substrate proteins.[40] Extrinsic apoptotic pathways are mediated via death receptors (e.g., TNF and Fas receptors) and activate caspase 8 and 3, while activators, such as many anti-cancer drugs, trigger intrinsic mitochondrial pathways that induce apoptosis through Apaf-1 and recruitment of caspase 9 (Fig. 2). However, other diverse proteases such as the cysteine proteases calpain and cathepsin B, the threonine protease of the proteosome and other serine proteases have more recently also been linked to cell death. Cellular serine proteases that can initiate apoptotic responses include granule enzymes (granzymes) that interact with intracellular caspase pathways and others that associate with the bcl-2 pathways.[41] Granzyme B is a serine protease secreted by Cytotoxic T Lymphocytes (CTLs) or Natural Killer (NK) cells. Granzyme B is reported to have the strongest apoptotic activity of all the known granzymes. Granzyme B enters cells through pores formed on the cell membrane caused by perforin, which is also secreted by activated CTLs and NKs, and initiates apoptosis through direct cleavage and activation of caspases and also cleavage of the anti-apoptotic Bid. The cleaved form of Bid inserts into the outer mitochondrial membrane to induce the intrinsic apoptotic/death pathway. Granzyme has also been reported to enter cells through pathways separate from perforin pore structures. Another serine protease in the thrombotic pathway, thrombin, also activates apoptosis.[42] Once inside the cell, granzyme B cleaves the "pro" form of caspase 3 which then activates the apoptotic machinery (Fig. 2). The caspase enzymes are proven activators of the inflammasome complex through activation of the pro-forms of the interleukin-1β and IL18 by interleukin converting enzyme (ICE, also named caspase 1). Caspase 1 (ICE) drives inflammation through the activity of a cellular structure named the inflammasome.[43,44] Other researchers have reported direct-protease driven inhibition of poly-ADP-ribose polymerase (PARP) and iCAD that can act to protect cells against DNA damage and apoptosis.

The first apoptotic inhibitor identified was a viral cross-class cysteine and serine protease inhibitor isolated from Cowpox virus, called Cytokine response modifier A (CrmA) or Serine protease inhibitor-2 (Spi-2).[45] CrmA will be described in greater detail in a later section of this

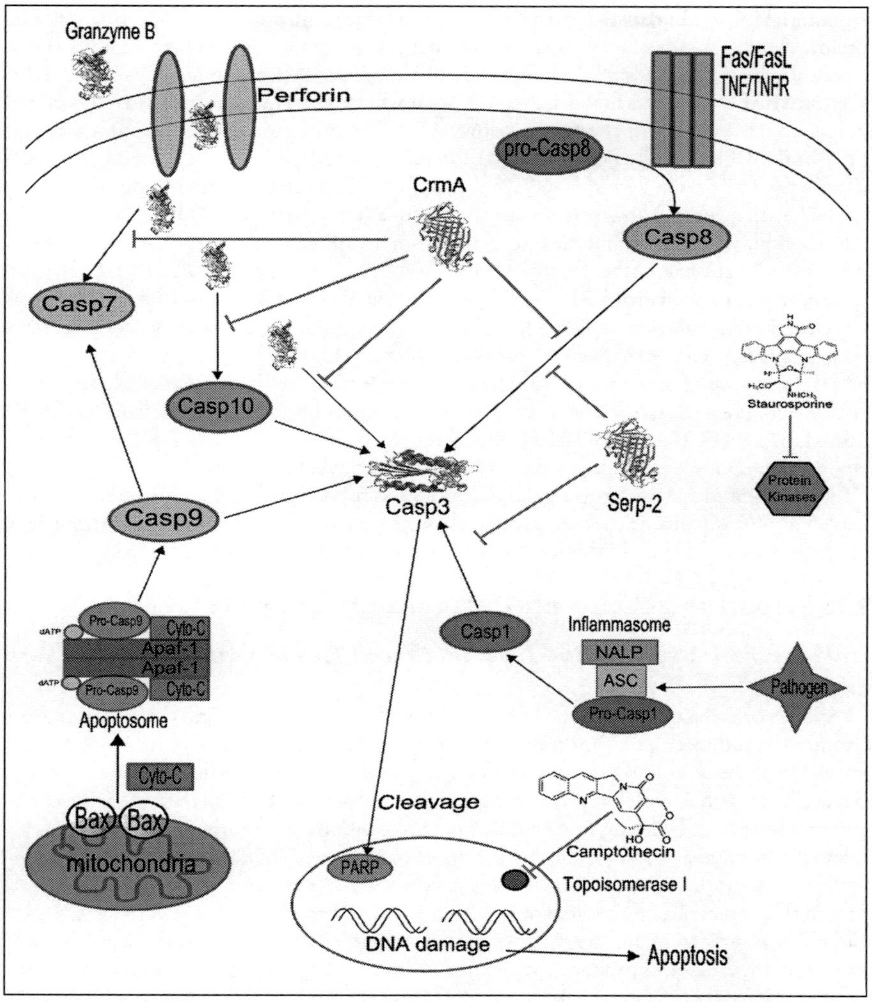

Figure 2. Apoptotic pathways-serine and cysteine proteases. Intrinsic and extrinsic Fas and granzyme B apoptotic pathways that are potentially targeted by Serp-2 and Crm A are illustrated. Staurosporine activates mitochondrial apoptotic pathways, Fas ligand and Granzyme B activate extrinsic apoptotic pathways; Camptothecin inhibits topoisomerase and blocks DNA repair.

chapter. Several mammalian serpins have more recently been identified as playing a part in the regulation of apoptotic responses.[46] CTLs also synthesize their own serpin inhibitors of granzyme B. Among these serpins are Protease Inhibitor-9 (PI-9, SERPIN B9), Plasminogen Activator inhibitor-2 (PAI-2) and maspin. Each of these cellular serpins has proven unusual, having both known extracellular actions as well as in many cases intracellular and even nuclear localization and function. PI-9 is a mammalian serpin that binds to and inhibits the actions of granzyme B effectively blocking cellular apoptosis mediated by Granzyme B.[47] PAI-2 preferentially binds uPA but is also localized in intracellular compartments and is poorly secreted. PAI-2 has been shown to alter apoptotic pathways and has been reported to have similarity in structure to Bcl-2 with potential for overlapping functions in this cell suicide pathway.[48] This potential overlap in function with Bcl-2 has potential impact in that Bcl-2 itself represents a larger family of anti-apoptotic intracellular proteins. Neuroserpin is a mammalian serpin that preferentially binds to and inhibits tPA and has now also been reported to alter apoptotic cell responses.[49] Maspin (mammary serine protease inhibitor, SERPIN B5) was first discovered as a breast cancer suppressor and also has both intracellular and extracellular functions by binding widely divergent proteins that include uPA, uPAR, HSp79 and 90, glutathione peroxidase, β-1 integrin and collagens I and III, among others.[50] In addition, the trypsin-like serpin PI-6 (SPI 3, SERPIN B6) in mouse brain can bind granzyme B and block cathepsin and downstream caspase 7 activation in neurons.[51-53] PI-6 also incidentally binds thrombotic and thrombolytic proteases, thrombin, uPA and plasmin. Three serpins that are best known as regulators of the thrombotic and thombolytic pathways, AT-III, PAI-1 and protein C inhibitor (PCI also named PAI-3, SERPIN A5), have also been found to alter apoptosis in human cells.[54] Nitric oxide nitrosylation of α-1 anti-trypsin (AAT) serpin which targets the trypsin protease modifies AAT such that it becomes an inhibitor of cysteine proteases.[55] Serpin activity is blocked through specific serine protease cleavage, nonspecific metalloproteinase (MMP) cleavage, oxidation and polymerization. Some of these cleaved serpins have additional activities as has been reported for AT-III. The precise mechanisms through which some of these serpin-driven changes in apoptotic responses are under investigation and are not yet fully defined.

These serine and cysteine protease cellular pathways are both preferentially targeted by virus-engineered serpins which are now known to block protease activity in the thrombotic/thrombolytic and apoptotic pathways and exhibit amazingly potent anti-inflammatory activities.

Viral Serpins and Their Anti-Inflammatory Activities

Viral Serpins That Target the Thrombotic and Thrombolytic Pathways; Serp-1 and Spi3

Myxoma virus is a member of the leporipoxvirus family of poxviruses that infects only rabbits. Myxoma virus induces a lethal infection in European rabbits with over 99% mortality.[56] The initial observation of the profound pathogenicity of this virus in rabbits was made in the late 1800s by Dr G Sanarelli in South America who had imported the European rabbit to his lab where they became inadvertently infected with myxoma virus.[56] Infection with myxoma virus in the European rabbit was lethal, causing a rapidly disseminated infection, immune dysfunction and overwhelming sepsis with over 99% mortality in the animals (European rabbits). In the early 1990s the McFadden lab reported that targeted genetic knockout of the myxoma virus Serp-1 gene resulted in a virus that could only cause a benign infection in normally susceptible European rabbits.[57] This Serp-1 gene knockout of myxoma virus produced only mild local dermal lesions that were eradicated within 1-2 weeks, similar to what is observed in rabbit infections with the closely related Shope fibroma virus, which has a naturally inactivated Serp-1 gene and also produces a benign, self-limiting infection in rabbits. Each of these infections, rabbits infected with wild type myxoma virus, myxoma virus with engineered Serp-1 knockout, or the Shope fibroma variant, differed greatly in lesion pathology. The Serp-1 knockout myxoma virus and the natural variant Shope fibroma virus, exhibit a more effective host inflammatory response to viral infection, while with the wild type myxoma infection, inflammatory cells do not properly migrate to tissue sites of viral infection.

In later studies, McFadden and Lomas demonstrated that the myxoma virus-encoded Serp-1 has sequence similarity to serpins, and the protein inhibits tPA, uPA and plasmin in the thrombolytic pathways as well as factor Xa in the thrombotic pathway.[58] However, despite the marked effects of Serp-1 on viral pathogenesis, the K_{ass} for Serp-1 is a lower affinity reaction at $7\text{-}8.6 \times 10^4$ $M^{-1}s^{-1}$, whereas the mammalian serpin PAI-1 has higher affinity and activity with K_{ass} on the order of $1.1\text{-}2.3 \times 10^7$ $M^{-1}s^{-1}$.[59] Serp-1 also binds human thrombin but acts as a protease target and is cleaved by thrombin under physiological conditions.[60] In the presence of heparin, however, Serp-1 becomes a more potent thrombin inhibitor with potential to become more anti-thrombotic than anti-thrombolytic in character.[61]

Spi-3 is a related but distinct viral protein expressed by orthopoxviruses like vaccinia virus and rabbitpox, which unfortunately has a confusing nomenclature with similarity to the unrelated PI6/SPI3 mammalian serpin nomenclature. The poxviral Spi-3 binds to and inhibits uPA and tPA and plasmin.[62,63] Spi-3 is also capable of forming weaker complexes with thrombin and factor Xa as measured by gel shift assays.[63] Spi-3 from vaccinia/rabbitpox and Serp-1 from myxoma virus share only 30% sequence similarity despite targeting similar host protease pathways. The K_i's for Spi-3 was measured as 0.51, 1.9 and 0.64 nM for uPA, tPa and plasmin, respectively. The K_i's for Serp-1 were similar at 0.16, 0.14 and 0.44 nM, respectively for uPa, tPa and plasmin.[63] However, whereas Serp-1 is secreted into the surrounding environment, Spi-3 remains tethered to the cell surface and exhibits a secondary function in the inhibition of cell fusion that is independent of its serpin-based activities.

Preclinical Analysis of Serp-1

Pilot Studies—Rabbit Model

Our founding hypothesis was that native, virus-derived, immunomodulatory proteins can be developed as a new class of protein therapeutics to treat inflammatory-based diseases. Rather than using a live virus or a viral vector in animal models or in patients and thus risk reactions to the viral construct itself, we followed the lessons of natural evolution and the selection pressures exerted by the immune system. Large DNA viruses, like many parasites, have evolved highly effective defenses against the host and inflammatory systems and thus such viruses have already accomplished the necessary research and development that created these viral immune evasion proteins. In other terms, rather than relying on man's imperfect knowledge and understanding of the inflammatory and immune responses, we utilized viral proteins already discovered, engineered, developed and proven effective as immune modifying agents and used these naturally-derived reagents in the form of the expressed and purified protein. We directed our initial studies to express secreted proteins encoded and engineered by complex poxvirus DNA genomes. Indeed, native viral immunomodulatory proteins that are often highly potent and naturally function at very low concentrations to divert the host immune response away from the invading viral organisms.

For the first preclinical animal studies, the viral serpin, Serp-1 protein, was expressed from a vaccinia virus expression system and the secreted serpin protein was purified on FPLC columns (J Macen, McFadden lab). Using our first generation purification strategy, only submicrogram (picogram to nanogram) quantities of purified protein were isolated, and it is these early preparations of secreted Serp-1 that were used for the first studies in 74 cholesterol fed New Zealand white rabbits. It was reasoned that the intact virus only expressed and secreted very low levels (femtomolar amounts) of Serp-1 protein into the surrounding tissues and thus the viral serpins might be capable of anti-inflammatory action even at very low dosages. Lower concentrations of therapeutic protein would also reduce the risk of inducing antibody and immune responses as well.

As this serpin was derived from a rabbit virus, the first disease model we tested was a rabbit angioplasty injury model, where the timing of initial injury and inflammatory response activation would be known, e.g., initiated with specific angioplasty injury. In the first studies, one bolus of 30-3000 picograms of Serp-1 protein per rabbit was then infused locally at sites of balloon injury in studies performed by LY Liu and E Dai in the Lucas lab.[64] The Serp-1 protein was infused locally using a Wolinsky catheter, which is an angioplasty balloon catheter with small perforations on the balloon that allowed the protein to be sprayed on the arterial surface while inflating the balloon (Fig. 3A). Later analyses with Evans blue dye delivered with the same device demonstrated that the majority of this presumed local infusion was in fact delivered in both trans-arterial and systemic fashion, with 90% of the delivered dose spilling into the blood stream. Efficacy of intravenous and intra-arterial infusion of 0.3 to 300 ng of Serp-1 protein were then tested, again demonstrating effective inhibition of plaque growth at 4 weeks follow-up after a single bolus injection given at the time of balloon angioplasty.[64]

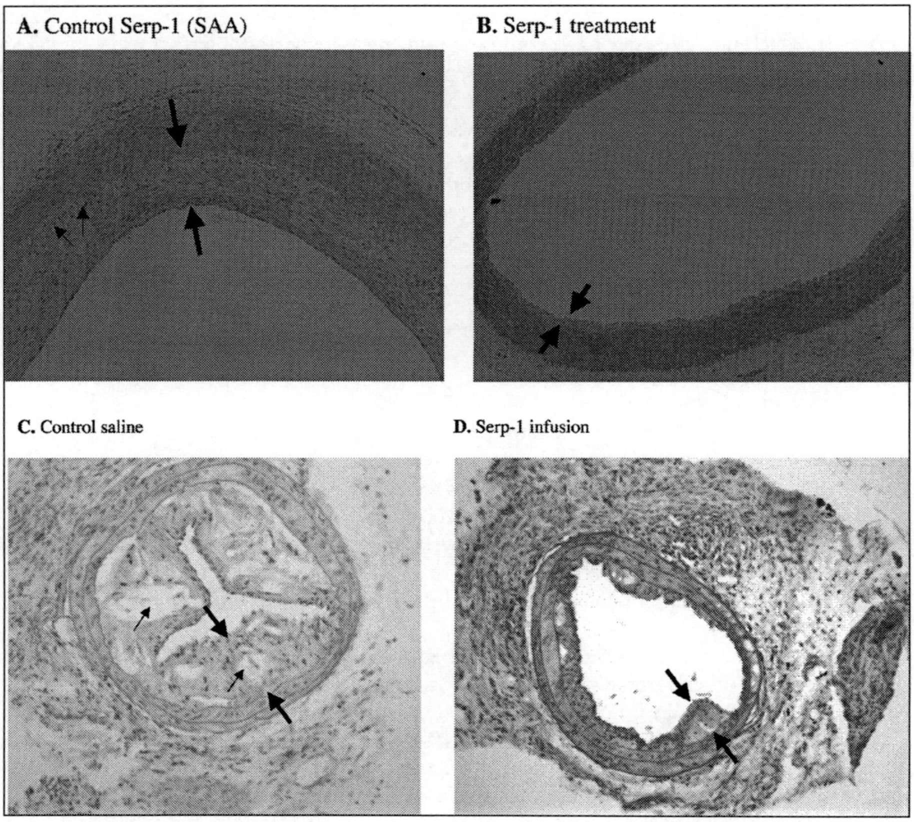

Figure 3E-F. In the first two panels representative histological cross sections of rabbit arteries taken at 4 weeks follow-up after balloon angioplasty injury and treatment with either inactive control Serp-1 (SAA) control (A) or active wild type Serp-1 (B) demonstrate the marked reduction in plaque area after a single bolus injection with Serp-1 (B) when compared to control (A). In this study rabbits were fed a high cholesterol diet and the drug was given as a single embolus injection immediately after balloon angioplasty. In the second two panels cross sections of ApoE null mouse carotid arterial histology sections are shown at follow-up after treatment with control saline treatment (C) or Serp-1 (D) infused by osmotic pump continuously. Again a marked reduction in plaque area was demonstrated with Serp-1 infusions. Large arrows bracket plaque growth areas in the intimal layer. Small arrows indicate areas of inflammatory cell invasion. Figure 3, continued on following page.

In these initial studies, early (i.e., 24 hrs to 7 days after angioplasty injury) inflammatory cell, macrophage and T-cell invasion was effectively blocked. Atherosclerotic plaque growth was significantly reduced at sites of angioplasty injury at 4 weeks follow-up, following injection of a single dose of Serp-1 protein (Fig. 3B). Smooth muscle cell and B cell invasion were not significantly altered in this model. Injection of a mutated Serp-1 protein, bearing a genetic replacement of the normal P1-P1′ Arg-Asn (R-N) scissile bond with Ala-Ala (A-A), produced an inactive serpin that was no longer able to block either inflammatory cell invasion or plaque growth.[64] In this original study there were no adverse effects observed; specifically no increased bleeding or clotting, infection or sepsis, no delay in wound healing, no increased mortality and overall no side effects. Inhibition of inflammation and plaque development was observed even after a single injection of picogram to nanogram doses of Serp-1 up to 4 weeks follow-up with no further bolus injections.[64] The half

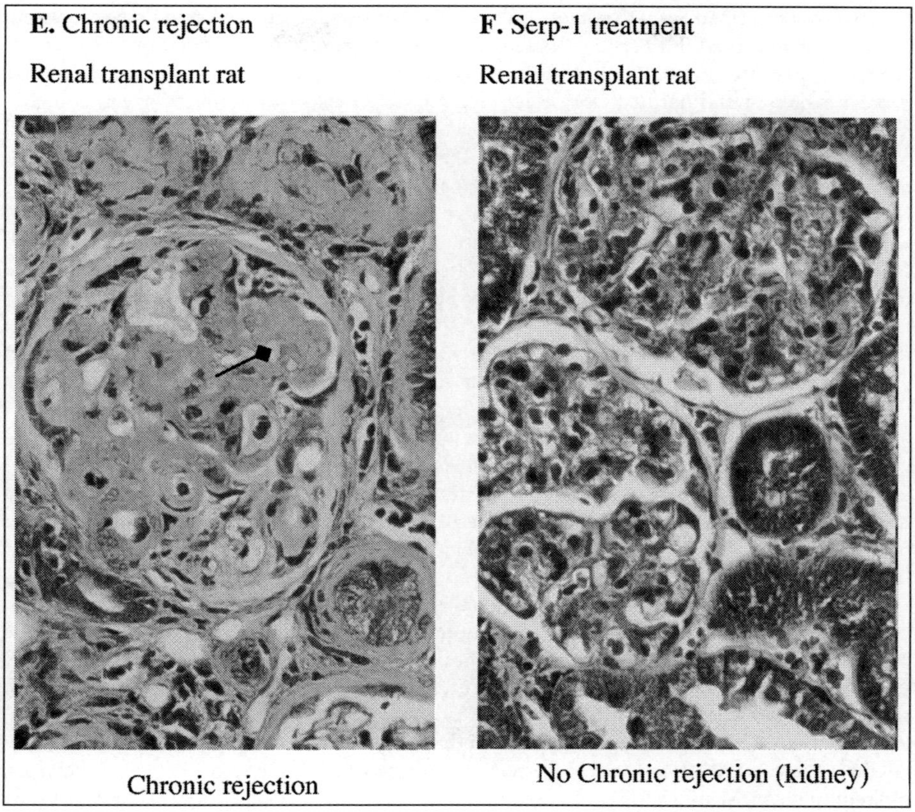

Figure 3, continued. Panels E and F illustrate treatment of rats after allograft renal transplant (F344 to Lewis rat) with either cyclosporine alone (E) or cyclosporine plus Serp-1 (F) at 5 months follow-up. Treatment was given as daily bolus injection i.v. for the first 10 days after renal transplant with no additional treatment. Areas of renal transplant scarring and loss of cellularity are indicated by the marker.

life of Serp-1 in normal rabbits was reported by Dr M Hatton (McMaster U, Hamilton, ON) to be less than 24 hours.[65]

Pre-Clinical Animal Models Studies of Angioplasty Injury

In subsequent work, the efficacy of Serp-1 therapy was tested across species and in differing models to assess the range of action and to confirm the anti-inflammatory and anti-atherogenic actions (blockade of atherosclerotic plaque growth) of Serp-1 in angioplasty injury models (Table 1). Both in the Lucas lab and in a collaborating lab (J Stroney, Case Western Reserve, OH) angioplasty-induced plaque was reduced at 4 weeks follow-up in cholesterol fed Yucatan microswine after peripheral arterial balloon angioplasty injury.[66] Plaque development was also significantly reduced after balloon angioplasty in rat iliofemoral and rooster aortic models (Lucas lab). In each case, early inflammatory mononuclear cell (macrophage and T-cell) invasion was significantly reduced with Serp-1 treatment at the site of injury. In these studies, doses of Serp-1 protein ranged from picogram/kg up to nanogram/kg body weight. Injections were given by either local Wolinsky perforated balloon injection (microswine) or by intravenous bolus (rabbit, rat and rooster), with Serp-1 infused immediately after angioplasty. No further injections were given until the time of follow-up at 4 weeks after angioplasty. While initial work was limited by the amount of purified Serp-1 protein available from the original small protein preps produced, for use later studies a CHO cell expression system was developed in collaboration with Biogen (Leona Ling, Boston, MA USA and Viron Therapeutics, Inc, London, ON, Canada). The Serp-1 protein expressed and purified after secretion from CHO cells was produced at much higher quantities and allowed for higher dose infusions. In the most recent work, Serp-1 protein used for preclinical animal models was expressed in CHO cells and secreted Serp-1 protein was purified under good manufacturing practice (GMP) conditions, in a format suitable for clinical use. In all of the initial studies in rabbits and Yucatan microswine models, the perforated Wolinsky balloon was used for Serp-1 delivery at the site of angioplasty injury, based upon the original rabbit studies.[64] However, later work demonstrated efficacy on an equivalent level with systemic intravenous (i.v.) or intra-arterial (i.a.) systemic blood infusions. Many of these studies assessing the local perforated balloon vs systemic i.v. injection routes proceeded in parallel such that local infusions were still utilized in the swine models while work was ongoing in rabbits demonstrated efficacy when Serp-1 was infused as an i.v. bolus infusion. Once proven effective, animal studies were in general performed using i.v., dosing of Serp-1 protein.

Subsequent to these initial studies, Serp-1 was tested for efficacy in preventing plaque growth after both balloon angioplasty injury and bare metal stent implant.[66] Plaque was reduced after 2, 3 or 11 daily injections of Serp-1 given i.v. and starting immediately after stent implant in cholesterol fed rabbit aorta.[66] However, Serp-1 was not effective when given as a single bolus after stent implant in either cholesterol fed New Zealand white rabbit or Yucatan microswine models.[66] With repeated balloon angioplasty injury in microswine there was a trend toward reduced plaque when Serp-1 was infused after the final angioplasty injury in a series of three angioplasty injuries over 1.5 months, but this trend did not reach significance.[66] In none of these cases was baseline preformed plaque reduced by subsequent Serp-1 treatment. Specifically, only plaque at the site of balloon angioplasty or stent implantation was reduced, but plaque caused by cholesterol feeding outside the range of the vascular injury was not affected by Serp-1 treatment.[66] Thus, with ongoing or recurrent injury, as with repeated balloon angioplasty or with indwelling arterial stent implants one observes less effective reductions in plaque with single i.v., bolus injections. Multiple Serp-1 injections, however, did reduce plaque significantly in the rabbit stent implant model.[66] With repeated Serp-1 dosing (2 up to 11 daily i.v. boluses) starting on the day of aortic stent implant in cholesterol fed New Zealand rabbits, there was effective plaque reduction at 4 weeks follow-up.[66] In each animal model assessed, when early histological specimens were taken, there was evidence for Serp-1 mediated reductions in monocyte and nonspecific T-cell invasion at sites of vascular injury.[66]

Spi-3 Treatment in Balloon Angioplasty Models

Spi-3 protein from Vaccinia virus binds uPA and tPA, with weaker binding to fXa and plasmin as described in preceding sections. Spi- 3 was also tested in parallel with Serp-1 in a limited study after balloon angioplasty injury in rat models. Spi-3, as for Serp-1, reduced plaque growth significantly when a single injection was given i.v. after balloon injury (A. Lucas and R. Moyer, unpublished observations).

Preclinical Animal Models Study of Inflamed Plaque in ApoE^{null} Mouse Carotid Cuff Injury

The capacity of Serp-1 to reduce plaque inflammation and growth after carotid cuff compression in hyperlipidemic ApoE^{null} mouse models was also assessed. In this study, Serp-1 was infused at 2 µg/kg/day subcutaneously by continuous osmotic pump over 4 weeks. A markedly significant 67.7% reduction in plaque size and a reduction in histological markers for plaque instability (Fig. 3C, D) were both detected following Serp-1 treatment starting one week after carotid cuff placement. Serp-1 plasma levels were measured at approximately ng/mL concentrations.[67] When Serp-1 treatment was started at 5 weeks post cuff placement, a nonsignificant trend toward a reduction in plaque size (30% reduction) was observed. The plaque development in Serp-1 treated mice at the site of carotid cuff compression displayed reduced numbers of invading macrophage with increased smooth muscle cells and increased collagen deposition suggesting improved stability of

Table 1. Viral anti-inflammatory serpins assessed for anti-inflammatory activity in preclinical animal models

Viral Serpin	RSL/P1-P1′ Sequence	Protease Target	Viral Origin	Route/X No Doses	Preclinical Model	Findings
Serp-1	EADERGTTAS SDTAITLIPRN ALTAIVANKP FMFLIYHKP	tPA, uPA, plasmin, FXa	Myxoma virus	Wolinsky local × 1	Balloon angioplasty-mouse, rat, rabbit, rooster, microswine	Inflammation-D, plaque-D
				V. X1, X2, X3 or X11	Balloon angioplasty and stent implant-rabbit, microswine	Inflammation-D, plaque-D plaque-NE
				Wolinsky Local × 1 Wolinsky Local × 1	Repeat balloon angioplasty injury-microswine	Plaque-D plaque-NE
				S.C. pump 30 days	Carotid cuff-mouse	Inflammation-D, plaque-D inflammation-D, plaque-D,
				I.V. X1	Renal artery vein bypass	Inflammation-D, plaque-D, scar-D
				I.V. X10	Aortic transplant-mouse, rat renal transplant-rat	Scar-D plaque-D chronic rejection-D
				I.V. X10	Cardiac transplant-mouse, rat	Prolonged allograft survival-I, acute rejection-D
				I.J. X1 I.V. X10	Antigen induced arthritis-rabbit,-rat	Inflammation-D, erosion-D
				Surface × 1	CAM-chicken	Angiogenesis-D
Spi-3	DVDEQGTVA EASTIMVAT ARSSPEQLEF NTPFIIRHDI	tPA, uPA, plasmin, thrombin, factor Xa	Vaccinia virus, cowpox virus	I.V. × 1	Balloon angioplasty-rat	Inflammation, plaque-D

continued on next page

Table 1. Continued

Viral Serpin	RSL/P1-P1' Sequence	Protease Target	Viral Origin	Route/X No Doses	Preclinical Model	Findings
Serp-2	... VTDFGG	Granzyme B, caspase 1 (ICE)	Myxoma virus	I.V. X1	Balloon angioplasty-rat	Inflammation, plaque-D
			Capripox virus,	S.C. pump × 30 days	Carotid cuff-mouse	Inflammation, plaque-D
			Yatapox virus	I.V. X1	Aortic transplant-rat, mouse	Inflammation, plaque-D
CrmA/ Spi2 Serpin B13R	ATCALVAD-CAST	Granzyme B, caspase	Cowpox virus,	I.V. X1	Balloon angioplasty-mouse, rat	NE
		1 (ICE), caspase 8, 10, 6, 3, 7	Vaccinia virus	S.C. pump × 30 days I.V. X1	Carotid cuff-mouse	NE
					Aortic transplant-mouse, rat	NE
					Prostate cancer leukemia	Cell growth-I
						Resistance to chemotherapy

uPA—urokinase type plasminogen activator, uPA receptor—uPAR, D—decrease, I—increase, NE-null or equivocal effects, Rb—rabbit, Rt—rat, Ms—mouse, MSw—microswine.

1. Srikanth S, Kraft AS. Inhibition of caspases by cytokine response modifier A blocks androgen ablation-mediated prostate cancer cell death in vivo. Cancer Res 1998; 58:834-839.
2. Antoku K, Liu Z, Johnson DE. Inhibition of caspase proteases by CrmA enhances the resistance of human leukemic cells to multiple chemotherapeutic agents. Leukemia 1997; 11:1665-1672.
3. Boomker JM, Luttikhuizen DT, Veniga H. The modulation of angiogenesis in the foreign body response by the poxviral protein M-T7. Biomaterials. 2005; 26: 4874-4881.
4. Richardson M, Liu L, Dunphy L, et al. Viral serpin, Serp-1, inhibits endogenous angiogenesis in the chicken chorioallantoic membrane model. Cardiovasc Pathol. 2007; 16: 191-202.

the plaque and reduced risk of plaque rupture.[67] Unstable atherosclerotic plaque is characterized by a necrotic lipid core with highly active macrophage and T-cell components that release factors that lead to thinning of the surface cap covering the inner plaque core. The core has activated macrophage and T-lymphocytes as well as smooth muscle cells, but can also contain apoptotic cells. The apoptotic macrophage cells in particular can act as small cytokine factories releasing large quantities of inflammatory activators. Proteases, such as the thrombolytic serine proteases (tPA, uPA and plasmin) and the matrix metalloproteinases (MMPs), are also released that breakdown the local connective tissue. In addition to these activated and apoptotic cells, layers of cholesterol deposits and connective tissue also can serve to activate cells and initiate inflammatory reactions.

With erosion, this thinned protective cap can rupture, exposing the underlying connective tissue, inciting thrombus formation and leading to heart attacks and strokes. Mice treated with Serp-1 in this study displayed reduced macrophage content and increased SMC and collagen content suggesting a more stable plaque phenotype.[67]

Preclinical Animal Model Studies of Transplant Rejection—Acute and Chronic Rejection

Aortic Transplant Models

Marked reductions were detected in both inflammatory cell invasion into the arterial wall along with associated reductions in late plaque growth in the balloon angioplasty and stent implant models, and thus the capacity of Serp-1 to reduce chronic inflammation in aortic allograft and renal transplant models was examined in rat models. In later studies mouse aortic transplant models were also utilized in order to analyze the effects of selected genetic 'knock out' on serpin mediated anti-inflammatory activities. Chronic transplant rejection of solid organ transplants is characterized by chronic vascular occlusion and associated end organ ischemic damage. Although inhibition of T-cell mediated rejection is efficacious in reducing chronic transplant vasculopathy and organ damage, it has been estimated that approximately 50% of chronic rejection responses are not adequately blocked and some of this chronic rejection and vasculopathy is now attributed to ongoing smoldering excesses in inflammation and scarring. Two collaborating labs, H Wong with R Zhong at the University of Western Ontario (ON, Canada) and R Morris at Stanford (CA, USA), examined the capacity of Serp-1 treatment to reduce the long term inflammation that contributes to chronic rejection.

The initial rat aortic transplant work was performed by Miller and Dai in the Lucas lab.[68] With a single i.v., bolus infusions of either ACI rat donor to Lewis rat recipient or Lewis rat donor to Sprague Dawley (SD) recipient rat aortic transplants, a significant reduction in early monocyte/macrophage and nonspecific CD2 positive T-cell invasion as well as a significant reduction in later plaque growth were detected.[68] Serp-1 was infused via i.v., bolus injection immediately after completion of an end-to-end aortic anastomotic connection once visible blood flow (aortic pulsation) was detectable. No further bolus doses of serpin were infused after this initial dose and no other drugs were given in this aortic transplant model study.

Renal and Heterotopic Cardiac Transplants

E Bedard in the Zhong lab subsequently examined F344 (RT^{lvl}) donor to Lewis (RT^{l}) recipient rat renal transplants at 5 months follow-up.[69] In this study Serp-1 i.v., plus cyclosporine A (0.75 mg/kg/day s.c.) treatment reduced both scarring and vascular stenosis at 5 months after giving Serp-1 treatment for only the first 10 days after transplant (Figs. 3E, F). Serp-1 treatment with and without cyclosporine A was compared to cyclosporine A alone.[69] When given together with cyclosporine A, Serp-1 at the higher 50 μg/kg i.v., dose markedly reduced all the classical histological markers for chronic rejection including tubular and glomerular atrophy, vascular hyalinization and cortical scarring in the renal transplant model at 5 months follow-up.[69] Additional bolus injections of Serp-1 at 1 month posttransplant did not produce further reductions in scarring and vasculopathy, indicating that the anti-inflammatory and anti-rejection activity of Serp-1 is realized during the first 7-10 days after transplant. Unlike the simple aortic transplant model, Serp-1 given alone, without concomitant cyclosporine injections, did not reduce transplant vasculopathy or scarring.[68] The work with the renal transplant rejection model provided several important advances in our work. First, treatment with Serp-1 together with cyclosporine for the first 10 days after transplant markedly reduced both scarring and vasculopathy at 5 months follow-up when compared to cyclosporine treatment alone.[68] Second, Serp-1 treatment could be limited to the first 10 days after transplant and result in continued protection of the allograft renal transplant even as late as 5 months follow-up with no further boluses. Finally, the renal transplant in this model was the only functioning kidney in the recipient rat and there was no evidence for toxicity, adverse events or adverse effects on renal function with Serp-1 treatment.

A similar study was performed by B Hausen in the Morris lab, examining the percentage of coronary vessels with evidence for vasculopathy and narrowing at 3 months follow-up in a rat cardiac transplant model following treatment with Serp-1 and cyclosporine treatment for the first 10 days post transplant.[70] The number of vessels displaying significant plaque and narrowing was significantly reduced with Serp-1 treatment at 3 months follow-up.

Acute rejection was more recently studied in a Brown Norway rat to Lewis rat heterotopic heart transplant model. In this work (H Wang in the Zhong lab) detected significantly reduced acute rejection following Serp-1 treatment.[71] Improved graft survival was also detected in rats treated with a lower dose of cyclosporin. Indefinite heterotopic heart allograft survival was demonstrated with greater than 100 days follow-up.[71] The Zhong lab also reported reduced xenograft rat to mouse cardiac transplant loss with Serp-1 treatment when given together with two other immunosuppressants. With this work one can propose that Serp-1 treatment together with low dose cyclosporine or other agents should reduce the toxicity of these acute rejection treatments.[70]

Preclinical Animal Model Study of Arthritis

In an early pilot study of 15 rabbits with ovalbumin antigen-induced arthritis, W Maksymovich and A Russell (University of Alberta, Edmonton, Canada) demonstrated reduced joint swelling and inflammation and reduced joint cartilage erosion.[72] In this study, Serp-1 protein was infused in nanogram doses via intra-articular injection. In a larger subsequent work, E Brahn (UCLA, CA, USA) detected marked reductions in joint swelling, erosions and an associated improvement in motility in a rat treated with Serp-1 prior to antigen challenge.[72]As this represented a preventative model and thus is an unlikely scenario in the clinic for arthritic patients, the Brahn group also proceeded to examine the effects of Serp-1 given after arthritis was already established in the rat model. In this follow-up study Serp-1 was infused either alone or together with cyclosporine. When given alone Serp-1 was not effective at reducing inflammation in this collagen induced arthritis model, whereas when dosed together with cyclosporine, Serp-1 protein treatment resulted in a synergistic reduction.[73]

Chicken Chorioallantoic Membrane (CAM) Model of Angiogenesis

In a chicken (*Gallus gallus*) chorioallantoic membrane (CAM) model of angiogenesis, Serp-1 treatment reduced new vessel creation significantly.[74]An inactive Serp-1 mutant Serp-1 (SAA), with an amino acid replacement of R-N to A-A at the P1-P1′ site, was ineffective. Treatment with the mammalian serpin, PAI-1, in the same model also did not reduce new vessel formation.[74] Treatment with Serp-1 reduced vascular endothelial growth factor and laminin gene expression at 6 hrs and 24 hrs after treatment in this model which has the potential to alter neovascular proliferation. Collagen IV expression was also altered but varied from reduced levels at 6 hours to increased at 14 hours follow-up. The role of Serp-1 in potentially controlling dysregulated angiogenesis merits further investigation.[74]

SERP-1 Mechanism of Action

Initial studies with Serp-1 used immunohistochemical analysis of tissue specimens from animal models of cellular invasion at early times after injury. In the rat and rabbit models there was a consistent early reduction in the invasion of macrophage and nonspecific CD2-positive, nonspecific lymphocyte (specifically NK cell) invasion from 24 hours up to 72 hours after Serp-1 treatment post angioplasty injury or aortic, cardiac and renal transplant. Similar reductions in macrophage cells and CD2-positive lymphocytes were detected at early follow-up after acute cardiac (<48 hours) and chronic renal transplant models. This reduction was greater than the effects of Serp-1 alone or cyclosporine treatment alone in these same models. This reduction in inflammatory cell invasion correlated closely with later reductions in plaque growth and vasculopathy development.

It should be noted that many viral proteins exhibit more than one function, often targeting two or more host response pathways. Serp-1 targets tPA, uPA, plasmin in the thrombolytic cascade and also factor Xa in the thrombotic cascade which represent more than one receptor and signaling pathway. It is certainly possible that this highly potent viral serpin may have acquired other functions during evolution as a host immunomodulator, functions that are as yet undisclosed. As the uPA/uPAR complex has been clearly demonstrated to play an important role in inflammatory cell responses after vascular injury, other cellular responses and changes in gene expression in human endothelial cells, monocytes and T-cells were examined in Serp-1 treated cultured human cells. These studies performed by K Viswanathan (Lucas lab) demonstrated that Serp-1 treatment was capable of reducing cell activation as measured by calcium content and membrane fluidity in all cell types tested. As one of the earliest cell responders to i.v. Serp-1 infusions, it is possible that endothelial cells mediate the first inhibitory actions of Serp-1 during vascular injury.[75] Gene expression was also noted to be altered in endothelial cells and monocytes after Serp-1 treatment and this activation pattern differed for Serp-1 and mammalian PAI-1 treatment. Of particular interest, in rat arteries tPA, PAI-1 and uPAR exhibited increased expression following Serp-1 treatments, while for human endothelial cells in culture, the genes for PAI-1 as well as the ITPR2 receptor for calcium underwent altered levels of expression.[28,75] In human monocytes, Serp-1 treatment, differed from PAI-1 as measured by microarray analysis, thereby causing an increase in an actin binding protein and a reduction in CD18, beta 2 integrin expression (unpublished observation).[108] In the CAM model of angiogenesis, Serp-1 treatment reduced VEGF and laminin expression and increased collagen IV. In the acute rejection model, when Brown Norway rat donor to Lewis rat recipient heterotopic heart transplants were performed, Serp-1 treatment significantly reduced TLR2, TLR 4 and myD88 gene expression.[71] Also, in the early stages after chronic renal allograft transplant with Serp-1 treatment, TGFβ gene expression was reduced. Associated with these changes in inflammatory gene expression, there was a reduction in macrophage and dendritic cell invasion. An ApoE^{null} mouse study similarly detected reduced macrophage invasion and associated increases in collagen and SMC in the plaque, all of which suggest increased plaque stabilization.[67] Serp-1 was also found to bind to the surface of endothelial cells, monocytes and T-cells in vivo using fluorescent microscopy and FACS (fluorescence activated cell sorting) analysis.

In mouse targeted gene knockout models, Serp-1 lost all inhibitory activity in uPAR deficient (uPAR$^{-/-}$) mouse model of aortic transplant. However, Serp-1 retained its potent anti-inflammatory activity in PAI-1 deficient (PAI-$^{1-/-}$) mice after aortic transplant, PAI-1 deficient C57Bl/6 background donor to PAI-1 expressing (PAI-$^{1+/+}$) Balb/C recipient mouse.[28] Conversely PAI-1 treatment only reduced plaque in PAI-1 knock out C57Bl/6 isograft transplants, but not in allograft transplants.[28] In the PAI-1 deficient allograft transplant model PAI-1 deficient C57Bl/6 background donor to PAI-1 expressing Balb/C recipient mouse, treatment with a single dose of PAI-1 caused a local excessive thrombotic activation and early mortality.[28] In stark contrast, a single injection of Serp-1 protein markedly reduced inflammation and plaque growth and caused no alteration in thrombosis or bleeding.[28] Treatment with an array of Serp-1/serpin RSL chimeras did not reduce inflammation or plaque growth indicating that Serp-1 requires an intact R-N sequence at the P1-P1′ site. Alteration of the adjacent P2-P7 amino acids, by replacing these residues with 6 alanines led to a loss of Serp-1 anti-inflammatory and anti-atherogenic activity.

Viral Serpins That Target Apoptotic Pathways: Preclinical Analysis of CRMA and SERP-2

Cowpox virus's cytokine response modifier A (CrmA) was the first viral serpin to be identified and has been intensely studied. CrmA is a stable intracellular protein synthesized early during cowpox viral infection that mediates the formation of characteristic red hemorrhagic pocks on the chorioallantoic membrane (CAM) of fertile hen eggs. A CrmA-like protein called Spi-2 is also expressed by certain vaccinia virus strains. This 38 kD intracellular protein is expressed early in viral infection and inhibits both caspases 1 (interleukin converting enzyme-1β or ICE) and 8 as well as Granzyme B, key players in apoptotic pathways. CrmA inhibits caspases-1 (Ki = 4-10 pM) and 8 (Ki = <340 pM) most effectively, but also inhibits caspase s10 (Ki = 4-17 nM), 6, 3 and 7 with decreasing effectiveness.[40] CrmA blocks apoptosis induced via death receptor signaling (Fas receptor and TNF), but is not effective at preventing cell death induced by stress or genotoxic damage induced through mitochondrial apoptotic (caspase 9) signaling. These inhibitions help the virus to replicate, as the primary way to eliminate a viral infection is for T-cells to induce apoptosis in infected cells. A chicken chorioallantoic membrane model demonstrated a strong reduction in inflammation when treated with CrmA; however, recent research in animal models of orthopox-virus infection has failed to confirm CrmA's anti-apoptotic capabilities in vivo.

CrmA (cytokine response modifier A) a cowpox viral protein and Serp-2 (an intracellular myxoma viral serpin) are two viral cross class serpins that bind and inhibit both granzyme B, a serine protease and caspase 1 (cysteine proteases). The presence of Asp (D) in the P1 site of the RSL of these two serpins allows these proteins to function both as serine and as cysteine protease inhibitors, which is why they are called cross class serpins (Fig. 2). Of interest, work by Moyer and Turner have demonstrated that while CrmA binds to both caspases 1 and 8 as well as granzyme B with greater affinity (Kass)[76,77] than Serp-2, whereas, Serp-2 displays greater effects on viral virulence in vivo during viral infections.[78] CrmA inhibits extrinsic apoptosis mediated through the Fas and TNF pathways, but does not block granzyme B mediated cell death after cytotoxic T-lymphocyte release of granules.[79] Serp-2 cannot block apoptosis in cowpox virus infected cells, but conversely Serp-2 deficiency in myxoma virus infection markedly attenuates virus infection in European rabbits with a reduction in mortality from 100% to 10% with inactivation of the Serp-2 gene.[76] CrmA cannot replace Serp-2 and results in only 70% mortality in myxoma virus rabbit infections.

Like CrmA, myxoma poxvirus 34kD Serp-2 is able to inhibit ICE. Serp-2 shares 35% similarity with CrmA. However, despite displaying a lower binding affinity in vitro for ICE and caspase 8, Serp-2 demonstrates a more robust anti-inflammatory activity in vivo during viral infections. Serp-2 is capable of inhibiting apoptosis in CAM models of infection but did not block inflammation in the CAM model.[79] Although both Serp-2 and CrmA are cross-class serpins and target some of the same proteases, in infected cells the insertion of Serp-2 in place of CrmA does not cause similar effects.[79] Thus these two cross class serpins are not functionally interchangeable and studies in animal models have confirmed a marked difference in their potential as anti-inflammatory agents.

Spi-1 is a less well described rabbit poxvirus intracellular protein with circumscribed anti-apoptotic activity. This viral serpin binds cathepsin G similar to the reports for the mammalian intracellular PI-6 protein. Spi-1 has been reported to inhibit a caspase independent form of apoptosis in selected cells.[80]

SERP-2 Preclinical Studies

In rat and mouse models of angioplasty injury and aortic transplant, effective reductions in inflammatory cell invasion and in plaque growth were observed following Serp-2 treatment.[81] CrmA conversely had no effect on plaque growth in these models, nor did two Serp-2 RSL mutants (D294A and E) provided by P Turner and R Moyer for these studies. Work by I Bot and E Biessen similarly detected significant reductions in plaque in the ApoE null mouse model with carotid cuff compression injury after Serp-2 treatment but not with CrmA.[82] While Serp-2 showed a trend toward reducing plaque at the site of carotid cuff compression injury, this trend did not reach significance. But, unlike Serp-1 treatment, Serp-2 was able to reduce the generalized increase in plaque detected at the aortic root of $ApoE^{null}$, mice suggesting that this protein can target and reduce systemic plaque buildup rather than plaque growth only at sites of vascular surgical injury.[82]

Other Mammalian Serpins

These studies with Serp-1 and other viral serpins provide a guide to the development of agents targeting pathways identified as having potential for high impact in the regulation of inflammatory and apoptotic pathways. Thus, mammalian serpins that inhibit the thrombolytic serine proteases, more specifically the uPA and tPA pathways, were studied to assess how close an analogy one might find for mammalian serpin inhibition of inflammation. PAI-1 and neuroserpin were tested in similar rodent models and compared to Serp-1 for anti-inflammatory and anti-atherogenic activity in the preclinical models. PAI-1 binds and inhibits tPA, uPA and thrombin as well as activated protein C (APC) with a higher K_{ass} than Serp-1 (often with an increase of K_{ass} on the order of two logs (10^7

versus 10^4)).[59] Neuroserpin binds and inhibits tPA and uPA, but with a greater predilection for tPA. As noted in prior reports, infusion of PAI-1 demonstrated anti-inflammatory and anti-atherogenic actions in some preclinical models, while other studies found a marked pro-inflammatory activity for PAI-1.[35-38] The work by Carmeliet demonstrated an increase in plaque after iliac injury in PAI-1 deficient mice,[34] while work by Plopis demonstrated reduced plaque in $FeCl_3$ injured arteries in PAI-1 deficient mice.[36] In our lab, PAI-1 infusion did reduce cell invasion and even plaque growth in PAI-1 deficient mice in both cell migration assays and in isograft aortic transplants where the donor and recipient mice both lacked PAI-1 expression.[28] In aortic allografts, however, with transplant of a PAI-1 deficient C57Bl/6 mouse aorta into a PAI-1 expressing Balb/c mouse there was excess local thrombosis and 100% mortality.[27] Thus, adding this natural mammalian serpin to a mouse already expressing background levels of PAI-1 leads to excess clotting and death. This pro-coagulant property is not seen following Serp-1 treatment.

Mammalian neuroserpin preferentially inhibits two chain tPA (Ki = 6.2×10^5 $M^{-1}s^{-1}$), but also inhibits single chain tPA, trypsin, uPA, nerve growth factor-γ, plasmin and thrombin (Ki ranging from 2.1×10^2 to 8.0×10^4 $M^{-1}s^{-1}$).[32] Neuroserpin has also been implicated in neurological diseases such as dementia with neuroserpin mutation and polymer formation, cerebrovascular diseases, epilepsy, multiple sclerosis and schizophrenia. In mouse cerebral ischemia models neuroserpin appeared to have a protective role as evidenced by the finding that delivery of neuroserpin to the affected area (or overexpression of neuroserpin) decreased the ischemic territory and reduced the numbers of apoptotic cells in the setting of cerebral infarction.[33] Neuroserpin has also been reported to play a central role in development of the visual cortex in axonal growth and in the regulation of synaptic plasticity. We thus tested neuroserpin in a collaborative study with D Lomas (U Cambridge, Cambridge, UK). In our rat model of angioplasty injury and in the mouse aortic transplant models, neuroserpin injection did significantly reduce plaque growth (unpublished observations).

In summary, we have assessed two mammalian serpins that target thrombolytic pathways, plasminogen activators, in preclinical vascular surgery models. One, PAI-1, induced excess thrombosis and mortality but the second, neuroserpin, reduced inflammation. Analysis of these cellular serpin proteins and discovery work on the mechanism of action of these viral serpins has therefore provided insights into new therapeutic targets for drug discovery and treatment of inflammation based disorders. This work in effect adds to the serpin chorus as potential therapeutic agents and new pathways to target in treating inflammation based diseases.

Other Parasite Derived Serpins

Investigation of diverse genomes has yielded a wide range of organisms that encode serpins, from poxviruses to mammals to more recently discovered bacterial and unicellular serpins. The function of the majority of these proteins are unknown, although most are thought to have inhibitory activity based on sequence homology of known serine protease inhibitors. In this section, serpins from various organisms are presented and their importance in possible immunotherapies discussed.

Insects are hosts to a wide variety of disease causing organisms and act as vectors of transmission to man. Serpins in haematophagous (blood-feeding) insects are expressed for two different purposes; immune evasion/invasion for feeding from their host and defense against infection.[83] Tsetse flies are responsible for harboring and transmission of African sleeping sickness in humans. These insects express the serpins Tsal1 and Tsal2 in their saliva, which are important for transmission of Trypanosomes by suppression of B and T-cell activity.[84]

Mosquitoes serve as an intermediary in the transmission of malaria-causing Plasmodium parasites into humans. There are mechanisms in the mosquito gut designed to inhibit the invasion of the parasites.[83] *Anopheles* mosquitoes express several serpins that have important actions for the killing and/or clearing of parasites from the insect. Thus the mosquito can carry and transmit the infecting parasite to other hosts such as man without succumbing to infection itself. For example, in *A. gambiae*, SRPN10 is expressed as four isotypes, with variations in the RSL.[85] Invasion of the midgut of *A. gambiae* mosquitoes by *Plasmodium berghei ookinetes* up-regulates the mRNA of a specific subset of isotypes, KRAL and RCM, possibly regulating the apoptosis of infected cells.[85] Progression of the timeline of infection can be observed just by the localization and expression of SRPN10. SRPN10 normally resides in the nucleus, but when invaded by *ookinetes*, SRPN 10 translocates to the cytoplasm.[86] Once a parasite is ready to exit from the gut basolateral membrane, SRPN10 protein expression is then increased.[86] Down-regulation of another mosquito serpin, SRPN2, by RNAi decreases oocyst formation after infection with *Plasmodium berghei* in the midgut,[87] however, in field isolates of *Plasmodium falciparum*, this did not reduce oocyst formation.[88] SRPN6, another *Anopheles* expressed serpin, is expressed in both *Anopheles gambiae* and *stephensi* and has an identical 28 residues for their RSL yet result in slightly different activities in the organisms when faced with Plasmodium infection.[89] In *A. stephensi,* SRPN6 is thought to have parasite killing activity whereas in *A. gambiae*, Plasmodium *berghei* SRPN6 possibly has a role in the clearance of parasites by delaying the progression of infection.[90] Interestingly, in *A. gambiae*, SRPN6 is also up-regulated upon exposure of the gut to *Escherichia coli*,[90] as observed in *Drosophila melanogaster.*[91]

Insects are not the only organisms to express serpins that allow these organisms to invade their hosts and evade the host inflammatory responses. Helminthes, such as *Schistosomes*[92,93] and *Brugia*[94,95] express serpins that dampen the immune responses of their hosts in not one, but several pathways.[96-102] In blood feeding, ticks also secrete serpins in their saliva to inhibit the local inflammatory response and blood clotting to prolong their ability to feed without detection.[103-107]

Thus these disease-causing organisms over time have developed an arsenal of proteins to facilitate successful invasion of their host organisms. By understanding the mechanism of action, serpins from parasitic organisms could be refined and utilized in therapeutics as anti-inflammatory and antithrombotic compounds. Serpins have potential uses other than specific protease activity. For example, serpin-coupled peptides, such as ovalbumin (a serpin that lacks classical serpin inhibitory functions) and coupled Tc52 peptides from *T. cruzi*, have been shown to inhibit T-cell activity.[108] By using ovalbumin as a carrier protein to reduce immunogenicity, specific host pathways can be targeted. Beneficial activities from proteins with more than one function can be dissected and isolated for coupling with ovalbumin, to minimize nonspecific activities. Utilizing already known mechanisms of immune modulation can be an efficient method of therapeutic development.

Clinical Study of SERP-1 Treatment in Acute Unstable Coronary Syndromes; Unstable Angina and Non-ST Elevation Myocardial Infarction (NSTEMI)

Like any new discovery in the drug field and clinical testing of a new class of protein therapeutics, this work could not have been done without the efforts of many investigators. A small biotech company was established in 1997 by Drs Lucas and McFadden, called Viron Therapeutics Inc, (London, ON, Canada). Viron was initially established around the Lucas and McFadden research laboratories and founded with the guidance of Dr M Ponansky, the then Director of the Robarts Research Institute, University of Western Ontario, London, ON, Canada. Viron coordinated the expertise and funding necessary to produce viral proteins according to good manufacturing practices (GMP) and to set up the preclinical animal toxicity screening according to good lab practice (GLP) mandates. With this foundation Serp-1 has been successfully taken through a Phase 1 safety trial in man and is being tested for safety and efficacy in a Phase IIa clinical trial conducted in the US and Canada.

For the Phase 1 study, single doses of Serp-1 protein were infused in normal volunteers as mandated by the FDA. This was the first trial in man with a new class of virus-derived native protein therapeutic. No changes in cardiac, renal or hepatic function, as well as no changes in clotting parameters were detected in this study. This Phase 1 safety study demonstrated that Serp-1 infusion was safe with no adverse events detected or reported. A Phase IIa study is currently ongoing in which the effects of a Serp-1 infusion is given for three days starting immediately after balloon angioplasty and stent implant in patients with acute unstable angina and non-ST elevation myocardial infarction (NSTEMI). This trial is still ongoing at 7 sites in Canada and the US. The results of this study have yet to be reported. This Phase 2a study represents a first in man clinical trial of a native anti-inflammatory viral serpin. Clinical efficacy and safety thus remain to be finally determined for this protein, but, if safe, this study potentially opens the door to testing of other viral anti-inflammatory proteins in inflammation driven diseases.

While taking Serp-1 to clinical trial is a first step in the path toward using viral anti-inflammatory proteins as a new therapeutic, this work really represents only the first step in our symphony and our first passage in the study of serpins as guides to new therapeutic modalities.

References

1. Sulek K. Prize in 1908 awarded to P Erlich and E Metchnikoff for their work in immunology. Wiad Leuk 1967; 20:1117-1118.
2. Opitz B, Hippenstiel S, Eitel J et al. Extra and intracellular innate immune recognition in endothelial cells. Thromb Haemost 2007; 98:319-326.
3. Mariathasan S, Monack DM. Inflammasome adaptors and sensors: intracellular regulators of infection and inflammation. Nature Rev Immunol 2007; 7:31-40.
4. Brown KL, Cosseau C, Gardy JL et al. Complexities of targeting innate immunity to treat infection. Trends Immunol 2007; 28:260-266.
5. Lin WW, Karin M. A cytokine mediated link between innate immunity, inflammation and cancer. J Clin Invest 2007; 117:1175-1183.
6. Ware C. Network communications: Lymphotoxins, light and TNF. Annu Rev Immunol 2005; 23:787-819.
7. Roy CR, Mocarski ES. Nature Imm 2007; 8:1179-1187.
8. Parker LC, Prince LR, Sabroe I. Translational minireview series on toll-like receptors: Networks regulated by toll-like receptors mediate innate and adaptive immunity. Clin Exp Immunol 2007; 147:199-207.
9. Methe H, Weis M. Atherogenesis and inflammation—was Virchow right? Nephrol Dial Transplant 2007; 22:1823-1827.
10. Schneider DF, Glenn CH, Faunce DE. Innate lymphocyte subsets and their immunoregulatory roles in burn injury and sepsis. J Burn Care and Research 2007; 28:365-379.
11. Opal SM, Esmon CT. Bench-to-bedside review: Functional relationships between coagulation and the innate immune response and their respective roles in the pathogenesis of sepsis. Crit Care 2003; 7:23-38.
12. Jancaiuskiene S. Conformational properties of serine proteinase inhibitors (serpins) confer multiple pathophysiological roles. Biochem Biophys Acta 2001; 1535:221-235.
13. Lomas DA, Belorgey D, Mallya M et al. Molecular mousetraps and the serpinopathies. Biochem Soc Transaction 2005; 33:321-330
14. Carrell RW, Travis J. α1-Antitrypsin and the serpins: Variation and countervariation. Trends Biochem Sci 1985; 10:20-24.
15. Irving JA, Pike RN, Lesk AM et al. Phylogeny of the serpin superfamily: Implications of patterns of amino acid conservation for structure and function. Genome Res 2000; 10:1845-1864.
16. Iwanaga S, Kawabata S. Evolution and phylogeny of defense molecules associated with innate immunity in horseshoe crab. Front Biosci 1998; 3:D973-984.
17. Law RHP, Zhang Q, McGowan S et al. An overview of the serpin superfamily. Genome Biol 2006; 7:216.2-216.11
18. Silverman GA, Bird PI, Carrell RW et al. The serpins are an expanding superfamily of structurally similar but functionally diverse proteins. J Biol Chem 2001; 276:33293-33296.
19. Marszal E, Shrake A. Serpin crystal structure and serpin polymer structure. Arch Biochem Biophys 2006; 453:123-129.
20. Huntington JA, Read RJ, Carrell RW. Nature 2000; 407:923-926.
21. Patston PA, Church FC, Olson ST. Serpin–ligand interactions. Methods 2004; 32:93-109.
22. Ye S, Cech AL, Belmares R et al. The structure of a michaelis serpin-protease complex. Nat Struct Biol 2001; 8:979-983
23. Lawrence DA, Olson ST, Muhammad S et al. Partitioning of serpin-proteinase reactions between stable inhibition and substrate cleavage is regulated by the rate of serpin reactive center loop insertion into β-sheet A. J Biol Chem 2000; 275:5839-5844.
24. Whisstock JC, Bottomley SP. Molecular gymnastics: serpin structure, folding and misfolding. Curr Opin Struct Biol 2006; 16:761-768
25. Li W, Johnson DJ, Esmon CT et al. Structure of the antithrombin-thrombin-heparin ternary complex reveals the antithrombotic mechanism of heparin. Nat Struct Mol Biol 2004; 11:857-862.
26. Johnson DJ, Li W, Adams TE et al. Antithrombin-S195A factor xa-heparin structure reveals the allosteric mechanism of antithrombin activation. EMBO J 2006; 25:2029-2037.

27. Dai E, Viswanathan K, Sun YM et al. Identification of myxomaviral serpin reactive site loop sequences that regulate innate immune responses. J Biol Chem 2006; 281:8041-8050
28. Dai E, Guan H, Liu L et al. Serp-1, a viral anti-inflammatory serpin, regulates cellular serine proteinase and serpin responses to vascular injury. J Biol Chem 2003; 278:18563-18572.
29. Ewa Marszal, Andrew Shrake. Serpin crystal structure and serpin polymer structure. Arch Biochem Biophys 2006; 453:123-129.
30. Blasi F. Proteolysis, cell adhesion, chemotaxis and invasiveness are regulated by the uPA-uPAR-PAI-1 system. Thromb Haemost 1999; 82:298.
31. Carmeliet P, Collen D. Molecular analysis of blood vessel formation and disease. Am J Physiol Heart Circ Physiol 1997; 273:H2091-H2104
32. Yepes M, Lawrence DA. Neuroserpin: a selective inhibitor of tissue-type plasminogen activator in the central nervous system. Thromb Haemost 2004; 91:457-464.
33. Yepes M, Sandkvist M, Wong MK et al. Neuroserpin reduces cerebral infarct volume and protects neurons from ischemia-induced apoptosis. Blood 2000; 96:569-576.
34. Carmeliet P, Moons L, Lijnen R et al. Inhibitory role of plasminogen activator inhibitor-1 in arterial wound healing and neointima formation. Circulation 1997; 96:3180-3191
35. Falkenberg M, Tom C, DeYoung MB et al. Increased expression of urokinase during atherosclerotic lesion development causes arterial constriction and lumen loss and accelerates lesion growth. PNAS 2002; 99:10665-10670.
36. Ploplis VA, Castellino FJ. Attenuation of neointima formation following arterial injury in PAI-1 deficient mice. Ann NY Acad Sci 2001; 936:466-468
37. Strauss BH, Lau HK, Bowman KA et al. Plasma urokinase antigen and plasminogen activator inhibitor-1 antigen levels predict angiographic coronary restenosis. Circulation 1999; 100:1616-1622
38. Vaughan DR. PAI-1 and atherothrombosis. J Thromb and Haemostasis 2005; 3:1879-1883
39. Munuswamy Ramunujam G, Lucas A. Mammalian serine protease inhibitor (serpin), neuroserpin, targets thrombolytic proteases to reduce inflammation, atherogenesis and T helper lymphocyte activation. Poster. American Heart Association. Orlando, FL 2007
40. Bortner CD, Cidlowski JA. Cellular mechanisms for the repression of apoptosis. Annu Rev Pharmacol Toxicol 2002; 42:259-281.
41. Trapani JA. Granzymes: a family of lymphocyte granule serine proteases. Genome Biol 2001; 2:3014.1-3014.7
42. Roemisch J, Gray E, Hoffmann JN et al. Antithrombin: a new look at the actions of a serine protease inhibitor. Blood Coagul Fibrinolysis 2002; 13:657-670.
43. Mariathasan S, Monack DM. Inflammasome adaptors and sensors: intracellular regulators of infection and inflammation. Nature Rev Immunol 2007; 7:31-40
44. Taniguchi S, Sagara J. Regulatory molecules involved in inflammasome formation with special reference to a key mediator protein, ASC. Semin Immunopathol 2007; 29:231-238.
45. Zhou Q, Snipas S, Orth K et al. Target proteases of the viral serpin CrmA: Analysis of five caspases. J Biol Chem 1997; 272:7797-7800.
46. Moffitt KL, Martin SL, Walker B. The emerging role of serine proteases in apoptosis. Biochem Soc Transactions 2007; 35:559-560.
47. Bird CH, Sutton VR, Sun J et al. Selective regulation of apoptosis: the cytotoxic lymphocyte serpin proteinase inhibitor 9 protects against granzyme B-mediated apoptosis without perturbing the Fas cell death pathway. Mol Cell Biol 1998; 18:6387-6398.
48. Medcalf RL, Stasinopoulos SJ. The undecided serpin: the ins and outs of plasminogen activator inhibitor Type 2. FEBS J 2005; 272:4858-4867.
49. Cinelli P, Maadani R, Tsuzuki N et al. Neuroserpin, a neuroprotective factor in focal ischemic stroke. Mol Cell Neurosci 2001; 18:443-457.
50. Khalkhali ellis Z. Maspin: The new frontier. Clin Cancer Res 2006; 12:7279-7283.
51. Kato K, Kishi T, Kamachi T et al. Serine proteinase inhibitor 3 and murinoglobulin I are potent inhibitors of neuroserpin in adult mouse brain. J Biol Chem 2001; 276:14562-14571.
52. Sun J, Ooms L, Bird CH et al. A new family of 10 murine ovalbumin serpins includes two homologs of proteinase inhibitor 8 and two homologs of the granzyme B inhibitor (proteinase inhibitor 9). J Biol Chem 1997; 272:15434-15441.
53. Charron Y, Madani R, Nef S et al. Expression of serpinb6 serpins in germ and somatic cells of mouse gonads. Mol Repro Develop 2006; 73:9-19.
54. Suzuki K, Hayashi T. Protein C and its inhibitor in malignancy. Sem Thromb Haem 2007; 33:667-672.
55. Miyamoto Y, Akaike T, Alam MS et al. Novel functions of human alpha(1)-protease inhibitor after S-nitrosylation: inhibition of cysteine protease and antibacterial activity. Biochem Biophys Res Commun 2000; 267:918-923.

56. Fenner F, Woodroffe GM. The pathogenesis of infectious myxomatosis: the mechanism of infection and the immunological response in the European rabbit (oryctolagus cuniculus). Br J Exp Pathol 1953; 34:400-411.
57. Upton C, Macen JL, Wishart DS et al. Myxoma virus and malignant rabbit fibroma virus encode a serpin-like protein important for virus virulence. Virology 1990; 179:618-631.
58. Macen JL, Upton C, Nation N et al. SERP1, a serine proteinase inhibitor encoded by myxoma virus, is a secreted glycoprotein that interferes with inflammation. Virology 1993; 195:348-363.
59. Xing Li, Schneider H, Peters A et al. Heparin binding alters viral serine protease inhibitor (serp-1) activity. Open Biochem J 2008; 2:6-15.
60. Nash P, Whitty A, Handwerker J et al. Inhibitory specificity of the anti-inflammatory myxoma virus serpin, serp-1. J Biol Chem 1998; 273:20982-20991.
61. Lucas A, Li X, Schneider H et al. Heparin alters viral serpin, serp-1, anti-thrombolytic activity to anti-thrombotic activity. Open Biochem J 2007; 2:6-15
62. Turner PC, Moyer RW. The cowpox virus fusion regulator proteins SPI3 and hemagglutinin interact in infected and uninfected cells. Virology 2006; 347:88-99.
63. Turner PC, Baquero MT, Yuan S et al. The cowpox virus serpin spi-3 complexes with and inhibits urokinase-type and tissue-type plasminogen activators and plasmin. Virology 2000; 272:267-280.
64. Lucas A, Liu LY, Macen J et al. Virus-encoded serine proteinase inhibitor serp-1 inhibits atherosclerotic plaque development after balloon angioplasty. Circulation 1996; 94:2890-2900.
65. Hatton MW, Ross B, Southward SM et al. Metabolism and distribution of the virus-encoded serine proteinase inhibitor serp-1 in healthy rabbits. Metabolism 2000; 49:1449-1452.
66. Lucas A, Dai E, Liu L et al. Transplant vasculopathy: Viral anti-inflammatory serpin regulation of atherogenesis. J Heart Lung Transplant 2000; 19:1029-1038.
67. Bot I, von der thusen JH, Donners MMPC et al. Serine protease inhibitor serp-1 strongly impairs atherosclerotic lesion formation and induces a stable plaque phenotype in ApoE−/− mice. Circulation 2003; 93:464-471.
68. Miller LW, Dai E, Nash P et al. Nation, Robert Zhong, Grant Mcfadden and Alexandra Lucas Inhibition of transplant vasculopathy in a rat aortic allograft model after infusion of anti-inflammatory viral serpin. Circulation 2000; 101:1598-1605.
69. Bedard ELR, Jiang J, Arp J et al. Prevention of chronic renal allograft rejection by serp-1 protein. Transplantation 2006; 81:908-914.
70. Hausen B, Boeke K, Berry GJ et al. Viral serine proteinase inhibitor (serp-1) effectively decreases the incidence of graft vasculopathy in heterotopic heart allografts. Transplantation 2001; 72:364-368
71. Jiang J, Arp J, Kubelik D et al. Induction of indefinite cardiac allograft survival correlates with toll-like receptor 2 and 4 downregulation after serine protease inhibitor-1 (serp-1) treatment. Transplantation 2007; 84:1158-1167.
72. Maksymowych WP, Nation N, Nash P et al. Amelioration of antigen induced arthritis in rabbits treated with a secreted viral serine proteinase inhibitor. J Rheumatol 1996; 23:878-882.
73. Brahn E, Do L, Lee S et al. Suppression of collagen-induced arthritis with a serin proteinase inhibitor cloned from a myxoma viral sequence. Arthritis and rheumatism 2000 annual scientific meeting, Philadelphia, PA. Abstract Supplement, American College of Rheumatology 2000; abstract no. 1000:43(9).
74. Richardson M, Liu L, Dunphy L et al. Viral serpin, serp-1, inhibits endogenous angiogenesis in the chicken chorioallantoic membrane model. Cardio Path 2007; 16:191-202.
75. Visinathan K, Liu L, Vaziri S et al. Myxoma viral serpin, serp-1, a unique interceptor of coagulation and innate immune pathways. Thromb Haemost 2006; 95:499-510.
76. MacNeill AL, Turner PC, Moyer RW. Mutation of the myxoma virus serp-2 P1-site to prevent proteinase inhibition causes apoptosis in cultured RK-13 cells and attenuates disease in rabbits, but mutation to alter specificity causes apoptosis without reducing virulence. Virology 2006; 356:12-22.
77. Turner PC, Sancho MC, Thoennes SR et al. Myxoma virus serp-2 is a weak inhibitor of granzyme B and interleukin-1beta-converting enzyme in vitro and unlike CrmA cannot block apoptosis in cowpox virus—infected cells. J Virol 1999; 73:6394-6404.
78. Messud Petit F, Gelfi J, Delverdier M et al. Serp-2, an inhibitor of the interleukin-1beta-converting enzyme, is critical in the pathophysiology of myxoma virus. J Virol 1998; 72:7830-7839.
79. Nathaniel R, MacNeill AL, Wang YX et al. Cowpox virus CrmA, myxoma virus serp-2 and baculovirus P35 are not functionally interchangeable caspase inhibitors in poxvirus infections. J Gen Virol 2004; 85:1267-1278.
80. Luttge BG, Moyer RW. Suppressors of a host range mutation in the rabbitpox virus serpin spi-1 map to proteins essential for DNA replication. J Virol 2005; 79:9168-9179.
81. Viswanathan K, Bot I. Liu L et al. Viral cross-class serpin inhibits apoptosis, inflammation and atherosclerosis. In: Modulation of Atherothrombotic Factors: Novel Strategies for Plaque Stabilization. Print Partners Ipstamp Enschede, The Netherlands 2005; pg 61-76.

82. Viswanathan K, Bot I, Liu L et al. Viral cross-class serpin inhibits apoptosis, inflammation and atherosclerosis. In modulation of atherothrombotic factors: Novel strategies for plaque stabilization. Print Partners Ipstamp Enschede The Netherlands 2005:61-76.
83. Abraham EG, Jacobs Lorena M. Mosquito midgut barriers to malaria parasite development. Insect Biochem Mol Biol 2004; 34:667-671.
84. Caljon G, Van Den Abbeele J, Stijlemans B et al. Tsetse fly saliva accelerates the onset of trypanosoma brucei infection in a mouse model associated with a reduced host inflammatory response. Infect Immun 2006; 74:6324-6330.
85. Danielli A, Kafatos FC, Loukeris TG. Cloning and characterization of four anopheles gambiae serpin isoforms, differentially induced in the midgut by plasmodium berghei invasion. J Biol Chem 2003; 278:4184-4193.
86. Danielli A, Barillas Mury C, Kumar S et al. Overexpression and altered nucleocytoplasmic distribution of anopheles ovalbumin-like SRPN10 serpins in plasmodium-infected midgut cells. Cell Microbiol 2005; 7:181-190.
87. Michel K, Budd A, Pinto S et al. Anopheles gambiae SRPN2 facilitates midgut invasion by the malaria parasite plasmodium berghei. EMBO Rep 2005; 6:891-897.
88. Michel K, Suwanchaichinda C, Morlais I et al. Increased melanizing activity in anopheles gambiae does not affect development of plasmodium falciparum. Proc Natl Acad Sci USA 2006; 103:16858-16863.
89. Abraham EG, Pinto SB, Ghosh A et al. An immune-responsive serpin, SRPN6, mediates mosquito defense against malaria parasites. Proc Natl Acad Sci USA 2005; 102:16327-16332.
90. De Gregorio E, Spellman PT, Rubin GM et al. Genome-wide analysis of the drosophila immune response by using oligonucleotide microarrays. Proc Natl Acad Sci USA 2001; 98:12590-12595.
91. Perlmutter DH, Joslin G, Nelson P et al. Endocytosis and degradation of alpha 1-antitrypsin-protease complexes is mediated by the serpin-enzyme complex (SEC) receptor. J Biol Chem 1990; 265:16713-16716.
92. Pleass RJ, Kusel JR, Woof JM. Cleavage of human IgE mediated by schistosoma mansoni. Int Arch Allergy Immunol 2000; 121:194-204.
93. Zang X, Atmadja AK, Gray P et al. The serpin secreted by brugia malayi microfilariae, Bm-SPN-2, elicits strong, but short-lived, immune responses in mice and humans. J Immunol 2000; 165:5161-5169.
94. Manoury B, Gregory WF, Maizels RM et al. Bm-CPI-2, a cystatin homolog secreted by the filarial parasite brugia malayi, inhibits class II MHC-restricted antigen processing. Curr Biol 2001; 11:447-451.
95. Newlands GF, Skuce PJ, Knox DP et al. Cloning and expression of cystatin, a potent cysteine protease inhibitor from the gut of haemonchus contortus. Parasitology 2001; 122:371-378.
96. Peanasky RJ, Bentz Y, Paulson B et al. The isoinhibitors of chymotrypsin/elastase from ascaris lumbricoides: isolation by affinity chromatography and association with the enzymes. Arch Biochem Biophys 1984; 232:127-134.
97. Schonemeyer A, Lucius R, Sonnenburg B et al. Modulation of human T-cell responses and macrophage functions by onchocystatin, a secreted protein of the filarial nematode onchocerca volvulus. J Immunol 2001; 167(6):3207-3215.
98. Pfaff AW, Schulz Key H, Soboslay PT et al. Litomosoides sigmodontis cystatin acts as an immunomodulator during experimental filariasis. Int J Parasitol 2002; 32:171-178.
99. Dainichi T, Maekawa Y, Ishii K et al. Nippocystatin, a cysteine protease inhibitor from nippostrongylus brasiliensis, inhibits antigen processing and modulates antigen-specific immune response. Infect Immun 2001; 69:7380-7386.
100. Rhoads ML, Fetterer RH, Hill DE et al. Trichuris suis: a secretory chymotrypsin/elastase inhibitor with potential as an immunomodulator. Exp Parasitol 2000; 95:36-44.
101. Zang X, Maizels RM. Serine proteinase inhibitors from nematodes and the arms race between host and pathogen. Trends Biochem Sci 2001; 26:191-197.
102. Limo MK, Voigt WP, Tumbo Oeri AG et al. Purification and characterization of an anticoagulant from the salivary glands of the ixodid tick rhipicephalus appendiculatus. Exp Parasitol 1991; 72:418-429.
103. Joubert AM, Crause JC, Gaspar AR et al. Isolation and characterization of an anticoagulant present in the salivary glands of the bont-legged tick, hyalomma truncatum. Exp Appl Acarol 1995; 19:79-92.
104. Mulenga A, Sugimoto C, Ingram G et al. Characterization of two cDNAs encoding serine proteinases from the hard tick haemaphysalis longicornis. Insect Biochem Mol Biol 2001; 31:817-825.
105. Prevot PP, Adam B, Boudjeltia KZ et al. Anti-hemostatic effects of a serpin from the saliva of the tick ixodes ricinus. J Biol Chem 2006; 281(36):26361-26369.
106. Leboulle G, Crippa M, Decrem Y et al. Characterization of a novel salivary immunosuppressive protein from ixodes ricinus ticks. J Biol Chem 2002; 277:10083-10089.
107. Borges M, Da Silva AC, Sereno D et al. Peptide-based analysis of the amino acid sequence important to the immunoregulatory function of trypanosoma cruzi Tc52 virulence factor. Immunology 2003; 109:147-155.
108. Richardson J, Viswanathan K, Lucas A. Serpins, the vasculature and viral therapeutics. Front Biosci 2006; 11:1042-56.

CHAPTER 12

Helminthic Therapy:
Using Worms to Treat Immune-Mediated Disease

David E. Elliott* and Joel V. Weinstock

Abstract

There is an epidemic of immune-mediated disease in highly-developed industrialized countries. Such diseases, like inflammatory bowel disease, multiple sclerosis and asthma increase in prevalence as populations adopt modern hygienic practices. These practices prevent exposure to parasitic worms (helminths). Epidemiologic studies suggest that people who carry helminths have less immune-mediated disease. Mice colonized with helminths are protected from disease in models of colitis, encephalitis, Type 1 diabetes and asthma. Clinical trials show that exposure to helminths reduce disease activity in patients with ulcerative colitis or Crohn's disease. This chapter reviews some of the work showing that colonization with helminths alters immune responses, against dysregulated inflammation. These helminth-host immune interactions have potentially important implications for the treatment of immune-mediated diseases.

Epidemiology of Immune-Mediated Disease and Worms

Pathogens and parasites have evolved mechanisms to modulate or evade host immune responses. These adaptations permit survival of the invading organism. Our work extends this concept. We hypothesized that our immune systems are adapted "to expect" confrontations with other organisms and that loss of these confrontations permits immune dysfunction that can cause disease.[1] People living in highly-developed industrialized countries enjoy increasingly hygienic environments, which restricts exposure to infectious agents including helminths. We have focused on loss of helminth exposure as the major environmental change permitting the current epidemic of immune mediated disease.

It is apparent that environmental change underlies the increasing prevalence of immune-mediated disease. Highly regulated immune responses protect us from pathogens and malignancy. However, when dysregulated or mis-targeted, these immune responses produce inflammatory disease. Over the last 70 years, immune-mediated diseases like inflammatory bowel disease (IBD), multiple sclerosis (MS), Type 1 (autoimmune) diabetes (T1D) and asthma have become common in industrialized highly developed countries but remain rare in less-developed countries. Patients with IBD have chronic inflammation of the small and/or large intestine that results in bleeding, ulceration, strictures and perforation. IBD was uncommon prior to the 1940s, but now afflicts more than 3 million people in the United States and Europe.[2] As countries develop economically, IBD in the population increases.[2,3] Patients with MS have immune-mediated inflammatory destruction of the central nervous system. MS shares the same geographical and temporal distribution with IBD[4] and as countries develop improved sanitation, the prevalence of MS increases.[5] Patients with T1D have immune-mediated destruction of pancreatic β-cells that make insulin. Like MS and IBD,

*Corresponding Author: David E. Elliott—Division of Gastroenterology (4611 JCP), University of Iowa College of Medicine, 200 Hawkins Drive, Iowa City, Iowa 52242-1009, USA. Email: david-elliott@uiowa.edu

Pathogen-Derived Immunomodulatory Molecules, edited by Padraic G. Fallon.

the incidence of T1D has increased dramatically over time.[6] Patients with asthma have chronic airway inflammation worsened by allergic provocation. Over the last 40 years, asthma increased dramatically in developed countries and is becoming prevalent in cities of developing countries.[7] Children of Mexican immigrants born in the United States have more asthma symptoms than children born in Mexico and immigrating to the US at older ages.[8] Again, this suggests that the environment in developed countries promotes or permits asthma.[9]

Prior to the 1930s, most individuals were colonized with helminths.[1] Paleoparasitological specimens of prehistoric human waste from up to 10,000 years ago show that infections with nematodes (e.g., *Trichuris trichiura, Ascaris lumbricoides, Enterobius vermicularis, Trichinella spiralis),* cestodes (e.g., *Taenia* spp., *Diphyllobothrium* spp., *Hymenolepis nana*) and trematodes (e.g., *Fasciola* spp., *Schistosoma* spp.) were common.[10] Infections with these organisms remain prevalent in lesser developed countries.[11,12] Generational exposure to helminths would select genetic traits in their host that are optimized for the presence of these organisms.[1]

In the nineteenth century, highly promoted but ineffectual "patent" medications promised eradication of these unwanted guests (Fig. 1). While these medications failed; modern sewage treatment, cement sidewalks and regulated food industries in developed countries eliminated lifecycle pathways required to maintain helminth colonization. Even helminths with simple "hand to mouth" lifecycles such as *Enterobius vermicularis* (pinworm) have declined in the United States[13] and Europe.[14]

It is possible that eradication of these organism has unforeseen consequences. Multiples sclerosis is exceedingly rare where *T. trichiura* carriage rates are greater than 10%, creating a strong dichotomy ($p < 0.0001$) between areas where *T. trichiura* and multiple sclerosis are prevalent.[15] Furthermore, patients with MS who carry helminths have a milder disease course compared to MS patients without helminths.[16] This is associated with decreased IL12 and IFNγ production and increased FoxP3 expression and IL10 and TGFβ production by peripheral blood mononuclear cells in helminth colonized patients.[16] People with *A. lumbricoides* or hookworm (*Necator*) infections report less wheezing (a sign of asthma) than people without these infections.[17] This protection may be due to hookworm rather ascarid exposure.[18] Fewer Gabonese school children have atopic skin reactions to dust-mite allergens if they are infected with *S. hematobium* compared to children without that helminth.[19] Children repeatedly treated for geo-helminths (*T. trichiura* and *A. lumbricoides*) had increased skin responses to allergen tests compared to untreated children.[20] People living in areas endemic for *S. mansoni* report less wheezing and use of asthmatic medications than individuals living in nonendemic areas.[21]

Animal Models of Helminth Exposure

Worm-exposure protects animals in experimental models of these diseases, which further implicates the loss of helminths as an important environmental change promoting immune-mediated disease. Helminth exposure can prevent or reverse colitis in animal models of IBD. Mice and rats develop colitis when rectally exposed to trinitrobenzenesulfonic (TNBS) acid in 50% ethanol. Exposure to helminths prevents TNBS-type colitis.[22-26] Mice exposed to *S. mansoni* eggs, *T. spiralis* or *Hymenolepis diminuta or Heligmosomoides polygyrus* make less pro-inflammatory IL12/23p40, IFNγ and IL17 but more IL4 and immunoregulatory IL10 and TGFβ[22,23,25-27] (Elliott unpublished results). Helminths also induce mucosal CD8+ T-cells that inhibit the proliferation of other T-cells.[28] Thus, colonization with helminths induces several immune regulatory circuits in the gut that impede excessive intestinal inflammation. Protection from intestinal inflammation in these models requires intact host IL4 and IL10 signaling.[22,25,26]

IL10-deficient mice develop chronic colitis driven in part by normal gut flora.[29] Colonization with helminths like *Trichuris muris* or *H. polygyrus* exposure inhibits development of colitis and can actually reverse established colitis in $IL10^{-/-}$ mice.[1,30] Like in wild type mice, helminth colonization of $IL10^{-/-}$ mice inhibits gut-associated immune cell production of IFNγ and IL12/23p40 and augments mesenteric lymph node T-cell FoxP3 mRNA. This shows that helminth exposure can modulate inflammation in the absence of IL10. TGFβ appears to be more critical. Helminths failed

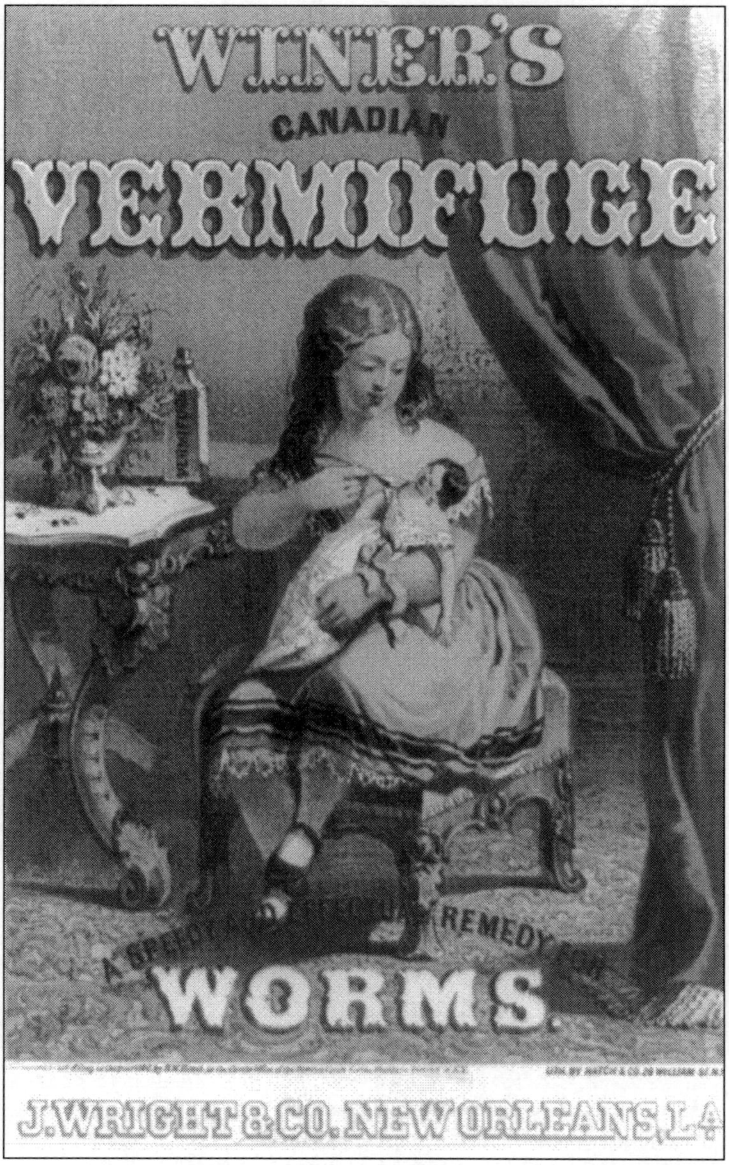

Figure 1. In the not so distant past, our predecessors were universally exposed to helminths. This is an advertisement showing a young girl treating her doll with a widely distributed circa 1850 herbal "de-worming medication" concocted by John Winer who was an enterprising former blacksmith from Hamilton Canada.

to regulate colitis or IFNγ production in mice that have a T-cell specific TGFβ signaling defect (Ince, unpublished results). Worms also induce CD8+ FoxP3+ T-cells in the gut that partially mediate this protection[28] (and Weinstock, unpublished results).

Exposing mice to *H. polygyrus* also changes the host's intestinal innate immune system. Once exposed to worms, the intestinal immune system is less able to support an adaptive immune response and becomes highly resistant to colitis. The gut T-cells display severely depressed IFNγ and IL17 responses (Elliott, unpublished results). The changes in the innate immunity and adaptive immunity persist even after elimination of the worms suggesting that the gut innate immune system is capable of prolonged 'memory' for this exposure.

IFNγ, IL17 and IL12p40 at least partly drive colitis in $IL10^{-/-}$ mice as demonstrated by antibody neutralization experiments. IL12p40 is the shared component of IL12 and IL23, which drive IFNγ and IL17, respectively. Worm infection blocks expression of all three in the gut. This could underlay the strong protection from colitis seen in worm colonization.

Exposure to helminths inhibits central nervous system inflammation in experimental autoimmune encephalitis (EAE), a rodent model of MS. Exposure of mice to viable *S. mansoni* or dead (freeze-thawed) *S. mansoni* eggs protects from EAE.[31,32] Schistosome exposure reduces proinflammatory IL12/23p40, IFNγ and TNFα and promotes regulatory TGFβ, IL10 and IL4 expression by splenocytes and CNS immune cells. Protection against EAE appears to require intact IL4 signaling.[31]

Colonization with helminths prevent onset of diabetes in the NOD murine model of T1D. NOD mice exposed to *S. mansoni* are protected from developing diabetes.[33,34] Schistosome egg or antigen exposure induces immune regulatory IL10 and augments NKT cell activity that may help prevent insulinitis. Colonization with *T. spiralis* or *H. polygyrus* also inhibits diabetes in NOD mice.[35]

Reactive airway disease improves with helminth exposure in murine models of asthma. Colonization of mice with male *S. mansoni* protects against airway hyperreactivity[36] but this protective effect is lost with productive (male and female worm) infection. Mice colonized only with male worms developed a modified immune response in the lung with decreased allergen-specific IL5 and increased IL10 production compared to uninfected mice. This decrease in allergen-specific IL5 release is similar to that reported in people with productive *S. mansoni* infections.[21]

Colonization with the intestinal helminth *H. polygyrus* inhibits airway inflammation and reactivity.[37,38] Helminth exposure is associated with decreased allergen-specific IL5 production. The regulatory cytokines induced by *H. polygyrus* differs by mouse strain showing parasite enhanced IL10 in C57BL/6 and TGFβ in BALB/c mice.[37] Protection requires intact IL10 signaling in C57BL/6[38] but not BALB/c[37] mice. *H. polygyrus* exposure increases the percentage of draining lymph node CD4+ T-cells that express regulatory cell markers (CD25 and FoxP3). Transfer of lymphocytes from colonized mice inhibits airway inflammation in recipient worm-free mice demonstrating that helminth exposure induced regulatory T-cell function.[37,38]

The effect of helminth infections on allergic-type diseases has prompted a search for helminth-derived factors with therapeutic potential. Cystatin is a filarial-derived molecule with antiallergic and anti-inflammatory properties. Treatment with cystatin protects mice from DSS colitis and lung hypersensitivity responses to OVA allergen challenge.[39] Cystatin induces macrophages to make IL10, which could mediate this protection.

Filarial nematodes also secrete a glycoprotein call ES-62 with anti-allergy properties. It directly inhibits FcεR1 induced release of mediators from human and rodent mast cells. It forms a complex with TLR4 resulting in cellular depletion of protein kinase C-α, a molecule important for FcεR1 signaling, thus preventing mast cell degranulation in response to IgE.[40]

Therapeutic Use of Helminths

These animal models of immune-mediated diseases demonstrate that helminth colonization induces multiple immune regulatory circuits that abrogate pathologic inflammation. Similar immune regulatory circuits have been shown in helminth-infected people.[19,41] A recent case report described a patient whose ulcerative colitis worsened after eradication of pin worm (*Enterobius vermicularis*). Mucosal biopsies taken while colonized showed enhanced FoxP3, IL10 and TGFβ expression in the mucosa compared to after eradication.[42] This suggests that helminth exposure may prevent or treat immune-mediated illness in patients through induction of regulatory T-cells in the gut.

Thus, helminths could have therapeutic application. If so, which helminths could be used therapeutically? There are many species of helminths that are divided taxonomically into nematodes (roundworms) and platyhelminths (flatworms). Platyhelminths are further divided into trematodes (flukes) and cestodes (tapeworms). Although nematodes and platyhelminths parasitize mammals and induce regulatory T-cell activity they are distant phylogenetically. Moreover, helminths have different mechanisms of transmission, different patterns of infection and dwell in different parts of the body. Some can cause significant disease that negates their therapeutic application. Others, with minimal or no known pathogenicity are being studied clinically.

Hookworm (*Necator americanus*) is being investigated for therapeutic use. Hookworms have pathogenic potential[43] but a light infection with *N. americanus* is likely to be safe.[44] Acute infection, even if light, can cause gastrointestinal symptoms.[45] People are colonized by applying infective larvae to the skin. The larvae migrate to the lungs, enter the bronchi and migrate up the trachea to the throat and are then swallowed. Once in the intestine they mature. Larvae are cultured from the stool of human volunteer donors that are actively colonized with *N. americanus*. These donors are extensively screened to reduce the risk of cotransmitting other infections. A feature of *N. americanus* that makes it an attractive candidate is that the worm establishes a chronic infection that can last at least 6 years. Therefore, repeat dosing may not be needed. In addition the systemic exposure created by larval migration may activate different immune compartments than a strictly intestinal exposure. However, exposure to strictly intestinal helminths does have systemic impact in mice.[46]

A small open-label trial tested *N. americanus* in 9 patients with Crohn's disease.[47] Two of these patients had moderately active disease when they received 50 larvae. Both showed improvement in their symptom scores. The other 7 patients in the initial trial had inactive or very mild disease which did not significantly change with helminth exposure. There was also a dose-ranging study of *N. americanus*[48] in preparation for a clinical trial in asthma.[49] This study enrolled 12 volunteers who were blinded to the dose of helminths they received. Two persons withdrew (one given 100 and one given 50 larvae) due to symptoms of diarrhea and vomiting or abdominal pain. However, the lowest (10 larvae) dose was well tolerated and resulted in patent colonization. Larger studies using patients with Crohn's disease and asthma are underway.

Most of the therapeutic helminth trials to date have used *Trichuris suis*, the porcine whipworm (Fig. 2). *T. suis* is closely related to *T. trichiura* (human whipworm) and can briefly colonize people.[50] *Trichuris spp*. have unique features that make them good candidates for clinical use. Ingesting microscopic embryonated parasite eggs initiates the colonization. Once ingested each egg releases one larva that matures into an adult worm. The larvae and adults do not migrate beyond the intestines. They do not multiply within their host. Freshly deposited ova require incubation in moist soil for one to two months to mature and become infective. Thus, normal hygienic practices preclude transmission of this agent from host to host.

Trichuris suis has additional characteristics that make it an attractive candidate. Human whipworms (*T. trichiura*) can be obtained only from colonized primates. *T. suis* is obtained from pigs raised in a specific pathogen-free environment. For the published studies, adult worms were collected from the colon, washed and cultured in vitro to collect freshly deposited ova. These ova were washed extensively and incubated for months to allow embryonation. This removed the risk of contaminating the *T. suis* ova with other infectious agents. Since *T. suis* is present in many pig herds, farmers must be exposed to *T. suis* and this organism has not been documented to cause human disease. This suggests that any risk from using this helminth is likely to be small.

Initially, the effect of *T. suis* colonization was studied in a small trial of 7 patients with IBD (4 Crohn's disease, 3 ulcerative colitis) under conditions where the patients and their care givers knew that they had received helminths ("open-label"). The patients ingested 2500 embryonated ova and were observed. All had improvement in their symptoms.[51] A second study tested repeated dosing of *T. suis* in 29 patients with active Crohn's disease.[52] The patients received 2500 *T. suis* ova every 3 weeks for 24 weeks. At week 24, 79% had responded with a significant reduction in symptoms. A third study was a double blind placebo-controlled trial of *T. suis* in 54 patients

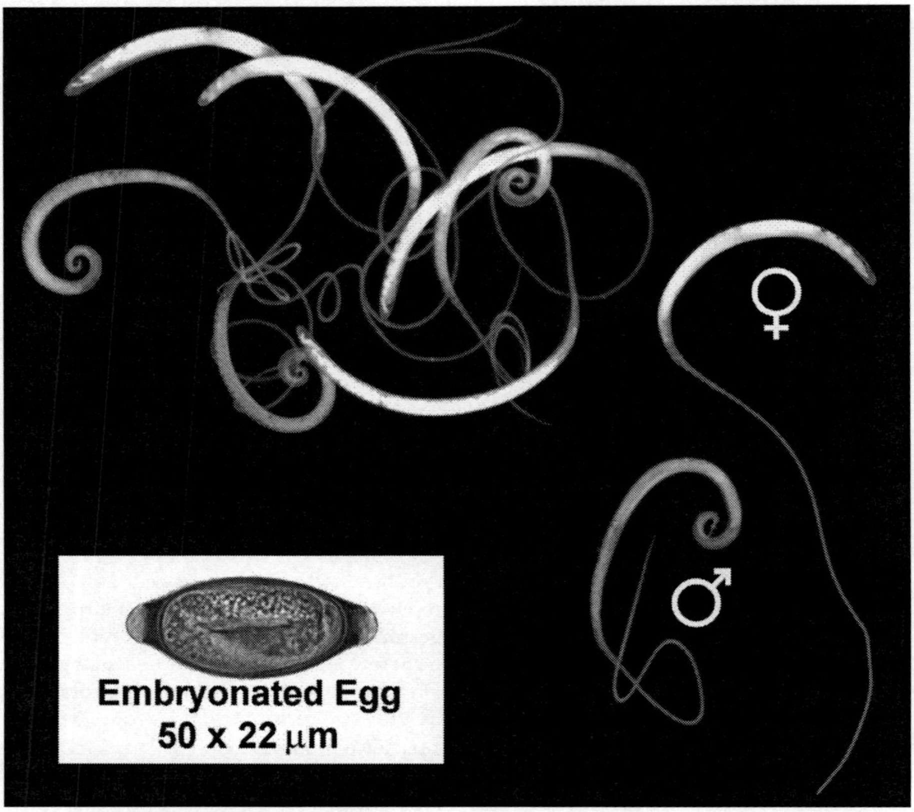

Figure 2. *Trichuris suis* adult male (♂) and female (♀)worms. The insert shows a microscopic embryonated egg which is the form ingested by patients.

with active ulcerative colitis.[53] Double-blind placebo-controlled trials remove a source of bias by preventing either the patient or the care givers from knowing if the patient receives the treatment or a placebo. The patients received either a placebo or 2500 *T. suis* ova every 2 weeks for 12 weeks. A significant percentage (43.3%) of the patients given *T. suis* improved compared to those given placebo (16.7% , $p < 0.04$). The study also included a blinded crossover limb where patients originally on placebo where switched to *T. suis* and those on *T. suis* were switched to placebo. In the crossover limb, 56.3% of the patients given *T. suis* improved compared to 13.3% of patients given placebo ($p = 0.02$).[54] Trials are now underway to investigate the safety and efficacy of *T. suis* in allergic rhinitis and multiple sclerosis. . Additional studies in Crohn's disease, ulcerative colitis and other immunological diseases are planned for the near future.

Controversy with Helminthic Therapy

The therapeutic application of helminths is not without controversy. At least a third of the world's population is actively infected with helminths and eradication programs struggle for resources. Some helminths can cause significant disease. This leads people to ask; "Isn't this dangerous?" or "Will patients actually want to have worms?"

We typically use medications with significant toxicities and side-effects to treat immune-mediated diseases. Most of these chronic life-threatening diseases require immune suppression mediated by glucocorticoids (e.g., prednisolone, hydrocortisone), anti-metabolites (e.g., methotrexate, azathioprine) and/or "biologics" (e.g., anti-TNFα antibodies, anti-integrin antibodies, anti-B-cell antibodies) that globally block or remove immune circuits. Each of these medications has adverse effects that severely limit their application. Each can be dangerous even when closely monitored. A decision analysis model investigating the use of infliximab (an anti-TNFα antibody) in Crohn's disease indicated that treating 100,000 patients for one year helps maintain remission, reduces the need for surgery and results in 33 fewer deaths from the disease, but results in 200 more lymphomas and 250 deaths from infliximab-related complications.[55] These medications have narrow "therapeutic windows", which is the dose range that improves disease without causing serious toxicity. Helminths can be closely monitored and discontinued (with an anti-helminthic if necessary). The therapeutic window for helminths is probably quite wide since most natural infections are asymptomatic.

One area of specific safety concern is whether helminth colonization may worsen certain pathogenic bacterial infections. For example, mice cocolonized with *H. polygyrus* and *Citrobacter rodentium* have increased susceptibility to this well-characterized bacterial pathogen.[56] This susceptibility requires Th2 signaling and results from induction of IL10-producing dendritic cells[57] and alternatively activated macrophages.[58] A similar effect is seen in pigs cocolonized with *T. suis* and *Camplyobacter jejuni.*[59] In both of these examples, helminths increase susceptibility to known pathogens but did not make commensal or nonpathogenic bacteria dangerous. Pathogenic bacteria, by definition, have the ability to evade host immune barricades. The immune suppressive medications presently used to treat immune-mediated disease (glucocorticoids, etc ...) can dramatically increase susceptibility to pathogenic and opportunistic organisms. Unlike immunosuppressive drugs, helminths do not increase susceptibility to opportunistic infections. Patients easily can be tested and treated for infection with pathogens. This is a common procedure ("standard of practice") when using immune suppressive drugs.

Another area of specific safety concern is whether helminths will become pathogenic in patients with impaired immunity like those treated with immune suppressive medications. For most helminths, host immune responses promote expulsion and retard re-infection. Therefore, patients with impaired immunity may be more easily colonized and may "keep" a helminth for a longer period of time. This would merely affect dosing schedules. Most helminths are biologically unable to multiply in their host. There is no alteration in host immunity that would permit an increase in helminth (e.g., *T. suis*) number without re-exposure to the infective stage. Only two helminths are known to circumvent this rule, *Strongyloides stercoralis* and *Paracapillaria philippinensis*. *S. stercoralis* can multiply and the infection can become fulminant when hosts are treated with glucocorticoids. *P. philippinensis* can multiply and become fulminant in hosts with intact immunity. Neither *S. stercoralis* nor *P. philippinensis* are proposed for therapeutic use. Other helminths neither multiply nor expand their niche in patients with HIV-AIDS.[60]

Patients are not averse to trying helminths. People are familiar with therapeutic virus exposure (vaccines). The concept of taking a helminth to alter immunity is not shocking. Helminth eggs and larvae are microscopic. Their appearance does not elicit a phobic response. Neither hookworm nor whipworm are easily seen in the stool. Whipworm cannot be passed to others in the household. Patients with life-threatening chronic diseases take medications with long lists of dangerous adverse effects. In this context, helminths are welcome. Many medications require daily ingestion or frequent injection. Episodic helminth exposure is far less challenging. If helminths prove to be as efficacious as early experience suggests, patient acceptance is unlikely to be a limiting factor.

Helminths alter immune responses by releasing factors. Many of these specific molecules are discussed in this book. Perhaps treatment with modified helminth-derived components or small molecule mimics could obviate the need for actual helminth exposure. On the other hand, each helminth may exude many immune modulating factors. Helminths are complex organisms with large diverse genomes.[61] They adapt to their environments. It is quite probable that the immune effects of a viable helminth will be greater than that provided by one or more of its molecular components. Moreover, a viable helminth will continuously interact with the host immune system. Administered components would require frequent administration. If viable helminths prove exceedingly safe for therapeutic application, there may be little benefit to administering only isolated components. On the other hand, different helminths could utilize different circuits. In this case, adding purified factors from different helminths could create an immune regulatory cocktail not achievable with natural infection.

Helminths may not inhibit or modulate all types of inflammation. Mice with oxazalone colitis, another model of intestinal inflammation, develop worse disease if colonized with the tape worm *Hymenolepis diminuta*.[62] Yet, exposure to this helminth protects from TNBS-like colitis[25] and improves some aspects of dextran sulfate sodium (DSS) colitis.[63] Similarly, mice with mixed-sex *S. mansoni* infections are more susceptible to DSS-induced colitis relative to uninfected mice, while schistosome male worm-only infections are refractory to DSS colitis.[64]

Conclusions

There is a spectrum of over 40 different immune-mediated diseases which together affect greater than 10% of the population in highly developed countries. Each immune-mediated disease is unique. However, Th1/Th17-type diseases like multiple sclerosis or Crohn's disease and Th2-type diseases like asthma appear to be suppressed by helminths. As we learn more about the specific immune dysregulation that occurs in each disease and the different immune regulatory pathways helminths exploit, we may be able to predict which patients will benefit from helminth exposure.

Acknowledgements

Supported by grants from the National Institutes of Health (AI49382, DK38327, DK58755, DK02428, DK25295), the Department of Veterans Affairs Office of Research and Development, the Crohn's and Colitis Foundation of America and the Broad Medical Research Program of the Eli and Edythe L. Broad Foundation.

References

1. Elliott DE, Urban JFJ, Argo CK et al. Does the failure to acquire helminthic parasites predispose to Crohn's disease? FASEB Journal 2000; 14(12):1848-1855.
2. Loftus EV Jr. Clinical epidemiology of inflammatory bowel disease: Incidence, prevalence and environmental influences. Gastroenterology 2004; 126(6):1504-1517.
3. Lakatos L, Mester G, Erdelyi Z et al. Striking elevation in incidence and prevalence of inflammatory bowel disease in a province of western Hungary between 1977-2001. World Journal of Gastroenterology 2004; 10(3):404-409.
4. Marrie RA. Environmental risk factors in multiple sclerosis aetiology. Lancet Neurol 2004; 3(12):709-718.
5. Cabre P, Signate A, Olindo S et al. Role of return migration in the emergence of multiple sclerosis in the French West Indies. Brain 2005; 128(Pt 12):2899-2910.
6. Onkamo P, Vaananen S, Karvonen M et al. Worldwide increase in incidence of Type I diabetes—the analysis of the data on published incidence trends. Diabetologia 1999; 42(12):1395-1403.
7. Braman SS. The global burden of asthma. Chest 2006; 130(1 Suppl):4S-12S.
8. Eldeirawi KM, Persky VW. Associations of acculturation and country of birth with asthma and wheezing in Mexican American youths. J Asthma 2006; 43(4):279-286.
9. Asher MI, Montefort S, Bjorksten B et al. Worldwide time trends in the prevalence of symptoms of asthma, allergic rhinoconjunctivitis and eczema in childhood: ISAAC phases one and three repeat multicountry cross-sectional surveys. Lancet 2006; 368(9537):733-743.
10. Goncalves ML, Araujo A, Ferreira LF. Human intestinal parasites in the past: new findings and a review. Memorias do Instituto Oswaldo Cruz 2003; 98 Suppl 1:103-118.
11. Bethony J, Brooker S, Albonico M et al. Soil-transmitted helminth infections: ascariasis, trichuriasis and hookworm. Lancet 2006; 367(9521):1521-1532.
12. Crompton DW. How much human helminthiasis is there in the world? J Parasitol 1999; 85(3):397-403.
13. Vermund SH, MacLeod S. Is pinworm a vanishing infection? Laboratory surveillance in a New York City medical center from 1971 to 1986. Am J Dis Child 1988; 142(5):566-568.
14. Gale EA. A missing link in the hygiene hypothesis? Diabetologia 2002; 45(4):588-594.
15. Fleming JO, Cook TD. Multiple Sclerosis and the Hygiene Hypothesis. Neurology 2006; 67(11):2085-2086.

16. Correale J, Farez M. Association between parasite infection and immune responses in multiple sclerosis. Ann Neurol 2007; 61(2):97-108.
17. Scrivener S, Yemaneberhan H, Zebenigus M et al. Independent effects of intestinal parasite infection and domestic allergen exposure on risk of wheeze in Ethiopia: a nested case-control study. Lancet 2001; 358(9292):1493-1499.
18. Leonardi-Bee J, Pritchard D, Britton J. Asthma and current intestinal parasite infection: systematic review and meta-analysis. Am J Respir Crit Care Med 2006; 174(5):514-523.
19. van den Biggelaar AH, van Ree R, Rodrigues LC et al. Decreased atopy in children infected with Schistosoma haematobium: a role for parasite-induced interleukin-10. Lancet 2000; 356(9243):1723-1727.
20. van den Biggelaar AH, Rodrigues LC, van Ree R et al. Long-term treatment of intestinal helminths increases mite skin-test reactivity in Gabonese schoolchildren. Journal of Infectious Diseases 2004; 189(5):892-900.
21. Araujo MI, Hoppe BS, Medeiros M Jr et al. Schistosoma mansoni infection modulates the immune response against allergic and auto-immune diseases. Mem Inst Oswaldo Cruz 2004; 99(5 Suppl 1):27-32.
22. Elliott D, Li J, Blum A et al. Exposure to schistosome eggs protects mice from TNBS-induced colitis. American Journal of Physiology 2003; 284:G385-G391.
23. Khan WI, Blennerhasset PA, Varghese AK et al. Intestinal nematode infection ameliorates experimental colitis in mice. Infection and Immunity 2002; 70(11):5931-5937.
24. Moreels TG, Nieuwendijk RJ, De Man JG et al. Concurrent infection with Schistosoma mansoni attenuates inflammation induced changes in colonic morphology, cytokine levels and smooth muscle contractility of trinitrobenzene sulphonic acid induced colitis in rats. Gut 2004; 53(1):99-107.
25. Hunter MM, Wang A, Hirota CL et al. Neutralizing anti-IL-10 antibody blocks the protective effect of tapeworm infection in a murine model of chemically induced colitis. Journal of Immunology 2005; 174(11):7368-7375.
26. Setiawan T, Metwali A, Blum AM et al. Heligmosomoides polygyrus promotes regulatory T-cell cytokine production in normal distal murine intestine. Infect Immun 2007; 75(9):4655-4663.
27. Ince MN, Elliott DE, Setiawan T et al. Heligmosomoides polygyrus induces TLR4 on murine mucosal T-cells that produce TGFbeta after lipopolysaccharide stimulation. Journal of Immunology 2006; 176(2):726-729.
28. Metwali A, Setiawan T, Blum AM et al. Induction of CD8+ regulatory T-cells in the intestine by Heligmosomoides polygyrus infection. Am J Physiol Gastrointest Liver Physiol 2006; 291(2):G253-G259.
29. Sellon RK, Tonkonogy S, Schultz M et al. Resident enteric bacteria are necessary for development of spontaneous colitis and immune system activation in interleukin-10-deficient mice. Infection and Immunity 1998; 66(11):5224-5231.
30. Elliott DE, Setiawan T, Metwali A et al. Heligmosomoides polygyrus inhibits established colitis in IL-10-deficient mice. European Journal of Immunology 2004; 34(10):2690-2698.
31. Sewell D, Qing Z, Reinke E et al. Immunomodulation of experimental autoimmune encephalomyelitis by helminth ova immunization. International Immunology 2003; 15(1):59-69.
32. La Flamme AC, Ruddenklau K, Backstrom BT. Schistosomiasis decreases central nervous system inflammation and alters the progression of experimental autoimmune encephalomyelitis. Infection and Immunity 2003; 71(9):4996-5004.
33. Cooke A, Tonks P, Jones FM et al. Infection with Schistosoma mansoni prevents insulin dependent diabetes mellitus in nonobese diabetic mice. Parasite Immunology 1999; 21(4):169-176.
34. Zaccone P, Fehervari Z, Jones FM et al. Schistosoma mansoni antigens modulate the activity of the innate immune response and prevent onset of Type 1 diabetes. Eur J Immunol 2003; 33(5):1439-1449.
35. Zaccone P, Fehervari Z, Phillips JM et al. Parasitic worms and inflammatory diseases. Parasite Immunology 2006; 28(10):515-523.
36. Mangan NE, Van RN, McKenzie AN et al. Helminth-modified pulmonary immune response protects mice from allergen-induced airway hyperresponsiveness. J Immunol 2006; 176(1):138-147.
37. Wilson MS, Taylor MD, Balic A et al. Suppression of allergic airway inflammation by helminth-induced regulatory T-cells. Journal of Experimental Medicine 2005; 202(9):1199-1212.
38. Kitagaki K, Businga TR, Racila D et al. Intestinal helminths protect in a murine model of asthma. J Immunol 2006; 177(3):1628-1635.
39. Schnoeller C, Rausch S, Pillai S et al. A Helminth Immunomodulator Reduces Allergic and Inflammatory Responses by Induction of IL-10-Producing Macrophages. J Immunol 2008; 180(6):4265-4272.
40. Melendez AJ, Harnett MM, Pushparaj PN et al. Inhibition of Fc epsilon RI-mediated mast cell responses by ES-62, a product of parasitic filarial nematodes. Nat Med 2007; 13(11):1375-1381.
41. Doetze A, Satoguina J, Burchard G et al. Antigen-specific cellular hyporesponsiveness in a chronic human helminth infection is mediated by T(h)3/T(r)1-type cytokines IL-10 and transforming growth factor-beta but not by a T(h)1 to T(h)2 shift. International Immunology 2000; 12(5):623-630.

42. Buening J, Homann N, Von SD et al. Helminths as Governors of Inflammatory Bowel Disease. Gut 2008; 57(8):1182-1183.
43. Hotez PJ, Brooker S, Bethony JM et al. Hookworm infection. N Engl J Med 2004; 351(8):799-807.
44. Pritchard DI, Brown A. Is Necator americanus approaching a mutualistic symbiotic relationship with humans? Trends in Parasitology 2001; 17(4):169-172.
45. Maxwell C, Hussain R, Nutman TB et al. The clinical and immunologic responses of normal human volunteers to low dose hookworm (Necator americanus) infection. Am J Trop Med Hyg 1987; 37(1):126-134.
46. Mohrs K, Harris DP, Lund FE et al. Systemic dissemination and persistence of Th2 and Type 2 cells in response to infection with a strictly enteric nematode parasite. J Immunol 2005; 175(8):5306-5313.
47. Croese J, O'neil J, Masson J et al. A proof of concept study establishing Necator americanus in Crohn's patients and reservoir donors. Gut 2006; 55(1):136-137.
48. Mortimer K, Brown A, Feary J et al. Dose-ranging study for trials of therapeutic infection with Necator americanus in humans. Am J Trop Med Hyg 2006; 75(5):914-920.
49. Falcone FH, Pritchard DI. Parasite role reversal: worms on trial. Trends Parasitol 2005; 21(4):157-60.
50. Beer RJ. The relationship between Trichuris trichiura (Linnaeus 1758) of man and Trichuris suis (Schrank 1788) of the pig. Research in Veterinary Science 1976; 20(1):47-54.
51. Summers RW, Elliott DE, Qadir K et al. Trichuris suis seems to be safe and possibly effective in the treatment of inflammatory bowel disease. American Journal of Gastroenterology 2003; 98(9):2034-2041.
52. Summers RW, Elliott DE, Urban JF Jr et al. Trichuris suis therapy in Crohn's disease. Gut 2005; 54(1):87-90.
53. Summers RW, Elliott DE, Urban JF Jr et al. Trichuris suis therapy for active ulcerative colitis: a randomized controlled trial. Gastroenterology 2005; 128(4):825-832.
54. Elliott DE, Summers RW, Weinstock JV. Helminths and the modulation of mucosal inflammation. Current Opinion in Gastroenterology 2005; 21(1):51-58.
55. Siegel CA, Hur C, Korzenik JR et al. Risks and benefits of infliximab for the treatment of Crohn's disease. Clin Gastroenterol Hepatol 2006; 4(8):1017-1024.
56. Chen CC, Louie S, McCormick B et al. Concurrent infection with an intestinal helminth parasite impairs host resistance to enteric Citrobacter rodentium and enhances Citrobacter-induced colitis in mice. Infect Immun 2005; 73(9):5468-5481.
57. Chen CC, Louie S, McCormick BA et al. Helminth-primed dendritic cells alter the host response to enteric bacterial infection. J Immunol 2006; 176(1):472-483.
58. Weng M, Huntley D, Huang IF et al. Alternatively activated macrophages in intestinal helminth infection: effects on concurrent bacterial colitis. J Immunol 2007; 179(7):4721-4731.
59. Mansfield LS, Gauthier DT, Abner SR et al. Enhancement of disease and pathology by synergy of Trichuris suis and Campylobacter jejuni in the colon of immunologically naive swine. Am J Trop Med Hyg 2003; 68(1):70-80.
60. Karp CL, Auwaerter PG. Coinfection with HIV and tropical infectious diseases. II. Helminthic, fungal, bacterial and viral pathogens. Clin Infect Dis 2007; 45(9):1214-1220.
61. Mitreva M, Blaxter ML, Bird DM et al. Comparative genomics of nematodes. Trends Genet 2005; 21(10):573-581.
62. Hunter MM, Wang A, McKay DM. Helminth infection enhances disease in a murine TH2 model of colitis. Gastroenterology 2007; 132(4):1320-1330.
63. Reardon C, Sanchez A, Hogaboam CM et al. Tapeworm infection reduces epithelial ion transport abnormalities in murine dextran sulfate sodium-induced colitis. Infection and Immunity 2001; 69(7):4417-4423.
64. Smith P, Mangan NE, Walsh CM et al. Infection with a helminth parasite prevents experimental colitis via a macrophage-mediated mechanism. J Immunol 2007; 178(7):4557-4566.

Chapter 13

Chemokine Binding Proteins Encoded by Pathogens

Antonio Alcami* and Margarida Saraiva

Abstract

Chemokines are chemoattractant cytokines that play an important role in immunity. The role of chemokines against invading pathogens is emphasized by the expression of chemokine inhibitors by many pathogens. A mechanims employed by poxviruses and herpesviruses is the secretion of chemokine binding proteins unrelated to host receptors that bind chemokines with high affinity and block their activity. Soluble chemokine binding proteins have also been identified in the human parasite *Schistosoma mansoni* and in ticks. The binding specificity of these inhibitors of cell migration point at chemokines that contribute to host defense mechanisms against various pathogens. Chemokine binding proteins modulate the immune response and may lead to new therapeutic approaches to treat inflamatory diseases.

Modulation of the Chemokine System by Pathogens

The infection with pathogens triggers signals that initiate the immune response and the recruitment of immune cells to sites of infection. The migration of leukocytes is largely controlled by chemokines, a family of chemoattractant cytokines that play a key role in inflammation and host defence against infectious agents.[1-3] Members of the chemokine family share structural similarities and can be divided into four classes: CC, CXC, C and CX3C chemokines. The induction of particular chemokines together with the differential expression of specific seven-transmembrane-domain G-protein-coupled chemokine receptors by leukocyte subsets determines the immune cells that migrate towards sites of replication of pathogens within the animal host.

Chemokines interact with both their specific receptors and with cell surface glycosaminoglycans (GAGs) via distinct binding sites.[4,5] Receptor binding is the means by which chemokines transduce their biological signals and trigger leukocyte migration. It is believed that under physiological conditions chemokines do not act in solution but are presented to chemokine receptors on leukocytes as ligands immobilized to a solid phase via interaction with GAGs. Disruption of either chemokine-receptor or chemokine-GAG complex formation might therefore inhibit chemokine biological activity.[6,7]

Infectious agents that replicate in an animal host must actively evade host immune defences that would otherwise eliminate them.[8-10] The immune evasion strategies encoded by parasites, bacteria and viruses are diverse and illustrate an adaptation to their replication strategy either outside the cell or within cell compartments, ability to mutate, host cell and tissue tropism, transmission mechanism and genome coding capacity. The large DNA genome of poxviruses and herpesviruses allow these viruses, in contrast to the smaller RNA viruses, to encode many genes that control their interaction with the host cell and the immune system. One of the immune

*Corresponding Author: Antonio Alcami—Centro de Biología Molecular Severo Ochoa (Consejo Superior de Investigaciones científicas—Universidad Autónoma de Madrid), Nicolás Cabrera 1, Cantoblanco, 28049 Madrid, Spain. Email: aalcami@cbm.uam.es

Pathogen-Derived Immunomodulatory Molecules, edited by Padraic G. Fallon.

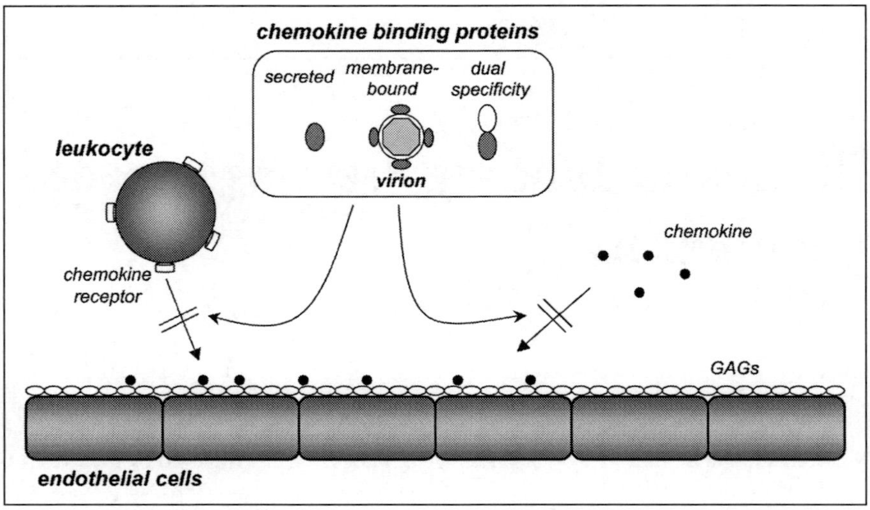

Figure 1. Mechanism of chemokine inhibition by the CKBPs encoded by pathogens. Chemokines are presented to the leukocytes on the surface of endothelial cells where they interact with GAGs. The chemokine receptors expressed at the surface of leukocytes interact with chemokines and intracelular signalling is triggered to induce cell migration. The CKBPs sequester chemokines and inhibit their interaction with either chemokine receptors or GAGs. The secreted CKBPs may be specific for chemokines, such as the gammaherpesvirus M3 or the poxvirus 35-kDa protein, or may be part of a protein with dual cytokine specificity, interacting with chemokines and other cytokines, such as TNF (poxvirus CrmB) or IFN-γ (MYXV M-T7). The CKBP gG encoded by alphaherpesviruses is expressed as a membrane protein at the virion surface that is proteolitically cleaved to produce a secreted protein.

evasion strategies employed by poxviruses and herpesviruses is the molecular mimicry of cytokines and their receptors to intercept the cytokine networks that control the immune response to infection.[11] Viral mechanisms that modulate the activity of chemokines include the expression of chemokine homologues, seven-transmembrane-domain chemokine receptor homologues or secreted chemokine binding protiens (CKBPs). More recently, some examples of CKBPs have been found in other pathogens different to viruses.

Here we review our current knowledge of pathogen-encoded CKBPs and describe their unique properties, mechanism of action and, if known, their contribution to pathogenesis.

The M-T7 Protein Encoded by Myxoma Virus (MYXV)

Myxomatosis is a severe disease of European rabbits that is caused by MYXV. Initial work identified the MYXV M-T7 gene encoding a 37-kDa glycoprotein that is abundantly secreted from infected cells.[12] The M-T7 ORF showed significant similarity to the extracellular binding domain of human and mouse interferon-γ receptors (IFN-γRs) and was shown to bind to and inhibit the biological activity of rabbit IFN-γ in a species specific manner.[12,13]

M-T7 was also found to bind a broad range of C, CXC and CC chemokines and studies with CXCL8 mutants suggested that chemokine binding to M-T7 is via the conserved C-terminal GAG binding domain found in a variety of chemokines.[14] This finding was unexpected and is a unique property of the IFN-γR encoded by MYXV since the IFN-γR orthologue encoded by VACV does not bind chemokines.[14,15] The interaction of M-T7 with the chemokine GAG binding domains led to the suggestion that M-T7 might prevent the correct localization of chemokines and the formation of a chemokine gradient, rather than the blockade of chemokine binding to specific receptors.[14]

Table 1. CKBPs encoded by pathogens

Pathogen	Protein	Species	Binding Properties and Mechanism of Action	Reference
Poxvirus	M-T7	MYXV	Secreted, binds C, CC and CXC chemokines through GAG binding domain and IFN-γ	14
	35-kDa	VAVC, CPXV, ECTV, MYXV	Secreted, binds CC chemokines, prevents interaction of CC chemokines with specific receptors, inhibits cell migration	15,17,18
	A41	VACV, ECTV	Secreted, binds CC and CXC chemokines, blocks interaction of chemokines with GAGs	34,35
	SECRET domain	VARV, ECTV, CPXV	Secreted, protein domain fused to the TNFRs CrmB and CrmD, or expressed independently as three SCPs encoded by different genes, binds a limited set of CC and CXC chemokines, inhibits cell migration	42
Herpesvirus	M3	MHV-68	Secreted, binds CC, CXC, C and CX3C chemokines, blocks the interaction of chemokines with receptors and GAGs, and cell migration	52,53
	gG	EHV-1, BHV-1, FeHV-1	Membrane-anchored at the surface of virus particles and infected cells, secreted after proteolytic cleavage, binds CC and CXC chemokines, blocks the interaction of chemokines with cellular receptors and GAGs, and cell migration	60,62
	pUL21.5	HCMV	Binds CCL5, mRNA packaged into virions, inhibits binding of CCL5 to cellular receptors	68
Parasites	smCKBP	Schistosoma mansoni	Binds some CC, CXC and CX3C chemokines, expressed by schistosome eggs, block migration induced by chemokines	72
Ticks	Evasins	Rhipicephalus sanguineus	Family of proteins (Evasins 1, 2 and 4) of restricted chemokine binding specificity, block binding of chemokines to cellular receptors and cell migration	73,74

Infection of rabbits with a MYXV mutant with an inactivated *M-T7* gene demonstrated de contribution of M-T7 to MYXV pathogenesis in European rabbits.[16] Marked differences were seen in the size and progression of skin lesions, the onset and severity of secondary bacterial infections and clearance of the virus. M-T7 was implicated in the control of migration of inflammatory cells to sites of infection. However, these results are difficult to interpret because M-T7 targets IFN-γ and chemokines and both have important roles in inflammatory responses.

The 35-kDa CKBP Encoded by Poxviruses

Poxvirus genomes encode a second class of CKBP that inhibits CC chemokines by interfering with the interaction of chemokines with their receptor.[15,17,18] The major secreted protein of 35 kDa encoded by VACV strains Lister and rabbitpox was identified as a CKBP. This protein is not expressed by the most-commonly used VACV strains Western Reserve (WR) and Copenhagen. In the WR strain the protein is truncated and its promoter, known as the p7.5 early-late promoter, has been widely used in VACV vectors driving the expression of genes of interest. Parallel experiments with the variola virus (VARV) orthologue indicated that the virus which caused smallpox in humans also expressed this activity.[18] The 35-kDa CKBP is also expressed by a wide variety of poxviruses including MYXV (M-T1 protein), cowpox virus (CPXV) and ectromelia virus (ECTV).[18,19] Orf virus is a parapoxvirus that causes infections in sheep and cattle and sporadic infections in humans. It was shown that orf virus encodes a protein related to the 35-kDa CKBP but with a broader chemokine binding specificity that included the C chemokine lymphotactin in addition to CC chemokines.[20] The binding of members of the 35-kDa CKBP family to numerous chemokines has demonstrated that it binds to almost all human and mouse CC chemokines with high affinity.[21] Low-affinity binding to CXCL1 and CXCL8 has been detected, but it has not been possible to demonstrate binding to a range of other CXC, C or CX3C chemokines.[15,18,21,22]

The mechanism of action of the 35-kDa CKBP is competitive inhibition of CC-chemokine binding to cellular receptors, inhibiting as a consequence the induction of transient increases in calcium concentrations and the migration of cells along chemotactic gradients.[15,18,22] The M-T1 protein of MYXV has the unique ability to interact with GAGs via a GAG binding domain at its C-terminus that is not present in other 35-kDa family members.[23] This unique property of M-T1 would retain the protein in the vicinity of infected cells and may enhance its ability to protect the sites of infection from chemokine-mediated anti-viral responses.

The structure of the CPXV 35-kDa CKBP was determined by X-ray crystallography and it was shown to be a compact globular protein composed of two large parallel β-sheets, two short α-helices and several large connecting loops.[24] The β sandwich topology of the protein is thought to be unique. By looking for exposed charged residues that have no apparent structural role and which are conserved among different members of the 35-kDa CKBP family, potential chemokine binding sites were suggested on the exposed face of β sheet II and at the edge of β sheet II. Further structural studies in solution by nuclear magnetic resonance demonstrated that the interface of interaction of the 35-kDa protein from VACV strain rabbitpox and human chemokine CCL2 involved the domains predicted from the initial crystalographic structure.[25] The N-terminal residues of CCL2, as well as residues in the 20′s region and 40′s loop of the chemokine are involved in binding to the CKBP and established the structural basis for the ability of this vCKBP to promiscuously recognize CC chemokines. These results confirmed previous binding studies with human CCL2 mutants that identified the amino acid residues required for high affinity interaction with the VACV 35-kDa CKBP which were similar to those involved in CCL2-CCR2b chemokine receptor binding.[26,27] The 35-kDa protein encoded by ECTV was also crystalized and shown to have a folding nearly identical to that of the CPXV and VACV proteins.[28]

The expression of an abundantly secreted CKBP would be expected to make a significant contribution to virus virulence in vivo by inhibiting CC-chemokine-mediated host inflammatory responses. Surprisingly, the experimental evidence in mouse and rabbit models of infection indicated that the 35-kDa protein encoded by VACV rabbitpox and MYXV inhibit the chemokine-mediated infiltration of immune cells into primary sites of infection but have little influence on the progression of disease.[17,29,30] The expression of the 35-kDa protein from VACV WR, a strain that does not encode the CKBP, caused a slight attenuation of the virus associated with reduced inflammatory pathology in the lungs, suggesting that this CKBP may attenuate the immune-mediated pathology caused by VACV infection.[31]

The A41 Family of Poxvirus CKBPs

The VACV A41 protein has immunomodulatory activity and sequence similarity to the 35-kDa CKBP from VACV and other poxviruses. Deletion of the *A41L* gene from VACV strain WR enhanced virulence slightly and showed an altered inflammatory response to infection in a dermal model.[32] Clark et al[33] showed that a VACV strain modified virus Ankara lacking the *A41L* gene induced better protection than control virus. However, none of these studies identify the ligand(s) for A41. Two different groups carried out an extensive screening of chemokines by Surface Plasmon Resonance (SPR) for their potential binding to recombinant A41 from ECTV and VACV and recently reported the identification of a set of CC and CXC chemokines that interact with the A41 protein.[34,35] In contrast to the poxvirus 35-kDa CKBP, the interaction of A41 with chemokines is inhibited in the presence of GAGs and the A41 protein does not block migration induced by chemokines. By using mutant forms of the chemokines CXCL10 and CXCL12α, which have reduced ability to interact with chemokine receptors or GAGs, Ruiz-Argüello et al[35] showed that the A41 CKBP interacts with the GAG-binding domain of chemokines, suggesting that the property of this protein to inhibit leukocyte migration in vivo may be related to its ability to block the correct interaction of chemokines with GAGs, which is required for appropriate chemokine function in vivo.[35]

The crystal structure of the A41 protein from VACV was determined and found to be related to the 35-kDa CKBP, but has notable structural differences particularly in surface loops and electrostatic charge distribution.[34] Although the A41-chemokine complex was not crystalized, structural modelling suggested that the interaction of A41 and 35-kDa proteins with chemokines involves the same domains. Based on the structural data, Bahar et al[34] suggested a revised model in which the functional distinction between A41 and 35-kDa proteins arises from the differences in binding affinity. The binding site on chemokines for GAGs and chemokine receptors frequently overlap and the high affinity of the 35-kDa protein is sufficient to interfere with receptor binding while the lower affinity of A41 for chemokines can block GAG binding but not receptor binding.

The A41 and 35-kDa CKBPs from poxviruses are structurally related but block the chemokine system in different but complementary ways, probably being advantageous for the virus to control the host anti-viral responses.

A Family of Poxvirus Proteins Containing the Smallpox Virus-Encoded Chemokine Receptor (SECRET) Domain

The finding of the SECRET domain was related to the study of the VARV-encoded Cytokine response modifier B (CrmB). CrmB is one of the four tumor necrosis factor receptor (TNFR) homologues encoded by poxviruses, named CrmB, CrmC, CrmD and CrmE and the only one encoded by VARV, the causative agent of smallpox.[36-39] A fifth homologue of the TNFR superfamily encoded by the poxviruses ECTV and CPXV is the viral homologue of CD30.[40,41] Sequence comparison of the poxvirus TNFRs showed that the N-terminal region of each molecule shares amino acid sequence similarity with the cysteine-rich domain of the mammalian counterparts that interact with TNF. However, it was evident that CrmB and CrmD have an additional C-terminal extension that contributes to half of the molecule size and for which no similarity was found. The C-terminal domain is not required for TNF binding and no function was predicted.

To explore the immunomodulatory activity of VARV CrmB, Alejo et al[42] expressed the protein in the baculovirus system to circumvent the direct manipulation of VARV, which is currently retricted to two high security laboratories.[43] The purified recombinant CrmB was tested in a screening with different cytokines by SPR and it was found that CrmB not only bound TNF but also various chemokines. Independent expression of the N-terminal and C-terminal CrmB domains demonstrated that the C-terminal domain interacts with chemokines. An extensive screening of all human and mouse chemokines by SPR showed that CrmB interacted with a restricted set of chemokines with binding affinities similar to that of TNF.[42] The human chemokines that best bound to VARV CrmB were CCL25, CCL28, CXCL12β, CXCL13 and CXCL14. Two additional chemokines (CCL27 and CXCL11) also interact with CrmB in the mouse system.

As the other CKBPs identified so far, the SECRET domain has no amino acid sequence similarity to host chemokine receptors or previously described CKBPs. However, the restricted chemokine binding specificity of the SECRET domain differs from the broad binding specificity of the previously identified CKBPs.

Given the sequence similarity between VARV CrmB and ECTV CrmD, this molecule was also tested for chemokine binding activity and shown to interact with the same limited set of human and mouse chemokines.[42] The number of molecules belonging to the SECRET family increased following the analysis of poxviral genomes and identification of other gene products encoding SECRET domain-containing proteins (SCPs), which were predicted to be secreted and of previously unknown function. The CPXV strain Brighton Red protein V218 and the ECTV strain Naval proteins E12 and E184 were shown to bind chemokines and named SCP-1, 2 and 3, respectively.[42] The fact that all members of the SECRET family bind the same set of chemokines, despite their relatively low sequence similarity, suggests that the SECRET domain has a specific folding, allowing it to bind chemokines with high affinity, either independently or fused to TNFRs.

The identification of the SECRET domain in five different poxvirus proteins is intriguing. This distribution may explain, in part, the variety of genes encoding TNFR homologues in poxvirus genomes, some of which (CrmB and CrmD) encode this additional chemokine-inhibitory activity. It may also provide the virus the ability to differentially block chemokines involved in controlling distinct antiviral responses, inhibit chemokines at different stages of infection in the animal host or simultaneously inhibit chemokines and TNF. It is likely that as poxviruses with narrow host species specificity adapted to particular hosts (i.e., VARV to humans or ECTV to mice), genes were selected to facilitate viral replication and transmission in each host.

The ability of VARV CrmB, CPXV CrmB, ECTV CrmD and CPXV SCP-1 to inhibity chemokine-induced migration in vitro has been demonstrated,[42] but the immunomodulatory activity of the SECRET domain has yet to be defined in vivo. Deletion of the SCP B7 encoded by VACV strain WR had no major effects on virulence.[44] It is interesting that the SECRET domain binds chemokines that are likely to be relevant in anti-viral defense: (i) chemokines mediating T- and B-cell recruitment that are expressed by epithelial cells in mucosal surfaces (CCL25 and CCL28) or the skin (CCL27),[45-47] which constitute the sites of virus entry; (ii) CCL25 and CCL28 recruit IgA-producing B-cells to mucosal sites;[47,48] (iii) CXCL14 is involved in dendritic cell migration to epidermal tissues;[49] and (iv) CXCL13 attracts B-cells to the spleen and lymph nodes.[46]

The CPXV CrmB protein, which blocks both TNF and chemokines in vitro,[42] has anti-inflammatory potential in vivo and increased approximately 50-fold the LD50 in infected mice.[50] However, the CPXV strain used still contained other active vTNFRs and the model of intracranial infection used can hardly mimick the natural route of infection. The MYXV M-T2 protein, a secreted vTNFR sharing the same domain organization as CrmB and CrmD, acts as a virulence factor in european rabbits causing a slight reduction of mortality.[51] However, none of these studies have defined the contribution of the anti-TNF vs anti-chemokine domains present in CrmB and CrmD to immune regulation and virulence.

The M3 Protein Encoded by Murine Gammaherpesvirus 68 (MHV-68)

MHV-68 is a pathogen of wild rodents that is related to the primate gammaherpesviruses Epstein-Barr virus, Kaposi's Sarcoma associated herpesvirus and Herpesvirus Saimiri. MHV-68 provides a useful mouse model of gammaherpesvirus pathogenesis. The M3 protein, unique to MHV-68 and not encoded by primate gamaherpesviruses, is a major secreted protein unrelated to host chemokine receptors that was identified as a CKBP.[52,53] M3 was identified as a CKBP in chemokine binding assays but, unlike the poxvirus 35-kDa CKBP that is specific for CC chemokines, M3 is able to bind to CC, CXC, C and CX3C chemokines.[52,53] Although M3 binds to chemokines of all four classes in the human and mouse systems, it is not likely to be an effective inhibitor of some CXC chemokines such as murine CXCL1 and CXCL5 or human CXCL12α.

The mechanism by which M3 inhibits chemokines is similar to that of the poxvirus 35-kDa protein. Both proteins bind free chemokine with high affinity in a manner which prevents their interaction with chemokine receptors and the induction of intracellular signalling events leading to cell migration and activation.[52,53] M3 has been shown to display a distinct property of inhibiting the interaction of chemokines with GAGs.[54] Moreover, it is able to disrupt preformed chemokine-GAG interactions in vitro, suggesting that M3 has a dual anti-chemokine function: the inhibition of chemokine-receptor and chemokin-GAG interactions.

The crystal structure of M3 complexed to a P8A variant of CCL2 showed that the dimerization of M3 brings in close proximity the N-terminal domain of a monomeric M3 molecule to the C-terminal domain of the second monomer to generate a binding site for chemokines.[55] The crystal structure was in accordance with binding studies of mutant chemokines showing that M3 interacts with chemokine residues involved in receptor binding, providing the structural basis for the ability of M3 to inhibit chemokine binding to cellular receptors.[55,56] Recent crystalographic studies have determined the structure of M3 complexed with wild type CCL2 or XCL1, showing that M3 engages the different chemokine classes with the same overall binding geometry.[57] The M3 C-terminal domain engages conserved residues involved in receptor binding whereas the acidic N-terminal domain exhibits electrostatic complementarity contacting chemokine basic clusters involved in GAG association, providing the structural basis for the interference of chemokine-GAG interactions by M3.[54]

Studies on the role of the M3 protein in MHV-68 pathogenesis have provided conflicting results. After intranasal infection, MHV-68 replicates transiently in respiratory epithelial cells and spreads to lymphoid tissue where latency is established in B-lymphocytes, macrophages and dendritic cells. An initial report showed that targeted disruption of the *M3* gene had surprisingly little effect on lytic virus replication in the respiratory tract or the initial spread of virus to lymphoid tissues after intranasal inoculation.[58] However, the mutant virus failed to establish normal levels of latency in splenic B-cells. Interestingly, in vivo $CD8^+$ T-cell depletion largely reversed the phenotype, suggesting that chemokine neutralization by M3 may function to block $CD8^+$ T-cell recruitment into lymphoid tissue and to enable the establishment of MHV-68 latency. A second report found no effect of the deletion of the M3 gene in the intranasal model, but demonstrated a role for M3 in controling brain inflammation in an intracranial model of infection.[59]

The Glycoprotein G (gG) Encoded by Alphaherpesviruses

The screening of supernatants from cultures infected with various alphaherpesviruses identified chemokine binding activity encoded by equine herpesvirus-1 (EHV-1), bovine herpesvirus 1 (BHV-1) and other alphaherpesviruses.[60] The activity was mapped to gG, which is expressed as a membrane-anchored protein at the surface of the enveloped virus particles of alphaherpesviruses. The chemokine binding activity identified in the supernatants was a secreted version of gG generated after proteolytic cleavage of the membrane form. Chemokine binding activity was also observed at the surface of insect cells infected with a recombinant baculovirus expressing the full-length gG. This is the first CKBP known to be expressed both as a membrane-anchored protein and as a secreted polypeptide.

gG encoded by EHV-1, BHV-1 and BHV-5 were characterized in more detail and found to bind a variety of CC and CXC chemokines.[60] These proteins block chemokine activity by interfering with chemokine interaction with cellular receptors and the subsequent activation of cell migration. EHV-1 gG was shown to inhibit chemotaxis of equine neutrophils by equine CXCL8.[61] As described for the MHV-68 M3 protein, it also prevents the interaction of chemokines with GAGs, suggesting that gG blocks chemokine activity at two different levels.[60]

Felid herpesvirus 1 (FeHV-1) gG was also found to bind a variety of chemokines with high affinity and to inhibit chemokine activity.[62] Studies on FeHV-1 demonstrated that the gG present at the surface of the virion binds chemokines.[63]

The contribution of gG to virus virulence has been addressed in several models of alphaherpesvirus infection. The infection of mice with low doses of an EHV-1 gG mutant led to an exacerbation of respiratory disease, with higher virus titers and a more pronounced inflammatory response in the lungs, compared to wild type infections.[61] EHV-1 gG was also found to reduce the infiltration of mouse neutrophils and macrophages into the lungs of infected mice and the chemotactic function of CCL3 in mice.[64] Infectious laryngotracheitis virus (ILTV) is an alphaherpesvirus that causes acute respiratory disease in chicken. A gG-deficient mutant of ILTV was attenuated compared to wild type virus and caused an increased tracheal mucosal thikness that reflects increased inflammatory cell infiltration at the site of infection.[65] These results suggest immunomodulatory activity of gG in this virus system, but the interaction of ILTV gG with chemokines has not beed formally demonstrated yet.

Chemokine binding activity has not been reported in the gG encoded by the human pathogens herpes simplex virus 1 (HSV-1) and HSV-2.[60] Interestingly, in contrast to gG encoded by HSV-2, HSV-1 gG is not secreted and suggests different functions of these proteins. A mutant HSV-1 lacking gG expression was tested in a mouse ear model but marginal attenuation and no effect on the ability of the virus to establish latency in neurons was observed.[66] Varicella zoster virus is an important human pathogen that causes a systemic disease but the gene encoding gG is not present in the varicella zoster virus genome.[67] The ability of gG encoded by HSV-1 and HSV-2 to bind chemokines and to contribute to virus-mediated pathology remains to be elucidated.

The Secreted CKBP from Human Cytomegalovirus (HCMV)

The pUL21.5 protein encoded by HCMV is a small secreted glycoprotein that has been demonstrated to bind CCL5 with high affinity and to block the interaction of CCL5 with specific cellular receptors.[68] This CKBP was reported to be highly specific for one chemokine, in contrast to the broad binding specificity of previously described virus-encoded CKBPs, but only three chemokines were tested in this study and the possibility that pUL21.5 binds some other chemokines of this complex family of chemoattractant cytokines cannot be ruled out. The mRNA encoding the pUL21.5 protein is packaged into virions and it was proposed that pUL12.5 may be expressed and secreted to modulate the host anti-viral response even before the newly infecting viral genome becomes transcriptionally active.[68]

A *Schistosoma mansoni*-Encoded Secreted Chemokine Inhibitor

S. mansoni is a trematode parasite that infects humans, causing schistisomiasis, a disease that is common in the developing world and that cause severe disease in 10% of the infected individuals.[69] In most cases, however, schistosomes are able to achieve chronic infections that cause high morbility. Schistosomes are particularly adept at manipulating the host's immune system to the benefit of the parasite. For example, the granulomatous inflammation around parasite eggs trapped in various organs, which, though a major cause of pathology, is evoked by the parasite to facilitate the expulsion of its eggs from the host. In situations where the formation of the granuloma is compromised, such as in immuno-supressed individuals, the excretion of eggs is diminished.[70,71]

A CKBP secreted by *S. mansoni* was identified in a cross-linking assay with radiolabelled chemokines and named smCKBP.[72] This protein was found to be expressed only in schistosome egg secretions but not in the other life cycle stages (cercariae, schistosomular, worms) of *S. mansoni*. Moreover, the smCKBP was also produced by eggs from the two other major schistosome species that infect man, *S. haematobium* and *S. japonicum*. smCKBP was the first CKBP identified in a human pathogen and is the only one identified in a parasite to date.

The gene encoding smCKBP was identified following a proteomic approach.[72] Despite having chemokine binding activity, smCKBP shares no amino acid sequence similarity to known viral CKBPs or mammalian proteins. Characterization of the binding properties showed that smCKBP interacts with several chemokines including CCL2, CCL3, CCL5, CXCL8 and CX3CL1. Furthermore, smCKBP was shown to prevent the interaction of chemokines with their specific cellular receptors and therefore chemokine-mediated cell activation and migration.

The study of the in vivo role of smCKBP might prove important to the understanding of the schistosomiasis pathology. It will be of interest to determine, in the murine model, the course of infection of a *S. mansoni* mutant lacking smCKBP. This, however, awaits the development of methodologies for the construction of deletion mutants in this parasite. It has been shown that, in an experimental granulomatous inflammation model, secretion of smCKBP by live eggs profoundly modulated the differential recruitment of cells and the size of the egg granuloma,[72] suggesting that smCKBP may ultimately facilitate granuloma formation and the propagation of the *S. mansoni* eggs.

Evasins, a Family of CKBPs in Ticks

A novel family of CKBPs, termed Evasins, has recently been described in ticks.[73,74] The Evasin family comprises four members and are small proteins of 7-12 kDa that are produced in the tick saliva. Ticks are blood-sucking parasites that feed on their hosts for several days but cause no inflammatory response. It has been suggested that Evasins may help ticks to inhibit chemokine-mediated innate responses that protect from parasites. In contrast to the broad binding specificity of the first CKBPs identified in viruses (M-T7, 35-kDa or M3 proteins), evasins show a restricted chemokine binding specificity. Evasin-1 binds CCL3, CCL4 and CCL18, Evasin-3 binds CXCL8 and CXCL1 and Evasin-4 binds CCL5 and CCL11.[73,74] Interestingly, it appears that ticks have evolved a family of CKBPs with narrow binding specificity as an alternative to a single CKBP with broad binding activity. Evasins block the interaction of chemokines with their cellular receptors and inhibit in this way chemokine-induced recruitment of leukocytes in vitro and in vivo.

The crystal structure of Evasin-1 and Evasin-3 has been determined and reveal novel protein folds.[73] Both proteins are unrelated in amino acid sequence and in their secondary and terciary structure and interact with host chemokines presumably in different ways. The determination of the structure of Evasins complexed with chemokines will define the structural basis for the Evasin selectivity of chemokines.

The Evolutionary Origin of CKBPs and Their Potential Therapeutic Applications

The production of secreted versions of cytokine receptors to control the activity of cytokines is a mechanism employed by large DNA viruses and other pathogens. The same strategy is used by the immune system to limit the activity of cytokines in order to avoid immune pathology. For example, it is well documented the proteolytic cleavage of TNFRs to release the ectodomain that retains TNF binding activity, the secretion of Type II IL-1 receptors to neutralize the activity of IL-1 or the production of secreted IL-18 binding proteins of structure different to that of the membrane-bound IL-18 receptor.[75,76]

Due to the structural nature of the seven-transmembrane-domain chemokine receptors, the production of secreted versions of these receptors is not feasable and alternative mechanisms have evolved to limit chemokine activity. The host immune system uses decoy receptors, such as the Duffy antigen receptor for chemokines (DARC), D6 or CCX-CKR, that bind chemokines but do not transduce signals and function as chemokine scavengers.[77,78] Alternatively, the anti-inflammatory cytokine IL-10 decouples chemokine receptors from intracellular signaling processes, thereby allowing them to sequester chemokines without inducing biological responses.[78] In contrast, viruses and other pathogens encode secreted proteins of unique structure not found in host proteins that are able to bind chemokines with high affinity and to block their activity.[9,11]

The chemokine family is complex, with more than 40 chemokine ligands[3] and the binding properties of the CKBPs reflects the targeting of specific immune functions mediated by particular sets of chemokines. Some CKBPs, such as the gammaherpesvirus M3 or the poxvirus 35-kDa proteins,[21,52,53] bind a broad range of chemokines, whereas a narrow binding specificity has been recently described for the poxvirus SECRET domain and the Evasins encoded by ticks.[42,73] Future in vivo studies in animal models of infection will be highly relevant to understand the immune modulatory activity of the different CKBPs and will shed light into the physiological role of the target chemokines in immunity.

Another interesting property of some virus-encoded CKBPs is that they may inhibit the interaction of chemokines with specific receptors, blocking chemokine-induced signalling, or may interfere with chemokine-GAG interactions. It is likely that the ability of gammaherpesvirus M3 and alphaherpesvirus gG to inhibit the binding of chemokines to both receptors and GAGs may enhance the anti-chemokine activity of these CKBPs in vivo.[54,57,60]

The chemokine network is a major target for the development of drugs useful in the control of inflammatory diseases.[6,7,79] The identification of CKBPs with no amino acid sequence similarity among themselves or to host receptors is providing us with novel protein structures capable of neutralizing chemokines and have great potential as new medicaments.[80,81] Many studies have demonstrated the therapeutic value of CKBPs in different animal models of human diseases. For example, treatment with the poxvirus 35-kDa CKBP significantly reduced inflammation of the airway and lung parenchyma and improved the physiological function of the lungs during airway hyperreactivity in a mouse model of allergen-induced asthma.[82] In another study, a single injection of MYXV M-T7 given to rats or rabbits caused a significant attenuation of restenosis, a response to vascular injury that leads to recurrent atherosclerotic plaque growth.[83] The generation of transgenic mice expressing the gammaherpesvirus M3 protein has been an important tool to demonstrate the efficient blockade of chemokine-mediated cell migration in vivo by M3 and its ability to block intimal hyperplasia in response to arterial injury and the development of diabetes.[84-87] The anti-inflammatory activity of the CKBPs encoded by *S. mansoni* and ticks has been shown in models of contact hypersensitivity model, pulmonary inflammation or arthritis.[72,73] Future studies will no doubt further assess the utility of pathogen-encoded chemokine inhibitors in other models of inflammation. It is interesting that products derived from pathogenic organisms hold substantial promise for the treatment of human inflammatory diseases.[81]

The vCKBPs identified in the genome of pathogens were not predicted from sequence analysis to bind chemokines and it is possible that the secretion of CKBPs of unrelated structure is a strategy used by other pathogens. Further studies on chemokine inhibitors from pathogens will provide interesting information on the contribution of these immune evasion proteins to pathogenesis and the function of chemokines in immunity.

Acknowledgements

We acknowledge the support from the Wellcome Trust, Spanish Ministry of Science and Innovation, Comunidad de Madrid and European Union.

References

1. Baggiolini M. Chemokines and leukocyte traffic. Nature 1998; 392(6676):565-568.
2. Mackay CR. Chemokines: immunology's high impact factors. Nat Immunol 2001; 2(2):95-101.
3. Zlotnik A, Yoshie O, Nomiyama H. The chemokine and chemokine receptor superfamilies and their molecular evolution. Genome Biol 2006; 7(12):243.
4. Johnson Z, Proudfoot AE, Handel TM. Interaction of chemokines and glycosaminoglycans: a new twist in the regulation of chemokine function with opportunities for therapeutic intervention. Cytokine Growth Factor Rev 2005; 16(6):625-636.
5. Handel TM, Johnson Z, Crown SE et al. Regulation of protein function by glycosaminoglycans—as exemplified by chemokines. Annu Rev Biochem 2005; 74:385-410.
6. Proudfoot AE. Chemokine receptors: multifaceted therapeutic targets. Nat Rev Immunol 2002; 2(2):106-115.
7. Wells TN, Power CA, Shaw JP et al. Chemokine blockers—therapeutics in the making? Trends Pharmacol Sci 2006; 27(1):41-47.
8. Finlay BB, McFadden G. Anti-immunology: evasion of the host immune system by bacterial and viral pathogens. Cell 2006; 124(4):767-782.
9. Seet BT, Johnston JB, Brunetti CR et al. Poxviruses and immune evasion. Annu Rev Immunol 2003; 21:377-423.
10. Alcami A, Koszinowski UH. Viral mechanisms of immune evasion. Immunol Today 2000; 21(9):447-455.
11. Alcami A. Viral mimicry of cytokines, chemokines and their receptors. Nat Rev Immunol 2003; 3(1):36-50.

12. Upton C, Mossman K, McFadden G. Encoding of a homolog of the IFN-gamma receptor by myxoma virus. Science 1992; 258(5086):1369-1372.
13. Mossman K, Upton C, McFadden G. The myxoma virus-soluble interferon-gamma receptor homolog, M-T7, inhibits interferon-gamma in a species-specific manner. J Biol Chem 1995; 270(7):3031-3038.
14. Lalani AS, Graham K, Mossman K et al. The purified myxoma virus gamma interferon receptor homolog M-T7 interacts with the heparin-binding domains of chemokines. J Virol 1997; 71(6):4356-4363.
15. Alcami A, Symons JA, Collins PD et al. Blockade of chemokine activity by a soluble chemokine binding protein from vaccinia virus. J Immunol 1998; 160(2):624-633.
16. Mossman K, Nation P, Macen J et al. Myxoma virus M-T7, a secreted homolog of the interferon-gamma receptor, is a critical virulence factor for the development of myxomatosis in euroapean rabbits. Virology 1996; 215(1):17-30.
17. Graham KA, Lalani AS, Macen JL et al. The T1/35 kDa family of poxvirus-secreted proteins bind chemokines and modulate leukocyte influx into virus-infected tissues. Virology 1997; 229(1):12-24.
18. Smith CA, Smith TD, Smolak PJ et al. Poxvirus genomes encode a secreted, soluble protein that preferentially inhibits beta chemokine activity yet lacks sequence homology to known chemokine receptors. Virology 1997; 236(2):316-327.
19. Smith VP, Alcami A. Expression of secreted cytokine and chemokine inhibitors by ectromelia virus. J Virol 2000; 74(18):8460-8471.
20. Seet BT, McCaughan CA, Handel TM et al. Analysis of an orf virus chemokine-binding protein: shifting ligand specificities among a family of poxvirus viroceptors. Proc Natl Acad Sci USA 2003; 100(25):15137-15142.
21. Burns JM, Dairaghi DJ, Deitz M et al. Comprehensive mapping of poxvirus vCCI chemokine-binding protein. Expanded range of ligand interactions and unusual dissociation kinetics. J Biol Chem 2002; 277(4):2785-2789.
22. Lalani AS, Ness TL, Singh R et al. Functional comparisons among members of the poxvirus T1/35 kDa family of soluble CC-chemokine inhibitor glycoproteins. Virology 1998; 250(1):173-184.
23. Seet BT, Barrett J, Robichaud J et al. Glycosaminoglycan binding properties of the myxoma virus CC-chemokine inhibitor, M-T1. J Biol Chem 2001; 276(32):30504-30513.
24. Carfi A, Smith CA, Smolak PJ et al. Structure of a soluble secreted chemokine inhibitor vCCI (p35) from cowpox virus. Proc Natl Acad Sci USA 1999; 96(22):12379-12383.
25. Zhang L, Derider M, McCornack MA et al. Solution structure of the complex between poxvirus-encoded CC chemokine inhibitor vCCI and human MIP-1beta. Proc Natl Acad Sci USA 2006; 103(38):13985-13990.
26. Beck CG, Studer C, Zuber JF et al. The viral CC chemokine-binding protein vCCI inhibits monocyte chemoattractant protein-1 activity by masking its CCR2B-binding site. J Biol Chem 2001; 276(46):43270-43276.
27. Seet BT, Singh R, Paavola C et al. Molecular determinants for CC-chemokine recognition by a poxvirus CC-chemokine inhibitor. Proc Natl Acad Sci USA 2001; 98(16):9008-9013.
28. Arnold PL, Fremont DH. Structural determinants of chemokine binding by an ectromelia virus-encoded decoy receptor. J Virol 2006; 80(15):7439-7449.
29. Lalani AS, Masters J, Graham K et al. Role of the myxoma virus soluble CC-chemokine inhibitor glycoprotein, M-T1, during myxoma virus pathogenesis. Virology 1999; 256(2):233-245.
30. Martinez-Pomares L, Thompson JP, Moyer RW. Mapping and investigation of the role in pathogenesis of the major unique secreted 35-kDa protein of rabbitpox virus. Virology 1995; 206(1):591-600.
31. Reading PC, Symons JA, Smith GL. A soluble chemokine-binding protein from vaccinia virus reduces virus virulence and the inflammatory response to infection. J Immunol 2003; 170(3):1435-1442.
32. Ng A, Tscharke DC, Reading PC et al. The vaccinia virus A41L protein is a soluble 30 kDa glycoprotein that affects virus virulence. J Gen Virol 2001; 82(Pt 9):2095-2105.
33. Clark RH, Kenyon JC, Bartlett NW et al. Deletion of gene A41L enhances vaccinia virus immunogenicity and vaccine efficacy. J Gen Virol 2006; 87(Pt 1):29-38.
34. Bahar MW, Kenyon JC, Putz MM et al. Structure and function of A41, a vaccinia virus chemokine binding protein. PLoS Pathog 2008; 4(1):e5.
35. Ruiz-Arguello MB, Smith VP, Campanella GS et al. An ectromelia virus protein that interacts with chemokines through their glycosaminoglycan binding domain. J Virol 2008; 82(2):917-926.
36. Hu FQ, Smith CA, Pickup DJ. Cowpox virus contains two copies of an early gene encoding a soluble secreted form of the Type II TNF receptor. Virology 1994; 204(1):343-356.
37. Loparev VN, Parsons JM, Knight JC et al. A third distinct tumor necrosis factor receptor of orthopoxviruses. Proc Natl Acad Sci USA 1998; 95(7):3786-3791.
38. Saraiva M, Alcami A. CrmE, a novel soluble tumor necrosis factor receptor encoded by poxviruses. J Virol 2001; 75(1):226-233.

39. Smith CA, Hu FQ, Smith TD et al. Cowpox virus genome encodes a second soluble homologue of cellular TNF receptors, distinct from CrmB, that binds TNF but not LT alpha. Virology 1996; 223(1):132-147.
40. Panus JF, Smith CA, Ray CA et al. Cowpox virus encodes a fifth member of the tumor necrosis factor receptor family: a soluble, secreted CD30 homologue. Proc Natl Acad Sci USA 2002; 99(12):8348-8353.
41. Saraiva M, Smith P, Fallon PG et al. Inhibition of Type 1 cytokine-mediated inflammation by a soluble CD30 homologue encoded by ectromelia (mousepox) virus. J Exp Med 2002; 196(6):829-839.
42. Alejo A, Ruiz-Arguello MB, Ho Y et al. A chemokine-binding domain in the tumor necrosis factor receptor from variola (smallpox) virus. Proc Natl Acad Sci USA 2006; 103(15):5995-6000.
43. Smith GL, McFadden G. Smallpox: anything to declare? Nat Rev Immunol 2002; 2(7):521-527.
44. Price N, Tscharke DC, Hollinshead M et al. Vaccinia virus gene B7R encodes an 18-kDa protein that is resident in the endoplasmic reticulum and affects virus virulence. Virology 2000; 267(1):65-79.
45. Homey B, Alenius H, Muller A et al. CCL27-CCR10 interactions regulate T-cell-mediated skin inflammation. Nat Med 2002; 8(2):157-165.
46. Kunkel EJ, Butcher EC. Chemokines and the tissue-specific migration of lymphocytes. Immunity 2002; 16(1):1-4.
47. Lazarus NH, Kunkel EJ, Johnston B et al. A common mucosal chemokine (mucosae-associated epithelial chemokine/CCL28) selectively attracts IgA plasmablasts. J Immunol 2003; 170(7):3799-3805.
48. Bowman EP, Kuklin NA, Youngman KR et al. The intestinal chemokine thymus-expressed chemokine (CCL25) attracts IgA antibody-secreting cells. J Exp Med 2002; 195(2):269-275.
49. Schaerli P, Willimann K, Ebert LM et al. Cutaneous CXCL14 targets blood precursors to epidermal niches for langerhans cell differentiation. Immunity 2005; 23(3):331-342.
50. Palumbo GJ, Buller RM, Glasgow WC. Multigenic evasion of inflammation by poxviruses. J Virol 1994; 68(3):1737-1749.
51. Upton C, Macen JL, Schreiber M et al. Myxoma virus expresses a secreted protein with homology to the tumor necrosis factor receptor gene family that contributes to viral virulence. Virology 1991; 184(1):370-382.
52. Parry CM, Simas JP, Smith VP et al. A broad spectrum secreted chemokine binding protein encoded by a herpesvirus. J Exp Med 2000; 191(3):573-578.
53. van Berkel V, Barrett J, Tiffany HL et al. Identification of a gammaherpesvirus selective chemokine binding protein that inhibits chemokine action. J Virol 2000; 74(15):6741-6747.
54. Webb LM, Smith VP, Alcami A. The gammaherpesvirus chemokine binding protein can inhibit the interaction of chemokines with glycosaminoglycans. FASEB J 2004; 18(3):571-573.
55. Alexander JM, Nelson CA, van Berkel V et al. Structural basis of chemokine sequestration by a herpesvirus decoy receptor. Cell 2002; 111(3):343-356.
56. Webb LM, Clark-Lewis I, Alcami A. The gammaherpesvirus chemokine binding protein binds to the N terminus of CXCL8. J Virol 2003; 77(15):8588-8592.
57. Alexander-Brett JM, Fremont DH. Dual GPCR and GAG mimicry by the M3 chemokine decoy receptor. J Exp Med 2007; 204(13):3157-3172.
58. Bridgeman A, Stevenson PG, Simas JP et al. A secreted chemokine binding protein encoded by murine gammaherpesvirus-68 is necessary for the establishment of a normal latent load. J Exp Med 2001; 194(3):301-312.
59. van Berkel V, Levine B, Kapadia SB. Critical role for a high-affinity chemokine-binding protein in gamma-herpesvirus-induced lethal meningitis. J Clin Invest 2002; 109(7):905-914.
60. Bryant NA, Davis-Poynter N, Vanderplasschen A et al. Glycoprotein G isoforms from some alphaherpesviruses function as broad-spectrum chemokine binding proteins. EMBO J 2003; 22(4):833-846.
61. Van de Walle GR, May ML, Sukhumavasi W et al. Herpesvirus chemokine-binding glycoprotein G (gG) efficiently inhibits neutrophil chemotaxis in vitro and in vivo. J Immunol 2007; 179(6):4161-4169.
62. Costes B, Ruiz-Arguello MB, Bryant NA et al. Both soluble and membrane-anchored forms of felid herpesvirus 1 glycoprotein G function as a broad-spectrum chemokine-binding protein. J Gen Virol 2005; 86(Pt 12):3209-3214.
63. Costes B, Thirion M, Dewals B et al. Felid herpesvirus 1 glycoprotein G is a structural protein that mediates the binding of chemokines on the viral envelope. Microbes Infect 2006; 8(11):2657-2667.
64. Van de Walle GR, Sakamoto K, Osterrieder N. CCL3 and viral chemokine-binding protein gg modulate pulmonary inflammation and virus replication during equine herpesvirus 1 infection. J Virol 2008; 82(4):1714-1722.
65. Devlin JM, Browning GF, Hartley CA et al. Glycoprotein G is a virulence factor in infectious laryngotracheitis virus. J Gen Virol 2006; 87(Pt 10):2839-2847.

66. Balan P, Davis-Poynter N, Bell S et al. An analysis of the in vitro and in vivo phenotypes of mutants of herpes simplex virus Type 1 lacking glycoproteins gG, gE, gI or the putative gJ. J Gen Virol 1994; 75(Pt 6):1245-1258.
67. Gomi Y, Sunamachi H, Mori Y et al. Comparison of the complete DNA sequences of the Oka varicella vaccine and its parental virus. J Virol 2002; 76(22):11447-11459.
68. Wang D, Bresnahan W, Shenk T. Human cytomegalovirus encodes a highly specific RANTES decoy receptor. Proc Natl Acad Sci USA 2004; 101(47):16642-16647.
69. Pearce EJ, MacDonald AS. The immunobiology of schistosomiasis. Nat Rev Immunol 2002; 2(7):499-511.
70. Fallon PG, Richardson EJ, Smith P et al. Elevated Type 1, diminished Type 2 cytokines and impaired antibody response are associated with hepatotoxicity and mortalities during Schistosoma mansoni infection of CD4-depleted mice. Eur J Immunol 2000; 30(2):470-480.
71. Karanja DM, Colley DG, Nahlen BL et al. Studies on schistosomiasis in Western Kenya: I. evidence for immune-facilitated excretion of schistosome eggs from patients with Schistosoma mansoni and human immunodeficiency virus coinfections. Am J Trop Med Hyg 1997; 56(5):515-521.
72. Smith P, Fallon RE, Mangan NE et al. Schistosoma mansoni secretes a chemokine binding protein with antiinflammatory activity. J Exp Med 2005; 202(10):1319-1325.
73. Deruaz M, Frauenschuh A, Alessandri AL et al. Ticks produce highly selective chemokine binding proteins with antiinflammatory activity. J Exp Med 2008; 205(9):2019-2031.
74. Frauenschuh A, Power CA, Deruaz M et al. Molecular cloning and characterization of a highly selective chemokine-binding protein from the tick Rhipicephalus sanguineus. J Biol Chem 2007; 282(37):27250-27258.
75. Dinarello CA. The IL-1 family and inflammatory diseases. Clin Exp Rheumatol 2002; 20(5 Suppl 27):S1-13.
76. Feldmann M. Development of anti-TNF therapy for rheumatoid arthritis. Nat Rev Immunol 2002; 2(5):364-371.
77. Graham GJ, McKimmie CS. Chemokine scavenging by D6: a movable feast? Trends Immunol 2006; 27(8):381-386.
78. Mantovani A, Bonecchi R, Locati M. Tuning inflammation and immunity by chemokine sequestration: decoys and more. Nat Rev Immunol 2006; 6(12):907-918.
79. Luster AD, Alon R, von Andrian UH. Immune cell migration in inflammation: present and future therapeutic targets. Nat Immunol 2005; 6(12):1182-1190.
80. Lucas A, McFadden G. Secreted immunomodulatory viral proteins as novel biotherapeutics. J Immunol 2004; 173(8):4765-4774.
81. Fallon PG, Alcami A. Pathogen-derived immunomodulatory molecules: future immunotherapeutics? Trends Immunol 2006; 27(10):470-476.
82. Dabbagh K, Xiao Y, Smith C et al. Local blockade of allergic airway hyperreactivity and inflammation by the poxvirus-derived pan-CC-chemokine inhibitor vCCI. J Immunol 2000; 165(6):3418-3422.
83. Liu L, Lalani A, Dai E et al. The viral anti-inflammatory chemokine-binding protein M-T7 reduces intimal hyperplasia after vascular injury. J Clin Invest 2000; 105(11):1613-1621.
84. Jensen KK, Chen SC, Hipkin RW et al. Disruption of CCL21-induced chemotaxis in vitro and in vivo by M3, a chemokine-binding protein encoded by murine gammaherpesvirus 68. J Virol 2003; 77(1):624-630.
85. Martin AP, Alexander-Brett JM, Canasto-Chibuque C et al. The chemokine binding protein M3 prevents diabetes induced by multiple low doses of streptozotocin. J Immunol 2007; 178(7):4623-4631.
86. Martin AP, Canasto-Chibuque C, Shang L et al. The chemokine decoy receptor M3 blocks CC chemokine ligand 2 and CXC chemokine ligand 13 function in vivo. J Immunol 2006; 177(10):7296-7302.
87. Pyo R, Jensen KK, Wiekowski MT et al. Inhibition of intimal hyperplasia in transgenic mice conditionally expressing the chemokine-binding protein M3. Am J Pathol 2004; 164(6):2289-2297.

Index